POWER PLAY

POWER PLAY

TESLA, ELON MUSK, AND THE BET OF THE CENTURY

TIM HIGGINS

RANDOM HOUSE
LARGE PRINT

Cover illustration: Hitandrun Media @ Début Art
Cover design by John Fontana

The Library of Congress has established a
Cataloging-in-Publication record for this title.

ISBN: 978-0-5934-1430-9

www.penguinrandomhouse.com/large-print-format-books

FIRST LARGE PRINT EDITION

Printed in the United States of America

10 9 8 7 6 5 4 3 2 1

This Large Print edition published in accord with
the standards of the N.A.V.H.

To my parents

CONTENTS

PROLOGUE

THE BEGINNING

On a breezy night in March 2016 at the Tesla design studio, Elon Musk took the stage in front of a crowd of supporters. Dressed like a James Bond villain, in a black jacket with the collar up, he was on the cusp of achieving a decade-long dream, a goal the famed entrepreneur had spent years building toward: the grand reveal of his Model 3 electric car.

The design studio—near the Los Angeles airport and in the same complex as Musk's privately held rocket company, SpaceX—was the home of Tesla's creative soul. It was a magical place, where Franz von Holzhausen, the automotive designer who contributed to the re-imagined Volkswagen Beetle and Mazda's resurgence, headed a team that put into form the ideas that Musk envisioned. Together they aspired to build revolutionary and dazzling electric cars, eschewing the techie, nerdy look favored by competitors, who had long seen such vehicles as experimental novelties.

Hundreds of customers turned out for the occasion. A Musk party wasn't to be missed; whether it was for Tesla or SpaceX, his events attracted an eclectic mix of Silicon Valley entrepreneurs, Hollywood names, loyal customers, and car enthusiasts. Until now Tesla had been a niche luxury brand—a fantasy for California environmentalists that had morphed into a whim for the rich, a must-have among those wealthy enough to have garages full of BMWs, Mercedes, and other gas-powered status symbols.

The Model 3, with its pledged starting pricing of $35,000, held the promise of something different. It was the embodiment of Musk's ambition to bring a fully electric car to the masses. It was a gamble in the form of a four-door compact car: that Tesla could generate the sales volume and cash to take on the biggest of the big boys in the century-old automotive industry: Ford, Toyota, Volkswagen, Mercedes-Benz, BMW, and, of course, General Motors. The Model 3 would determine if Tesla was a real car company.

Musk, just a year younger than Henry Ford when the Model T was introduced 108 years earlier, stood onstage that night, greeted by the pounding bass of techno music and screams of his fans, to rewrite history. He came to usher in a new era.

It was his mission to change the world and maybe even save it (and presumably get rich while doing so) that had helped him attract a team of executives to put his vision into reality. In the crowd, those key deputies—who had been pulled from the automotive industry, tech, and venture capital (including Musk's trusted confidant, his brother, Kimbal)—reveled in the excitement.

Onstage, Musk fumbled through charts about rising CO_2 pollution as he lamented the damage done to the planet. "This is really important for the future of the world," he told the crowd to cheers.

A highly produced video gave the first glances of the Model 3. The car looked like a beacon of the future, outside and in. Sleek curves and lines encased an interior unlike anything on the market, the gauges of a typical car gone, replaced by a single large tablet-like screen in the center of the cockpit. The car zoomed along winding roads down the California coast. Again, the crowd cheered. One attendee screamed: "You did it!"

Musk commanded the stage, telling his audience that Tesla had already received more than 115,000 deposits of $1,000 each—a $115 million boost in cash to the company. Within weeks, Tesla would claim more than 500,000 reservations. It was a staggering figure: Thirty-two percent more than Toyota Motor Corp. sold that year in the U.S. of its popular family sedan archetype, the Camry. And these were reservations—people lining up two years before the car would even be produced.

The Tesla team had devised a plan to slowly start making cars, with the goal of having a few thousand ready by the end of 2017, then boosting their capacity week by week, through 2018, until finally reaching the goal of 5,000 cars a week by the middle part of that year.

That volume—5,000 cars a week, 260,000 cars a year—was a widely held benchmark, a defining figure of what it meant to be a viable factory in the stable of a major carmaker. If Elon Musk and Tesla hit it, they could become a new force in the automotive industry.

But even that wasn't enough for Musk. He was already boasting that by 2020 he could get production at the company's single assembly factory outside Silicon Valley to 500,000 vehicles in a single year—twice as many as most car factories in the U.S.

It is hard to overstate just how crazy this would sound coming from anyone other than Elon Musk.

Automakers typically take five to seven years from the start of designing a new vehicle to delivering it to customers. It's a tedious and complex process, refined over generations of experience. Before a new car hits dealer showrooms, it's tested in the desert, the Arctic, and the mountains. Thousands of suppliers contribute to the effort, building parts with astonishing precision for vehicles that will ultimately be pieced together in factory blitzes choreographed to the second.

But for all of his startup gumption, undeniable ambition, and vision, walking off the stage that day, even with pre-orders pouring in, Musk couldn't escape the inexorable financial logic learned over a century by the likes of GM, Ford, and BMW: the process of making cars is a brutal business—and an expensive one.

And Elon Musk's books were a disaster. Tesla was burning, on average, $500 million a quarter and had only $1.4 billion free cash on hand—meaning Tesla was on track to be out of money by the end of 2016 if something drastic didn't change.

But this was all part of the confidence game he always knew he'd have to play if he were ever going to create the world's most valuable automaker. Belief created the vision; the vision would create a market; the market would create cash; and cash would create cars. He just had to do it on an unimaginable scale, and do it fast enough to stay a car's length ahead of competitors, creditors, customers, and investors betting against the company through a process called short selling that could pay handsomely if Tesla's stock plunged in value.

It was, he knew all too well, a dangerous race.

Or, in his darker moments, the ultimate game of chicken.

—

In June 2018, a little over two years after Musk's glitzy reveal of the Model 3, I visited him deep inside the cavernous Tesla Inc. assembly factory half an hour outside Silicon Valley. Musk looked weary. In a cubicle on the factory floor, he was dressed in a black Tesla T-shirt and jeans, his six-foot-two frame hunched over an iPhone. His Twitter account streamed taunts from short sellers; some of the world's most powerful investors were wagering against him, predicting his imminent failure. His email in-box contained new messages from a recently fired employee accusing the CEO of cutting corners and putting lives at risk.

Over his shoulder, the body shop towered: It was the greatest expression yet of Musk's vision: a mechanical beast that ate raw parts at one end and spit out cars at the other. Two stories tall, with more than a thousand robot arms anchored to the floor and hung from the ceiling, it was a gauntlet for the car skeletons it forged. Sparks flew as the robot arms swooped in to weld pieces of sheet metal to the frame. The air filled with an acrid smell. A reverberation of metal clanged like a deafening metronome.

From the body shop, the car moved to the paint shop where pearl white, midnight silver, and Tesla's iconic race-car red were applied. Then to the general assembly line where the thousand-pound batteries were added, along with all of the touches that make a car a car—the seats, the dashboard, the display.

It was at this juncture where the problems

currently were, and why Musk had been sleeping alone on the factory floor. The assembly line was plagued by snags. He'd relied too much on robots to make the cars, he said. The 10,000 parts required from hundreds of suppliers had created an endless loop of complexity. Everywhere he turned, he found something working not quite right.

He apologized for his unkempt appearance—his brown hair hadn't been combed in a while and he hadn't changed his T-shirt in three days. In a few days, he'd turn forty-seven. He was a year behind schedule in cranking up production of the Model 3, the compact car that would make or break Tesla.

Musk sat at an empty desk. His pillow from a few hours' sleep rested on a chair next to him. A salad went half-eaten. A bodyguard stood nearby. The company teetered on the verge of bankruptcy.

But for all of this, he was in surprisingly good spirits. He assured me that everything would work out.

A few weeks later, he called me in a decidedly darker mood. The world was out to get him. "It's not like I desperately want this fucking role," he said. "I'm doing this because I believe in the god-damn mission, that I think that sustainable energy needs to prosper."

If it seemed like rock bottom for Elon Musk, it wasn't.

—

At the heart of Musk's fight and Tesla's history is a central question: Can a startup conquer one of the biggest and most entrenched industries in the global economy? The automobile changed the world. Aside from the autonomy and mobility it offered individuals, and the entire swaths of modern civilization it has helped incubate and connect, it has generated an economy unto itself. Detroit helped make the middle class, establishing wealth and stability for the communities it touched. It also became one of the nation's largest industries— creating almost $2 trillion in revenue annually in the U.S. and employing one in twenty Americans.

GM, Ford, Toyota, and BMW have grown into global icons that design, build, and sell tens of millions of vehicles each year. Those car purchases have come to represent more than just an appliance; they convey independence and status, a symbol of the American Dream and, more and more so, the Global Dream.

The downside is that as those dreams have spread throughout the world, those same cars, through their manufacture and use, have contributed to unprecedented scales of congestion, pollution, and climate change.

Enter Musk, a self-made multimillionaire by his twenties who dreamed of using his newfound wealth to change the world. His belief in electric cars was so resolute that he staked his fortune on its success, teetering on bankruptcy because of it and burning through three marriages—twice to the same woman—along the way.

It's one thing to create a social network when the incumbent is MySpace. Or to use an online platform to unlock excess inventories of cars and apartments, to take on taxi cartels or the hotel industry. It's something else entirely to stare down some of the biggest companies in the world and to challenge them on their own turf, with something they've been learning to make—often painfully—for over a century.

It is often a narrow-margin business. The average car may only generate about $2,800 in operating profit. To get to that point, you need to achieve extraordinary scale, including the ability to keep a factory humming to the tune of 5,000 cars a week. And even if you do that, you have to be damn sure that someone's going to **buy** them all.

Any snag in production or sales can very quickly lead to disaster. Costs mount every day that a factory isn't in use, or that cars aren't moving out to dealerships, or that consumers aren't driving them home. That cash flow, from consumer to dealers to manufacturers, is the lifeblood of the automotive industry; it in turn funds the development of a company's next vehicles, something that can require massive investments and sunk costs.

GM spent a total of $13.9 billion on developing new products in 2016 and 2017. And in a business where profits can swing wildly from one year to the next (GM was $9 billion in the black in 2016; $3.9 billion in the red in 2017), it's perhaps little surprise that the biggest carmakers can't get by without hoards of cash: in 2017 GM had $20 billion in cash on hand; Ford had $26.5 billion; Toyota and VW both finished fiscal 2017 with $43 billion in their bank accounts.

So great are the barriers to entering the car business that the last major new U.S. carmaker to emerge that is still around today was Chrysler. That was in 1925. Or as Musk likes to remind people, in playing up the outrageous long shot he's taking, only two U.S. car companies have **not** gone bankrupt: Ford—and Tesla.

So you'd pretty much have to be delusional to enter such a competition; which some think Elon Musk is. But he hasn't shrunk from the challenge. Instead he has willed himself and his company to where the lofty visions of Silicon Valley meet the harsh reality of Detroit. His big idea is that in Tesla he can make electric cars really work. That they can outperform their gas-guzzling cousins; that they can out-style them; out-tech them; save customers billions of dollars a year on gasoline; and in so doing save the world from itself.

But it's a promise that at times obscures the ruthless business ambition—and imperative—under which Musk and Tesla operate. Many of us might misunderstand or underestimate Tesla's endgame. They might see the car as a toy for the green-conscious family down the block with money to burn, or for the status-inclined hedge funder with the progressive air. Or else it's the new Ferrari for the walking midlife crisis who just parked next to you at the train station.

But these niche existences? These are decidedly not what Tesla is about. And that is why the fate of the company rides on the Model 3, the electric car for the masses. As one Wall Street banker lamented years ago, "Either they become a niche manufacturer like Porsche or Maserati and make 50,000 high-end cars annually, or they crack the code on a $30,000 car that would put them on the inflection point of a large industrial."

That inflection point is the Model 3.

Musk's relentless drive to create the Model 3, and the questionable tactics he has used to get there, have unsettled competitors and industry observers alike. Unlike most automotive executives, Musk's philosophy of decision-making flows from his California ecosystem, where it's better to make a fast, wrong choice that can be undone quickly than to spend time perfecting hypotheticals. For a startup, time is money, something even truer for a new car company that has been, largely from day one, burning through millions of dollars a day.

Musk is a firm believer in the power of momentum, that one win leads to another. And as he has developed and sold successive models of electric cars that have shattered preconceptions of what electric cars can do, he has undoubtedly strung together some Ws.

His success with Tesla's early luxury models has sprung his competitors into action. The world's largest automakers in 2018 were rushing to catch up with their own electric cars, investing more than $100 billion to build and bring out 75 all-electric and plug-in hybrid vehicles by the end of 2022, according to one study. By 2025, analysts were predicting then, almost 500 new types of electric cars could be on sale, representing one in five new car sales globally.

But Musk has carved out a decided brand advantage. He has almost singlehandedly created the contemporary electric car zeitgeist. He embodies it. To many, he is it.

And that is why, in 2018, investor enthusiasm for Musk's vision pushed the market value of Tesla higher than that of any other U.S. automaker—before having ever turned an annual profit and selling just a fraction of the cars. That increasing share price indicated that investors were betting on the potential of Tesla to lead the electric car revolution. Tesla's access to billions of dollars of capital had fueled its growth and allowed it to survive.

Investors had been valuing Tesla more like a tech company than a typical automaker, which gets judged harshly by quarterly performance and their low expectations for the future. That was good news for Musk in 2018. If Tesla had been valued by investors the way they valued GM, it would have been worth $6 billion—not $60 billion. If GM had been valued like Tesla, it would have been worth $340 billion, not $43 billion.

But despite all the hype, Tesla must abide by the same financial logic as any carmaker—each new product represents a stretch, and a possibly fatal stumble. This is, in fact, only more true for Tesla, given its minuscule lineup. The stakes grow larger as Tesla grows bigger, as the bets go from a few million dollars to billions.

And while Musk's vision, enthusiasm, and determination carry Tesla; his ego, paranoia, and pettiness threaten to undo it all.

His fans and detractors can't get enough of him. His face has appeared on magazine covers for a decade. He was the inspiration for Robert Downey Jr.'s portrayal of Tony Stark in the **Iron Man** movies. He's prolific on Twitter, sparring with government regulators he disagrees with, attacking short sellers betting against him, and joking with his fans about everything from Japanese anime to drug use. But increasingly people have seen another side of him. Frazzled. Stressed. Worried. Despairing. Insecure. In short: vulnerable.

Would Musk's naked ambition to upend the

automotive industry enable him to do what once was seen as impossible? Or would his hubris be his undoing?

Amid the controversial figures to have emerged from Silicon Valley in recent years, you couldn't help but wonder: Is Elon Musk an underdog, an anti-hero, a con man, or some combination of the three?

PART I

A REALLY EXPENSIVE CAR

CHAPTER 1

THIS TIME COULD
BE DIFFERENT

The idea for an electric car kept JB Straubel up late one summer night in 2003. His tiny, rented house in Los Angeles brimmed that evening with members of Stanford University's solar car team, who had just finished a race from Chicago. The biennial event was part of a growing movement to stoke interest among young engineers in developing alternatives to gas-powered vehicles. Straubel had offered to play host to his alma mater's team, and the grueling run left many sleeping on his floor.

Intensely focused on his own projects, Straubel had never joined the team himself during his six years at the Stanford engineering school. But his interests aligned with those of his guests: He too was obsessed by the idea of powering cars with electricity—an interest he had held since his

childhood in Wisconsin. After graduating, he had
floated between LA and Silicon Valley, struggling
to find his place. Straubel didn't look like a mad
scientist intent on changing the world; he had a
quietness about him and the bland good looks of
a midwestern frat boy. But inside, he had a gnaw-
ing desire to do more than take a job with friends
at a startup like Google or join the bureaucracy of
a Boeing or General Motors. He wanted to create
something that changed everything, whether it was
in a car or an airplane; he wanted to chase a dream.

Stanford's team, like its competitors, had designed
a car that ran on energy it collected from the sun
using solar panels. Small batteries stored some of
that energy—for use at night, or else when the sun
was obscured by clouds. It being a solar race, however, organizers placed limits on how batteries could
be used.

Straubel thought this prohibition was misguided.
Battery technology had improved dramatically in
recent years, with the rise of personal electronics.
He wanted to think beyond the arbitrary rules defined by competition organizers. Better batteries
meant a car could run longer without relying so
much on finicky solar panels and the whims of the
weather. Why not emphasize battery power, whatever the source, instead of fixating on the sun?

He'd been studying a promising new type of battery that used lithium ion, first made popular by
Sony in its camcorders a decade earlier before it

spread to laptops and other consumer electronics. Lithium-ion cells were lighter weight and packed more energy than most of the rechargeable batteries then on the market. Straubel knew the challenges posed by older batteries—those lead-acid, brick-shaped containers were heavy, and they held comparatively little energy. He might get twenty miles of driving range out of a car before needing to find a place to recharge. With the rise of lithium-ion batteries, however, he saw the potential for something more.

And he wasn't alone: Among those who stayed awake with him that night was one of the Stanford team's younger members, Gene Berdichevsky, who shared an interest in batteries. As they chatted, he grew excited about Straubel's idea. For hours they batted ideas back and forth. If they strung thousands of small lithium-ion batteries together to create enough energy to power a car, would they need to harvest the sun's energy at all? They did the math to figure out how many batteries they'd need to power a car on a single charge to go from San Francisco to Washington, D.C. They sketched out a torpedo-shaped vehicle designed for aerodynamics. With half a ton of batteries and a lightweight driver, they figured their electric car might get a range of 2,500 miles. Just imagine the attention that would get—it was precisely the kind of stunt that could spark worldwide interest in electric cars. Animated by their conversations, Straubel suggested the team shift gears from solar power to a long-range electric car. They could raise money from Stanford alumni.

With the sun rising in the backyard, Berdichevsky and Straubel were giddy as they began messing with lithium-ion batteries that Straubel kept around for experiments. They fully charged finger-length cells, then videotaped themselves as Straubel hit them with a hammer. The impact set off a reaction that ignited a fire, sending the battery tubes off like rockets. The future looked bright.

"This needs to be done," Straubel told Berdichevsky. "We've got to do this."

Jeffrey Brian Straubel had spent his childhood summers in Wisconsin rummaging through the dump to find mechanical devices to take apart. His parents indulged his curiosity, allowing their basement to be converted into a home lab. He built an electric golf cart, experimented with batteries, and became enthralled with chemistry. One evening, while in high school, he tried to decompose hydrogen peroxide to make oxygen gas, but he had forgotten that there was some leftover acetone in his flask, which resulted in an explosive mixture. It detonated into a fireball that shook the house and sent shards of glass flying. His clothes caught fire; the smoke detector blared and Straubel's mother rushed to the basement to find her son's face gushing blood, requiring

40 stitches. To this day, though Straubel looks the part of the earnest, baby-faced midwesterner, a scar down his left cheek hints at something a bit more mysterious.

Straubel learned a new respect for the dangers of chemistry, leading him in 1994 to Stanford University, where he kept an interest in how energy worked, realizing a passion for the juncture between lofty science and the real-world application of engineering. He became enamored, specifically, with energy storage and renewable energy generation, power electronics, and microcontrollers. Ironically, he dropped a class on vehicle dynamics—he found the details surrounding a car's suspension and the kinematics of tire movement boring.

Straubel wasn't so much a car guy as a battery guy. His engineering brain saw an inefficiency in the world of gas-powered cars. Petroleum was finite and burning it for energy dumped harmful carbon dioxide into the air. To him, engineering an electric vehicle wasn't about creating a new car, per se, but addressing a crappy solution to an engineering problem. It was like being cold, spying a table in the room, and burning it for warmth. Yes, it created heat, but you were left with a room full of smoke and no table. There had to be a better way.

During his third summer of college, a professor helped land him an internship at a startup car company in Los Angeles called Rosen Motors. The company had been founded in 1993 by legendary aerospace engineer Harold Rosen and his brother Ben Rosen, a venture capitalist and chairman of Compaq Computer Corp. They envisioned a car that was nearly pollution-free and were working to develop a hybrid-electric powertrain. They wanted to marry a gas-powered turbogenerator with a fly-wheel. Their flywheel, a spinning body that generates more and more energy the faster it spins, was designed to create the electricity needed to keep the vehicle going once the engine had started moving it.

It would be Straubel's introduction to the car business. Harold Rosen forged a connection with him and took him under his wing. Soon, Straubel was working on the magnetic bearings for the fly-wheel and helping with test equipment. The summer flew by; it made Straubel realize he needed to return to Stanford for his senior year to learn more about car electronics.

Back at school, he worked remotely for Rosen until he got a call with disappointing news: the company was shutting down. It was an early lesson for Straubel in the challenges of launching a car company from scratch. Rosen Motors had burned through almost $25 million. They had installed their system in a Saturn coupe as a kind of proof of concept. (They'd torn apart a Mercedes-Benz as well.) They promised a car that could do zero to sixty in six seconds, with the hope, ultimately, of partnering with a carmaker to implement their technology.

But even with glowing press, they couldn't see a

way forward. The joke in the auto industry has long been that to make a small fortune in the car business, start with a large one. In the company's obituary, Ben, whose fortune came in part from a highly successful investment in Compaq, was sanguine about their effort: "There are not many chances you have in a major industry to change it and do something that's good for society and clean up the air and reduce the use of petroleum," he said. "It was a chance to change the world."

Back at Stanford, Straubel rented an off-campus house with a half-dozen friends. Inspired by his experience that summer but suspecting that Rosen's flywheel idea would be too challenging to implement, he took over the garage to work on converting a used Porsche 944 into a purely battery-powered vehicle. He had some early success: His jerry-rigged car, powered by lead-acid batteries, was as quick as the devil, producing burnouts and a blazing quarter mile. Straubel didn't concern himself with the handling or suspension. Instead, he focused on the car's electronics and battery management system. That was key, trying to figure out how to juice enough power without blowing a motor or burning up the batteries. He began spending time with other like-minded engineers in Silicon Valley, who introduced him to electric car competitions. Similar to how Henry Ford had demonstrated his abilities at the track every weekend, a hundred years earlier, Straubel and his friends took to drag racing. The trick to these races, he found, was ensuring the batteries didn't get overheated and melt down.

As Straubel continued to tinker with electric cars, he got to know an engineer named Alan Cocconi, who had worked as a contractor on General Motor Corp.'s failed electric car called the EV1. In 1996, Cocconi's shop, about thirty miles from downtown Los Angeles in San Dimas, was working on ways to create excitement around the idea of electric cars. They took advantage of a kit car that was then favored by home-car enthusiasts, with a fiberglass frame for a low-slung, two-seat roadster. But instead of installing a gas engine, they powered the car with lead-acid batteries, which were packed into the doors. The result: a hot rod that could accelerate from zero to sixty miles per hour in 4.1 seconds, as good as a super car. The car could drive about 70 miles on a single charge—nowhere near what the average car could do on a tank of gas, but a promising start. More impressive still, he began beating Ferraris, Lamborghinis, and Corvettes in drag races. He called his bright-yellow car the tzero—a math symbol that marks a starting point (when elapsed time equals zero).

By late 2002, however, Cocconi's shop was going through tough times. Carmaker clients were less interested in converting cars into electric in order to impress regulators, who had themselves shifted their interests away from electric cars toward other zero-emissions technology. And the tzero proved costly and time-consuming to build. Undeterred, Cocconi, who had been tinkering with lithium-ion batteries to build remote-controlled airplanes, began working on converting the tzero's batteries away from lead-acid.

That idea caught Straubel's attention as he hung out at the shop after graduation, splitting his time between LA and Silicon Valley. He proposed to Cocconi the same idea for a cross-country car that he and the Stanford solar team had batted around that long night in the summer of 2003. He figured he'd need about 10,000 batteries strung together, and that it would cost about $100,000 to make the demo car. The team at AC Propulsion liked Straubel's enthusiasm and were eager to do it—if Straubel could find the money for it. In fact, Cocconi wanted to hire Straubel but the business couldn't afford him.

For his part, Straubel wasn't sure he was ready to settle down into a real job. He was also spending time with his old boss, Harold Rosen, then in his seventies, who wanted to pursue another wild idea: a high-altitude, hybrid-propulsion aircraft that could be used to create wireless internet access. Straubel thought lithium-ion batteries might be the solution Rosen needed, too.

As Rosen and Straubel looked for investors for their new aerospace venture, Straubel remembered a guy he'd heard of back in Palo Alto. At the time, Straubel had known of Elon Musk as a seemingly

eccentric member of the flying club at the local airport. After Musk was late in returning an airplane,
upsetting other members who were scheduled to
fly, he sent a giant bouquet of flowers to the front
desk. Lately, Musk had been in the news for his
involvement with a startup called PayPal that had
been acquired by eBay for $1.5 billion, and for
starting a rocket company with his new fortune.
He seemed like a person drawn to big, impossible
ideas. He could be just the investor they needed.

That October, Straubel settled in to a lecture series about entrepreneurship at Stanford University
to hear Musk, then thirty-two, speak. "If you like
space, you'll like this talk," Musk began. Before
getting into how he founded a company to make
rockets, called the Space Exploration Technologies
Corp., or SpaceX, he walked through his own
founding story. It had a Horatio Alger quality to it.
He'd grown up in South Africa, emigrated to Canada
at age seventeen by himself, then to the U.S. to finish his undergraduate studies at the University of
Pennsylvania. Shortly after finishing there, Musk
and his best friend Robin Ren drove across the U.S.
to study at Stanford. Musk wanted to go deep into
energy physics, convinced he could make radical
advances in battery technology, only to abandon his
studies—two days later—ahead of the gold rush era
of the late 1990s dot-com boom.

Straubel listened as Musk, dressed in black with his shirt unbuttoned like he was at a European nightclub, elaborated on his origin story. He said few venture capitalists on Sand Hill Road at the time had shared his vision for the internet. The fastest way to make money, Musk had figured, would be helping existing media companies convert their content for the World Wide Web. He and his younger brother, Kimbal, founded Zip2 to do just that, eventually attracting attention for a first-of-its kind Web program that gave turn-by-turn map directions between two locations—an idea that would later become ubiquitous. It was an attractive feature for newspaper companies, including Knight Ridder, Hearst, and **The New York Times,** that were looking to create city-directory-type websites. The two young men quickly sold the company for cash ("That's a currency I highly recommend," he joked wryly) and the newly rich Musk, with $22 million in the bank, had one goal: to start another company. His next bet, in early 1999, that he could replace the ATM with a safe online payment system—a company eventually known as PayPal— created the true fortune that he would use to fund his grander ambitions.

There was a question that had long nagged at him: Why had the space program stalled? "In the '60s, we went from basically nothing, not being able to put anyone into space to putting people on the moon and developing all the technology from scratch to do that, and yet in the '70s, the '80s and the '90s we've kind of gone sideways and we're currently in a situation where we can't even put a person into low-earth orbit," Musk said. It didn't align with other technologies, such as microchips and cell phones, which had only gotten exponentially better and cheaper with time. Why had space technology languished?

What Musk said struck a chord with Straubel, who had been thinking similar things about the auto industry. Afterward, Straubel rushed forward to talk with Musk, dangling as bait his connection to Rosen, known in aerospace circles for his role in helping develop modern communication satellite technology. Musk invited Straubel and Rosen to come see the SpaceX rocket factory near Los Angeles.

Straubel watched as Rosen walked through SpaceX's office in a former El Segundo warehouse seemingly unimpressed. He kept pointing out flaws in Musk's plans for a rocket that would supposedly cost a fraction of what was then being built. "This is going to fail," Rosen told Musk, to Straubel's dismay. Musk was no less critical of Rosen's idea for an aircraft to create wireless internet: "That idea is stupid." By the time they sat down for lunch, Straubel was convinced the entire visit was a disaster.

To keep the conversation going, Straubel turned

to his own pet project, an electric car that could cross the country. He described to Musk how he was working with a shop called AC Propulsion to use lithium-ion batteries, which could be just the breakthrough he needed. It was a pitch Straubel had given whenever he had the chance, one that most thought was nuts. But not Musk. It clicked, Straubel could tell just by looking at him. His face embraced the idea. His eyes darted up to seemingly process the information. He nodded in agreement. Musk simply got it.

Straubel left feeling like he had met somebody who shared his vision. After their lunch, he followed up with an email, suggesting Musk reach out to AC Propulsion if he was interested in seeing an example of a lithium-ion powered car. Musk didn't hesitate. He wrote back that he wanted to contribute $10,000 to Straubel's long-range demo vehicle, and he promised to give AC Propulsion a call. "This is really cool stuff and I think we are finally nearing the point where electric cars are a viable option," Musk wrote.

Straubel didn't know he'd soon be competing for Musk's attention.

CHAPTER 2

THE GHOST OF EV1

The idea for Tesla Motors began not with Elon Musk or JB Straubel, but with Martin Eberhard—a man walloped by middle age. As the new millennium began, in quick succession, he had sold his fledgling business and was divorcing his wife of fourteen years. She would receive much of the money he'd made, while he got to keep their house in the hills above Silicon Valley, the place his architect brother had helped him create and where, on a good day, he could see the Pacific Ocean. The lengthy drive to his new job at a tech incubator, a place to help startups get off the ground, took Eberhard along a road of hairpin turns through a canopy of redwoods, giving him plenty of time to think about what was next, personally and professionally. At forty-three, Eberhard wasn't sure what field to dive into, but he knew he wanted to start another company, something that mattered. Or else—maybe law school?

While stewing on that subject, Eberhard began dreaming about something more immediate, if clichéd: He wanted to buy a sports car. Something fast. Something cool.

At lunch each day, Eberhard, who with his salt-and-pepper hair and beard looked like the father in the 1980s sitcom **Family Ties,** debated with his longtime friend Marc Tarpenning which car to buy. The two had founded a company together five years earlier, in 1997, called NuvoMedia Inc., with the rather audacious goal of blowing up the book publishing business. Both read a lot, traveled often, and had grown tired of hauling their books around on long flights. The thought occurred to them: Why couldn't books be digital?

What followed was the Rocket eBook, the precursor to the Amazon Kindle and its ilk. In 2000, ahead of the dot-com bubble burst, they sold their business for $187 million to a company more interested in their patents than their digital revolution. Since they had relied so heavily on outside investors, such as Cisco and Barnes & Noble, the two had very little ownership stake left, which meant they didn't become super rich like Musk did when PayPal was acquired. And of what Eberhard did get, much was going to his soon-to-be ex-wife.

As Eberhard looked at all those fast cars, he complained to Tarpenning about fuel efficiency.

The 2001 Porsche 911 with a manual transmission was a helluva car to drive, but it sucked gas at the pump. It got just 15 miles per gallon in city driving; a little better on the highway with 23 mpg. Ferrari and Lamborghini meanwhile might average just 11 mpg. A more mainstream 2001 BMW 3 Series averaged about 20 mpg on combined city and highway driving.

Global warming in 2002 hadn't yet begun to occupy its central place in the cultural conversation, but Eberhard had seen studies supporting it. His mind was predisposed to believe the rational arguments of science. "It's kind of foolish to think that we can continue to spill carbon dioxide into the air and expect nothing to happen," he said. Beyond that, it was hard for him to turn away from the idea that the U.S.'s troubles in the Middle East were a byproduct of its dependence on oil—a sentiment that Tarpenning shared.

Ever the engineer, Eberhard began researching what kind of car, hypothetically speaking, would be the most efficient, electric or gasoline. He prepared elaborate spreadsheets to calculate well-to-wheel efficiency (the total energy used by a car compared to the greenhouse gas emissions created). He became convinced that electric was the way to go. The only problem was that he couldn't find an electric car that suited his needs—especially not anything as sexy as a Porsche.

Eberhard wasn't alone. Increasingly in California,

a vocal minority were clamoring for better electric options. His colleague at the startup incubator, Stephen Casner, was one of them. Casner had leased an EV1, GM's entrant into the still-nascent electric car market, which plunged him into the emerging subculture of electric vehicle enthusiasts. He attended the annual Electric Auto Association's rally, where he saw a Porsche converted to electric by a recent Stanford graduate (none other than JB Straubel). That car had garnered acclaim for posting a speed record at a drag race in Sacramento.

Sitting in an EV1 for a test ride, Eberhard took the car in. It certainly didn't look like a sports car, more like a weird, two-seat spaceship. For the sake of aerodynamic slickness, the EV1 sat low to the ground with a teardrop-shaped body. The rear wheels, partially covered by body panels, looked from the side like half-opened eyes. The design decisions helped give the car 25 percent less wind drag compared to other production vehicles, which in turn meant it was more energy efficient and required fewer batteries.

Weight and efficiency were a constant battle for electric-car makers. The EV1 battery pack weighed half a ton, a considerable amount at a time when the average sedan weighed a little more than 1.5 tons. Situated in the middle of the car, the pack formed a squat wall between the car's two seats, making the tiny vehicle feel even more claustrophobic. Atop the battery pack and surrounding the gear shifter at Eberhard's right hand were dozens of buttons, giving the appearance of a scientific calculator more than a typical sports car.

Still, Eberhard marveled at the acceleration. Stepping on the pedal led to a quick burst. GM claimed the car could go from zero to sixty miles per hour in less than nine seconds. And without a gas engine roaring, the ride was quiet, just the gentle whir of the motor.

It wasn't the car for Eberhard, wasn't the sexy sports car he sought. The EV1 would prove a tease anyway. GM was taking all of the cars back and killing off the line—it was deemed a money loser. But as they talked, Casner mentioned his neighbor, a man named Tom Gage, who worked for one of the original engineers on the EV1 project at a company in LA called AC Propulsion. There, shop owner Al Cocconi had concocted an electric car that they were calling the tzero.

Eberhard had read about it. Soon he was on a flight to Los Angeles to see it.

When Eberhard arrived at AC Propulsion, he learned that Cocconi and Gage had already sold two of the three tzeros that had been made. The selling price had been $80,000 each. Bright yellow, the car had a cartoonish, sloped front end and squat rectangular body. Twenty-eight lead-acid batteries were stacked inside panels where doors would normally have been, requiring Eberhard to climb in and out of the car's tight cockpit **Dukes of Hazzard**-style.

What it lacked in refinement and comfort, Eberhard discovered it made up for in acceleration. No clutch. No gear shifting. Pure adrenaline.

For all that, it shared a problem with the EV1: The tzero still depended upon big, expensive batteries that provided little range. As they talked, Eberhard recalled, he raised the idea of using lithium-ion batteries, which he was familiar with from his eBook. In his telling, the room grew uncomfortably silent when he shared his thoughts—almost as though he'd hit a nerve. Cocconi quickly wrapped up for the day.

When Eberhard returned, Cocconi had something to show him. It turned out the two had alighted on a similar idea. The remote-airplane buff had also noticed the benefits of lithium-ion as a replacement for the nickel metal hydride commonly found in such toys. The batteries were cheaper and had better performance. Using a small grant, Cocconi had set about to test a hunch that he could string a bunch of the laptop battery cells—sixty of them to start—together to create a battery pack that would put out more energy. He began talking about how they needed to convert the tzero to lithium ion.

If Cocconi could replace the tzero's lead-acid batteries with 6,800 of these cheap laptop batteries, in theory they could have a lighter-weight car with more range and greater performance. The only issue was that AC Propulsion was running out of money. Without California regulators pushing for mandated electric vehicles, one of its largest clients, Volkswagen AG, had pulled out of a contract (they had been working together to convert VW vehicles into electrics to help augment the company's California sales and avoid fines associated with vehicle emissions). Things looked bleak now for the shop, which was laying off employees as it tried to find a way to save the business.

Eberhard wanted a tzero and was prepared to pay for it. He agreed to pay $100,000 for the car and to give them an additional $150,000 to keep the lights on as they converted the tzero to lithium ion. Maybe he could help turn the shop into a real car company, Eberhard thought to himself. Solving a more mundane real-life problem, the problem of lugging around heavy books, had led him and Tarpenning to start their previous company. Maybe now he could address this new problem, start a new company, and resolve his midlife crisis in one fell swoop.

The idea of an electric car is as old as the car itself. Creators tinkered with battery-run vehicles going back to the mid-1800s. Henry Ford's wife had an electric car in the early 1900s when the technology was seen as especially appealing to women put off by the hand crank, the noise, and the smell of gasoline-powered horseless carriages. Her husband's success with the Model T, a gas-powered car for everyman, pretty much ended the gas vs. electric debate. Ford Motor Co.'s ability to mass-produce the vehicle at a cost that a growing middle class could afford created an industry of carmakers and a network of filling stations based around fossil fuels. An electric car that might cost on the order of three times more than a typical sedan and have a range of only fifty miles just wasn't an appealing proposition.

Not until the 1990s did it look like an all-electric car might make a comeback. General Motors Corp. surprised the industry when it introduced a concept car at the Los Angeles auto show in 1990, then rushed the idea into production with what would, in 1995, become the EV1.

The inherent problem for the EV1 was one of math. Among the challenges for electric-car makers was that the cost of batteries for any fully electric vehicle added tens of thousands of dollars to its price tag. That was simply never going to cut it for corporate bean counters at the big automakers, people who often dropped features to save mere dollars—if not cents. The idea of adding thousands to a car's cost was generally a nonstarter, especially because the electric car faced so many hurdles for customers. Most notable among them was how they would even go about charging them; electric cars didn't have the benefit of drawing from filling stations dotted across every city and town in the country, developed over nearly a century.

To make the EV1, engineers worked from the formula that fewer batteries would cut cost. But that in turn reduced range. It amounted to an unsolvable riddle.

Even if lithium-ion power could resolve the range problem, Eberhard faced the fact that those batteries, even less expensive ones, would still add cost compared to a comparable gas-guzzling car, which has no analogous expense. As he studied the issue, he concluded that too many car companies, including GM, had failed by emphasizing electric cars for everyday buyers. They had tried to reach scale, in hopes of driving down battery costs. The result was a car full of compromises—not thrilling enough to suit high-end buyers and too ineffective to compete at the lower and middle parts of the market.

From his days in consumer electronics, Eberhard saw that carmakers were taking the wrong approach. The latest tech always sold for a premium first, then came down to suit the everyday buyers. It was gobbled up by early adopters willing to pay more for the cool new thing—before the mainstream followed suit. Why should an electric car be any different?

He found inspiration from sales of the Toyota Prius, a hybrid that used an onboard battery to store power generated either from braking or from the car's gasoline-powered engine, reducing overall fuel consumption. The buyers for what was essentially a low-end Corolla with an expensive, green-power powertrain were skewing largely like those who bought Toyota's Lexus luxury brand. Movie stars, wanting to signal their environmental credentials, were showing up in Hollywood with them. As Eberhard drove the tony streets of Palo Alto, he took pictures of the driveways of the homes he saw, where BMWs and Porsches were lined up alongside Prius sedans. Those were, he wagered, his potential customers. They worried about the environmental impact of driving while still craving performance.

By developing a high-priced electric sports car, Eberhard wouldn't face as much pressure to get his battery costs down. And from personal experience, Eberhard knew that sports car buyers could be a forgiving lot, willing to overlook some things—even dependability—if the performance was there and the brand was considered cool.

As Eberhard became more excited about a potential market, he convinced Tarpenning to join him in starting what he christened Tesla Motors, after Nikola Tesla, who designed the alternating current electric system that powers homes around the world. They registered the company in Delaware on July 1, 2003—nine days ahead of Tesla's date of birth 147 years earlier—and began trying to figure out what they needed to know about being in the auto industry.

Admittedly, they knew little. They saw that as an asset. The industry was undergoing structural changes, as behemoths such as GM tried to adapt to changing customer tastes and legacies of debt, high labor costs, and shrinking market share. For generations, carmakers had pioneered vertically integrated operations, where in-house part suppliers fed the pieces needed to make a car to the assembly end of the business. To slash cost, those parts makers were being spun out to become third parties around the world. In theory, Eberhard bet, tiny Tesla could buy the same parts being used by the big players to make his sports car.

And why **build** the car themselves, for that matter? When the two were making eBook readers for NuvoMedia, they didn't actually assemble the gadgets. Rather, like most consumer electronics companies, they outsourced the work to third parties. Few car companies offered such services, they discovered, but one that did caught their eye: Lotus, the UK sports car maker.

Lotus had recently released a new version of its Elise roadster. What if Tesla could simply buy a few Elise cars, have Lotus tweak their designs enough to make them look singular, and swap out their combustion engines for the electric motors made by Cocconi? They'd have a high-end electric sports car on their hands.

Others had thought about engineering an electric car, but they had failed to make it profitable. Eberhard was thinking of ways to change the business of making cars altogether, bringing to a hundred-year-old industry the lessons learned from a career as an entrepreneur in Silicon Valley. Tesla would be asset-light, focused on brand and customer experience. He figured the timing was right.

By September 2003, AC Propulsion had completed the conversion car that Eberhard had commissioned and the results made him think they were on to something. The new tzero, with its lithium-ion batteries, had shed a remarkable 500 pounds from its weight. The 0-to-60 acceleration speed improved to a mind-melting 3.6 seconds, ranking it among the world's quickest cars. The conversion also improved the vehicle's range, allowing it to go 300 miles on a charge, a substantial improvement from the 80 miles that Cocconi and Gage had been mustering from their older tzeros.

Eberhard had his neighbor Ian Wright, an amateur race car driver, give it a test drive. It was unlike anything Wright had driven before. To attempt a super-quick launch of a gas-powered car requires an advanced vehicle and driver. A driver has to rev up to speed and engage the clutch at just the right moment. Engage the clutch too soon and you'll stall, since there's not enough torque to advance the car. Too late and the tires smoke. It all has to be timed just right, so the tires take off precisely when the engine has enough power to turn them, and no more. Automatic transmissions can help a driver, but those are dependent on gas-powered systems.

Not so with the tzero. Eberhard and Wright found that the electric car was already a full vehicle length ahead of where its competition would be by the time any gas-powered car could go through those steps. The acceleration didn't fade as the car sped up, either. "It felt like a racecar in first gear but a first gear that just kept going and going, all the way to 100 mph," Wright later wrote. He soon ditched his own startup idea and joined Eberhard and Tarpenning instead.

While all of this bolstered Eberhard's idea for a sports car, the tzero suggested a different possibility to Gage and Cocconi. They wanted to use the new battery technology to make an everyday EV. The lithium-ion batteries could give cars a range that took the technology from a plaything to a daily commuter.

They had made great strides together but it was clear: the two sides needed to go their own ways. Gage and Eberhard came to an understanding. If they couldn't be in business together, they could at least **do** business together. Eberhard would buy Gage's motors and electronics for the roadster he planned to develop. But it wouldn't be an exclusive deal: AC Propulsion would be free to do its own thing.

The challenge for both, however, was that neither team had a bank account large enough to create a car, no matter how good an idea it seemed. They would need to raise millions of dollars to get off the ground. A gentleman's agreement was reached that they wouldn't pursue the same investors.

Very quickly, one of those potential investors for AC Propulsion became Elon Musk—thanks, in part, to a recommendation from JB Straubel. Musk's name had also come up when Gage showed the tzero to Sergey Brin and Larry Page, co-founders of Google, the Web search engine that had launched just a few years prior. During a demonstration of the tzero, Gage had impressed them with the car, but they had demurred from investing; their company was still a few months away from going public. Brin suggested Musk. "Elon has money," he said.

"What do you know about building a car?" Judy Estrin asked Eberhard. The co-founder of Packet Design, the incubator where Eberhard had worked before quitting to work on his Tesla idea full-time, had listened intently to his pitch. She'd always been impressed with Eberhard's intelligence and risk-taking as an entrepreneur. She didn't doubt his theory that the time was right for an electric car. But creating a car company was hard to pull off, to put it mildly.

Eberhard was receiving similar pushback as he and his partners looked for money. Laurie Yoler, a friend and colleague at Packet Design who was well connected in the investor world, was setting them up with meetings around Silicon Valley. A startup typically began with some seed money—maybe from the founder's own pocket or scraped together from friends and family, to show that the company had strong early support—before raising increasingly large rounds of funding. At each round, the founder gave up some ownership stake of the company as others bought in, and current investors either had to up their investment or else face dilution. Venture capitalists (VCs), meanwhile, ran funds that raised millions of dollars with the goal of investing in such startups, then cashing out—either through acquisition of the startup by a bigger company or else through an initial public offering at some point during the fund's lifetime (typically eight to twelve years).

In those early days, a software-based startup, such as Facebook, was judged on its expanding user base rather than its profit. Mark Zuckerberg's social network didn't turn a profit until five years after he founded it in his Harvard University dorm room, still years before its IPO. Amazon went nearly a decade before turning an annual profit; meanwhile it spent its cash to grow its user base and build unparalleled logistics and digital infrastructure systems. And those are the successes. They bled and bled cash until at some point they turned a corner.

A car company just wasn't something most investors were conditioned to consider. First, there hadn't been a successful new car company in generations. And simple math led investors to quickly recognize that it would be hugely capital-intensive, meaning their chance of a large return was less likely. Even to become a supplier to a car company, as the team at AC Propulsion aspired to do, was fraught; it might take a decade or more before its widget turned up in a production vehicle, long past the time horizon of a normal VC's investment. But on a Tuesday afternoon in late March 2004, Eberhard opened his email to find a short message from Gage. Musk had been copied in with a note suggesting the two connect. "He would be interested in hearing about your activities at Tesla Motors," Gage wrote.

Eberhard looked at the email and immediately remembered listening to Musk speak years before at a lecture on the Stanford campus, about his vision for space travel and visiting Mars. Clearly, he wasn't allergic to thinking big.

What wasn't said in the email: AC Propulsion had failed to convince Musk that they had a sound business. Unknowingly echoing Eberhard's own sentiments, Musk told Gage that he thought the best way to make electric cars cool was to begin at the high-end and work down to the mainstream. What about converting the sports car made by UK carmaker Noble? He was thinking about importing one without an engine. That car was sexy for sure. But Gage and Cocconi weren't interested in going in that direction.

Given how closely their visions were aligned, a meeting between the small Tesla team and Musk was quickly arranged. Tarpenning had family plans, so that left just Eberhard and Wright to fly to LA with a business plan that they had been crafting for months. They had run the numbers; they figured it would cost them $49,000 to build each sports car, which they had taken to calling Roadster. The biggest part of that number—almost 40 percent— would come from the cost of the batteries, though they hoped they could get discounts as production ramped up. Surely battery suppliers would be thrilled at the prospect of a company whose products depended not on a single one of their cells, like a digital camera or cell phone, but thousands. About $23,000 would be needed to fabricate the car—buying the chassis from a third party such as Lotus, along with the components that make up your typical vehicle. They would then turn around and sell it to a small number of approved car dealers for $64,000, who would turn around and sell it to customers for $79,999. That math meant Tesla would make a gross margin of $15,000 per vehicle, allowing it to break even if it sold 300 cars per year, or just about 1 percent of the high-end U.S. sports car market.

They projected they'd need to raise a total of $25 million to fund the creation of the company through its breakeven, which they imagined they'd reach in 2006, at which point it would start turning a profit. They planned to first raise $7 million, money they'd use to begin hiring engineers and to pay for the creation of a handmade prototype. They then planned to raise an additional $8 million by that year's end, to pay for the vehicle's final development and the creation of additional prototypes. Nine months later they would raise $5 million more to pay for the factory tooling and to begin building inventory, capped by a final fundraising of $5 million in March 2006—two years after their first investment—to fund the production launch.

Their plan was straightforward enough: Make an awesome sports car that could beat the best in the world. Sell it for $79,999. Deliver 565 of them in four years. Make a profit. Change the world. What could be easier? The math seemed so simple. They didn't realize at the time that they were using the wrong numbers.

Musk was enthusiastic but skeptical. "Convince me you know what you're talking about," he told them as they sat in a glass conference room near his cubicle at the SpaceX offices in El Segundo, decorated with airplane models. He quizzed them on how much funding would be required. "Why won't it be twice that or five times that or ten times?"

The two men didn't have a good answer beyond being able to say they'd been in contact with Lotus, after hounding executives during the LA auto show, and that they had read that the Elise had been developed for less than $25 million. Musk looked unconvinced. The biggest risk, he told them—presciently, it turned out—was that they might need way more money than they anticipated. Still, Musk was intrigued.

They discussed their plans for after the Roadster. Eberhard called for either making a super version of the car, with a two-speed transmission and nicer interior, a car that could compete against six-figure competitors. Or going down-market a little and offering a four-seat coupe to target the buyers of the Audi A6. Or else maybe an SUV? Once the company reached volume production, which would get them to less expensive parts and cut manufacturing costs, they envisioned lower-priced "follow-on" cars. What Eberhard's Tesla represented was a road map to ushering in electric cars. Like an old treasure map, some parts of it were fuzzy and built on assumptions—rightly or wrongly. At its core, though, the path forward looked clear. The Roadster could be the start of something.

After two and a half hours, Musk said he wanted to do it, though he still wanted to talk to the absent Tarpenning. He had some conditions, too. He'd become chairman, he stipulated. Also, they needed to close the deal in ten days because his wife, Justine, was scheduled to have a C-section for the birth of their twin boys.

Eberhard and Wright left the meeting elated. "We just got the company funded!" Eberhard told his new business partner.

While Musk did his final due diligence, so did the Tesla team. Yoler, their VC connector, asked around about Musk. She reported back to Eberhard that he had a reputation for being difficult and stubborn. What wasn't widely known at the time was how Musk's time as CEO at PayPal had come to an end. Musk had been kicked out of his position by a board unhappy with his management style—while he was on his honeymoon. It was just ahead of the company being sold to eBay, a move that cemented him as wildly rich but also left him feeling like he'd lost control of his creation. It was a formative moment for Musk. He would never forget the feeling—or the financial knife fight—of losing a company.

Yoler was worried. Eberhard could be difficult and stubborn as well—a common trait among entrepreneurs who've got to trust their guts against a crowd of people telling them their ideas are too risky, unwise, unproven. "You gotta talk to him," Eberhard told her. She got on the phone with Musk and was quickly reassured. "He really impressed me," she said. "He said, 'I'm gonna be a very wealthy board member and investor, that's all I'm looking for.'"

And so they closed the deal swiftly. Musk put in $6.35 million of the $6.5 million initial investment. Eberhard contributed $75,000, with the rest coming from other small investors. He became CEO; Tarpenning, president; and Wright, chief operating officer. Yoler was put on the board, as was a longtime friend and mentor of Eberhard's, Bernie

Tse. On the night their new chairman's check was deposited, they all gathered, except for Musk, in a tiny Menlo Park office that Eberhard had rented. They passed around a bottle of champagne and toasted the start of their company. It was an auspicious beginning for a promising business, one that was truly their own.

CHAPTER 3

PLAYING WITH FIRE

Outside the three-bedroom home JB Straubel rented in Menlo Park sat a pile of used electric motors packed individually in large wooden crates—dozens of them, creating a stark contrast to the white fence and manicured lawn of the neighbor's house across the street. For years, Straubel had collected discarded motors from General Motors' failed EV1 cars.

First, his collection was driven by curiosity. When GM eventually canceled the EV1 program in 2003, the Detroit automaker recalled all of its cars, then shipped them off to be crushed at junkyards, to the collective horror and protests of customers. Prior to that, Straubel wondered if any spare EV1 parts survived, perhaps at the Saturn dealerships tasked by GM with servicing them. Those parts might be useful in his effort to retrofit gas cars with electric capacity, as he had done with his Porsche. His hunch proved correct; he found a scrapped motor sitting behind a local Saturn service center. Gobsmacked by his good fortune, Straubel steeled himself for a tough negotiation with the car dealer sitting on such a rare treasure. Instead, the dealer looked at him with disbelief: This is garbage. Take it!

Soon Straubel had mapped out all of the Saturn dealerships in the western U.S. that had serviced the EV1, and he began calling them all. By the time he was done, he had collected almost a hundred motors. He tore some apart to learn their secrets, sold others to fellow EV enthusiasts, and made plans to use the rest for his electric car projects—such as the one he had just sold Elon Musk on, during their otherwise unproductive lunch with Harold Rosen.

Straubel had moved back to Silicon Valley in early 2003, with Musk's $10,000 investment in hand, to begin work on a prototype electric car capable of traveling across the country on a single charge. He picked his rental house based on its proximity to Stanford's solar car team garage, where Straubel hoped to enlist eager students. As he settled in, he lined up his motor collection, set up his garage workshop, and began experimenting with how to wire together lithium-ion cells to make the battery pack that would produce energy for his car. He didn't yet know that Musk had recently funded an electric car company, to the tune of millions of dollars, based on a very similar technology to the one he was perfecting.

So he was surprised when he got a call in 2004 from Martin Eberhard, who claimed to have just founded an electric car company that he called Tesla Motors, and who wondered if Straubel was interested in a job. Straubel was incredulous. He knew everyone in the EV community, and yet Eberhard's name had never come up. Even more surprising, Tesla was located in an office building just a mile from his house. Stunned, Straubel biked over to learn more.

There he found Eberhard, Marc Tarpenning, and Ian Wright, who were looking for employees to help make **their** dream of an electric car into reality. They needed help—lots of it. They fancied themselves car guys—Eberhard had restored parts of a 1966 Ford Mustang in his dorm room at the University of Illinois years earlier, and Wright spent his free time racing cars—but their experience with car building was limited.

Tesla's challenge in those early days was to create a so-called mule, a prototype vehicle of what they thought they could later manufacture: specifically a marriage between the Lotus car frame and AC Propulsion's electric motor, which would be powered by stringing together thousands of laptop batteries. AC Propulsion's tzero had shown them that lithium-ion worked. Now they had maybe a year's worth of money to pull a mule together. If they were successful, they'd have something to show investors with the hope of raising the next round of funding, and to begin development of an actual car for Lotus to manufacture. If all went as planned, it would go into production in 2006, less than two years away.

Straubel, then twenty-eight, had come recommended to Tesla by the guys at AC Propulsion. As he and Eberhard talked in their first meeting, Straubel realized the two of them had been pursuing the same idea in parallel without knowing it. They knew many of the same people; both had crossed paths with Alan Cocconi and Tom Gage at AC Propulsion. Both had struck up arrangements with Elon Musk.

Straubel left the meeting with his head spinning. All of this was going on in his backyard, involving people he knew, and yet he had heard nothing of it? He called Musk. "This doesn't sound real," Straubel said. He peppered Musk with questions: Is this legit? Is this really what you're doing? Are you really committed to funding this? Are you going to be in it for the long haul?

Musk assured him it was true. "I'm super excited about this," he said. "We need to do it. . . . You either need to do that or you need to join SpaceX."

Straubel chose Tesla; he was hired as an engineer.

Other than Straubel, Eberhard had been staffing Tesla mainly through friends and former colleagues from NuvoMedia, which bolstered the car company's consumer electronics mindset. Tarpenning, technically the company's president, had been charged with heading up software development and serving as CFO until they could find someone suitable for the role. Wright was overseeing engineering. Eberhard's neighbor Rob Ferber would tackle battery development.

With the beginnings of a team in place, Straubel's first order of business was to hand-deliver a check from Tesla to AC Propulsion, as part of the deal to license its technology. In San Dimas, Straubel checked into a $40-a-night motel near the shop, where he would live while he reverse-engineered AC Propulsion's electric motor and other systems. In many ways, it was a dream come true for Straubel. He was working alongside his friends at AC Propulsion and getting paid to do it. Eberhard and Wright had already gone to the UK to finalize their agreement with Lotus, and the carmaker shipped its first Elise to San Dimas. Straubel and the team at AC Propulsion dug in. They began by ripping out the gas engine to make room for an electric motor and batteries. Straubel quickly ran into one of Tesla's first hurdles.

While the EV1 motors stacked up back at Straubel's home in Menlo Park illustrated the precision and uniformity that came from a big carmaker, AC Propulsion's motors were something else. Each was a jewel, he thought. Beautifully crafted—and unique. That was a problem. Eberhard's plan called for selling hundreds of Roadsters a year. Straubel couldn't return to his team with jewels; he needed cogs for a machine.

They'd cross that bridge when it came. For now, before they could even consider mass production, Tesla needed to make a mule, and with limited time and money. Uniform or not, Straubel needed to get a powertrain working. In theory, the mule should be easy; AC Propulsion had already built a battery pack for the tzero, after all. But AC Propulsion wasn't in the same rush as Tesla, being seemingly more interested in converting a Toyota Scion into an electric car, or else focused on other projects that had captured its interest. Perhaps it shouldn't have come as a surprise that AC Propulsion wasn't prioritizing the Tesla Roadster; it was in some ways their competition.

Eberhard wasn't having better luck. He and Wright had returned from the UK discouraged after receiving a painful wakeup call from Lotus. A key part of Eberhard's plan had been to simply use the Elise with essentially cosmetic changes, to distinguish it as a Roadster. Instead, they were blown away to learn how costly and time-consuming it would be to change the design of, for example, a door. **Everything** was going to cost them. And Lotus wanted way more money for the job than what Eberhard's business plan had called for. They left the meeting with the realization that they needed to raise money sooner than they'd planned.

If not, Wright concluded with understatement: "This is not going to work."

Tension was rising among the small team. By fall of 2004, a decision was made to ship the Elise body north from LA to Silicon Valley, where Eberhard had found a new office (a former plumbing warehouse) with room for a shop. At Straubel's recommendation, they hired Dave Lyons, a former employee at IDEO, the renowned design firm, to add some engineering muscle to the team.

The battery pack was turning out to be a larger project than expected, so Straubel, who had already been working on his own battery pack for a demo car, simply repurposed it for the mule. Already deeply involved in the electronics and motor, Straubel began taking on more of the battery work. He sought help from his friends at Stanford, who in turn looked to him for leadership. His home garage became the Tesla Annex: His workshop was outfitted with the tools they needed, his living room converted into an office. Gene Berdichevsky, the member of Stanford's solar team who had shared Straubel's early enthusiasm for an EV, dropped out of school (with a promise to his parents of returning to finish) and joined Tesla; he took daily cereal breaks in Straubel's kitchen while Straubel handglued battery cells near the backyard pool.

Early signs of interpersonal trouble were emerging. Some on the team felt Wright was threatened by Straubel's growing Stanford fraternity and

obvious clout. The two disagreed on technical issues, such as whether to cool the batteries with air
or liquid coolant. Wright went so far as to ask the
young engineer if he was after his job. Straubel was
perplexed by such a question. For him, working at
Tesla was simply a dream come true. He had no
time for politics.

Wright wasn't clashing just with Straubel. As they
settled into the engineering work, Wright's experience as a race car driver loomed large. Some members of the team began to feel as if he expected them
to be as proficient as a pit crew. When he ordered
them to calculate the car's center of gravity, for example, they struggled with how to respond—they
could only turn to the internet for guidance. The
engineering team was months old and already starting to see departures.

During coffee breaks, Straubel and Tarpenning
would wonder aloud about how long the big personalities might last together. Eberhard was clearly
having friction with Wright. An engineer at heart,
Eberhard enjoyed wading into the details, whereas
Wright felt Eberhard should stick to his CEO duties. He questioned Eberhard's vision, doubting
there'd be a market for anything other than a fully
electric super car.

Late that year, Straubel got a call from Musk.
He wanted to know how things were going, specifically with Eberhard and Wright. Naively, Straubel
told him that Wright wasn't doing so well, he was

rubbing the engineers the wrong way. Eberhard seemed fine, Straubel told him. He enjoyed working with him.

The next day Straubel realized what Musk's call was really about when he learned the news: Wright was out. He'd later learn that Wright had secretly flown to LA to speak with Musk about firing Eberhard and replacing him as CEO. The company was barely a year old and already undergoing internal drama—the kind that would be the hallmark of its corporate culture for years to come. The ousting demonstrated, to Straubel and others, that Musk wasn't afraid to reach into the engineering ranks to gather intel on what was really going on with his investment.

But on that day, with Wright gone, the team largely felt relieved. They celebrated the holiday season at Eberhard's home. He'd worked with a friend from IDEO to come up with themes for what the production car might ultimately look like. He presented them at the party and the team voted on their favorites. They eschewed designs that exuded a high-tech feel, like that of the Prius, settling instead on a car that looked more like a Mazda Miata, only with more angular front headlights.

There would be time for the production car, though. First things first: They needed to knock out that mule.

To make the battery pack, Eberhard secured roughly 7,000 batteries from LG Chem, a unit of the Korean tech giant. A pallet arrived with each cell individually wrapped in plastic. Colette Bridgman, Tesla's office assistant, ordered a stack of pizzas for lunch and the entire office gathered with X-Acto knives to extract the batteries from their casings, paying close attention not to puncture the cells, which could cause them to ignite.

By that spring, their collective efforts had paid off. The mule was nearly complete. They had successfully swapped out the Elise's engine for an electric motor and battery pack.

To Straubel, the day of its test drive at first represented something of an anticlimax. The car still looked like an Elise, one they had all stared at for months now. And Straubel had driven many electric vehicles before. But with Tesla's prospects hinging on its performance, there was a palpable weightiness to the moment, as the company gathered outside the former plumbing warehouse turned Tesla workshop for the car's first spin.

Straubel sunk into the low-slung roadster, windows down so he could talk to the engineers. When it came time to start, he touched the accelerator. The car tore forward down the warehouse-lined street. Its quickness stunned the assembled group, but so did its silence. Eberhard teared up as he looked at the car. "It was the first evidence that Tesla could happen," he said. When it came time for his turn to drive, he didn't want to let go of the steering wheel. The team was months behind schedule, and running low on cash. But **this**—this was a milestone, one that would enable them to unlock money from investors. More than elation, a sense of relief surged through the team.

Musk was delighted at the development. When Eberhard had returned from the UK months earlier with the discouraging news that they would need more money than planned, Musk was unhappy—but he also wasn't surprised. He had told them from the beginning that he thought their $25 million estimate was too little for the creation of a new car. (He had mentally given the team a 10 percent chance of success.) But stoked by the team's progress with the mule, he agreed to contribute much of the $13 million they sought in their second round of funding. That funding round also brought some new faces to the company, including Antonio Gracias, whose Chicago-based investing firm became the largest investor besides Musk.

Straubel's efforts were recognized as well; he was promoted to chief technology officer.

Their next challenge would be to develop the actual Roadster, the version that would begin production at Lotus's factory. But just as the team celebrated its windfall, a problem arose that threatened to kill the company before it went any further.

A panicked letter from LG Chem arrived at Tesla with a dire demand: Return its batteries.

Just as Tesla was proving it could craft a lithium-ion battery pack on its own, the battery industry was grappling with the danger that cells posed when they were handled incorrectly. AC Propulsion had learned this the hard way months earlier, in one of an increasing number of incidents that sent shudders through the battery industry. En route from Los Angeles to Paris, a shipment of AC Propulsion's batteries caught fire as it was being loaded onto a FedEx airplane while it refueled in Memphis, triggering an investigation by the National Transportation Safety Board and prompting concerns about how to transport batteries in the future. Personal electronics companies, such as Apple Inc., were recalling devices with lithium-ion batteries out of concern they could overheat and catch fire. In 2004 and 2005, Apple recalled more than 150,000 laptops—with batteries made by LG Chem.

When LG Chem realized it had sold a large number of its batteries to a Silicon Valley startup that planned to use all of them for a single device—a car, as it happened—its legal department sent a letter demanding the cells be returned. The battery maker didn't want to be associated with a potentially fiery experiment.

Eberhard ignored the request. He had little choice. His bet that Tesla would be able to find a ready battery supplier was proving harder to cover than expected. Without these batteries, there might not be a second chance to get more.

Amid all of the attention on lithium-ion batteries, Straubel thought back to his former house in LA, where he and Berdichevsky celebrated the idea of an electric car by setting cells afire. If struck with a hammer, they put on quite a show. Cars were always in danger of that kind of impact, but there was also a more insidious threat. He began to wonder what might happen if one of the cells in the tightly packed cluster that would form a car's battery pack got too warm.

One day in the summer of 2005, he and Berdichevsky decided to find out. With the office cleared out for the day, they went to the parking lot with a brick of cells—a cluster of batteries glued together. They wrapped one of the cells with a wire that would allow them to heat it remotely. Then, from a safe distance, they flicked the heater on. The individual cell quickly rose to more than 266 degrees Fahrenheit (130 degrees Celsius), causing the battery to flash into a blinding flame as the temperature spiked to 1,472 degrees, then explode altogether, sending the remaining skin of the battery into the sky like a rocket. Then another cell in the pack caught fire, launching into the air. Soon all of the cells were on fire. **Bang. Bang. Bang.**

Straubel recognized the implications of his amateur pyrotechnics. If an incident like the one he cooked up were to happen in the wild, it could spell the end for Tesla. The next day, after they disclosed their experiment to Eberhard, they showed him the scorched pavement, pitted with holes from the night before. Eberhard urged them to be more careful, but he couldn't deny that more testing was needed. He gathered the team at his rural house on the hills above Silicon Valley for more experiments. This time, they dug a pit and put a brick of cells in it, then covered it with plexiglass. They heated one of the cells and again the batteries ignited, causing a chain of explosions. Straubel had been right: this wasn't good. They needed outside help to understand exactly what they were dealing with—the team needed battery experts.

Days later, a small group of battery consultants were gathered with what at first seemed like a manageable message: Yes, even the best battery manufacturers produced a random cell that would have a defect, causing it to short and catch fire. But the odds were remote. "It happens really, really infrequently," one of the consultants said. "I mean like between one in a million and one in ten million cells."

But Tesla planned to put about 7,000 cells in a single car. Sitting near Straubel, Berdichevsky pulled out his calculator and computed the likelihood that a cell in one of their cars might catch fire by chance. "Guys, that's like between one in 150 and one in 1,500 cars," he said.

And not only would they be churning out cars with defective batteries that, if ignited, could set off a chain reaction, but their cars could be detonating in the garages of the richest of the rich—burning down mansions and lighting up local TV news. The mood in the room shifted. The questions became more urgent: Was there anything that could be done to avoid defective cells?

Nope. Random cells were always going to get too hot and spark thermal runaway—basically an explosion sparked by overheating.

Straubel and the team returned to their work deflated. The stakes couldn't have been higher for Tesla. This wasn't just about solving a hard problem, one that threatened to drain limited resources and derail development of the Roadster. If they forged a solution that seemed to work, only to see Tesla vehicles catch fire in years to come, the company would be doomed. And it would be a failure not just for Tesla; their dreams of the electric car could be set back a generation. They could not only cause injury or death, they might kill the electric car in the process.

If they wanted to truly become an automaker, they had to face the challenge that GM, Ford, and others had been dealing with for a hundred years: They had to ensure they were putting safe cars on the roadway. A solution to thermal runaway could amount to a true breakthrough, one that would set Tesla apart from the auto industry for years to come. Using lithium-ion batteries had seemed like a smart idea, one that a number of thinkers had alighted on. But figuring out how to use them without turning the car into a ticking time bomb could be their greatest innovation.

They stopped work on all aspects of the Roadster project and formed a special committee to find a solution. The team set up whiteboards, listing what they knew and what they needed to learn. They began running daily tests. They'd configure a battery pack with the cells spaced differently, to see if there was an ideal distance for containing chain reactions. They tried different methods of keeping the batteries cool, such as having air flow over them or tubes of liquid brush past them. They'd take the packs to a pad used by local firefighters for training and ignite one of the cells to better understand what was taking place.

The danger of the situation was driven home while en route to one of those tests. Lyons, their recruit from IDEO, began to smell smoke coming from the back of his Audi A4, where he had loaded a pack of test batteries. It was a sign that a cell was heating up and approaching thermal runaway. He immediately stopped and yanked the batteries out of the car and threw them to the ground before his car could catch fire—a close call.

Eventually, Straubel began to narrow in on a solution. If they couldn't keep a cell from warming, maybe they could keep it from reaching the point where it set off a chain reaction. Through trial and error, the team realized that if they had each cell lined up a few millimeters from its neighbor, snaked a tube of liquid between them, and dumped a brownie-batter-like mixture of minerals into the resulting battery pack, they could create a system that contained overheating. If a defective cell within began to overheat, its energy would dissipate to its neighboring cells, with no individual cell ever reaching combustibility.

Where just months earlier they had been struggling to set up a workshop, now they were on to something utterly new. Straubel was thrilled. Now he just needed to figure out how to convince the battery suppliers to trust them. Straubel was hearing from Eberhard that the established manufacturers weren't interested in their business. As one executive at a supplier told Eberhard: You guys are a shallow pocket. We're a deep pocket. If your car blows up, we'll probably get sued.

That supplier was right, and he wasn't alone. Eberhard's original business plan had assumed that suppliers would be willing to supply them. In fact, there was little interest. Even with thousands of batteries in a single car, Tesla's volume was just too small compared to those of other battery buyers. The likelihood of them being around to pay their bills was too low, the probability of failure too high.

The one thing Tesla had, however, was its Roadster. If they could attract attention, maybe they could convince suppliers that they were the real thing. They needed to show the world that electric cars weren't a fantasy.

CHAPTER 4

A NOT-SO-SECRET PLAN

As a student at Queen's University in Ontario, Canada, Elon Musk spotted a young woman named Justine Wilson as she walked to her dorm. Eager for a date, he approached her with a made-up story about having met at a party. He invited her to get ice cream. She accepted the invitation but blew him off. Several hours later, he found her in the student center hunched over a book studying Spanish. He coughed politely. She looked up to see him holding two melting ice cream cones.

"He's not a man who takes no for an answer," she later wrote.

Ultimately, he would transfer to the University of Pennsylvania to finish his undergraduate education. But they kept in touch and the two eventually became a couple. She followed him to Silicon

Valley, where after dropping out of Stanford after two days, he quickly found success. With the money he'd made from selling Zip2, he bought an 1,800-square-foot condo and a $1 million McLaren F1 sports car. It was a rare indulgence for a young man who had until recently been sleeping on his office floor and showering at the YMCA. The day the car arrived in 1999, Musk was giddy. A camera crew from CNN captured the occasion for a story about the enormous wealth being created in the Valley. In their footage, a balding Musk talks awkwardly with a reporter. He discussed his dreams, including appearing one day on the cover of **Rolling Stone** magazine.

He rolled most of his new wealth into a company called X.com, which by 2000 was preparing to merge with a competitor called Confinity Inc. to form what would eventually be PayPal. Consumed with the deal that January, he arrived in St. Maarten the day before he was set to wed Wilson. Among the many last-minute details to address, the two spent several hours wandering the island failing to find a notary to witness the signing of a prenup. As they danced during their reception, Musk whispered to his new bride: "I'm the alpha in this relationship."

With the newly merged companies to manage, Musk delayed their honeymoon until September of that year, only to arrive in Australia to news that the PayPal board was ousting him from his role as CEO. Immediately, he returned to California.

The couple tried another go at a honeymoon a few months later. This time Musk contracted malaria in South Africa—a bout that nearly killed him. It left him in a state of mind that had him re-evaluating the meaning of his life.

With his new bride, Musk decamped to Los Angeles—to escape Silicon Valley and for a fresh start. There, the idea of SpaceX would take root in his conviction that he could create reusable rockets and cut the cost of space travel to a fraction of what the industry was spending. Musk threw himself into the world of aeronautics, developing a reputation as a fantastically wealthy eccentric willing to put his money on unusual bets, the kinds that the venture capitalists of Silicon Valley's famed Sand Hill Road shied away from.

Beyond SpaceX and Tesla, Musk would later encourage his cousins, Lyndon and Pete Rive, to create a company to sell solar panels, an idea that melded well with his idea of where he thought Tesla might go. He imagined a world where a customer's solar panels would charge their electric Tesla cars—a combination that would create a truly zero-emissions system.

His ambitions for Tesla only grew as Martin Eberhard, JB Straubel, and the team showed success in early 2006. With the mule completed, they had turned to developing the next milestone in making their Roadster, their engineering prototype. The EP1, as it was called internally, would be the car they aimed to eventually take to production. This would be the rough draft that was completed before the final draft, the production car. Importantly, the EP1 would give them a prototype that they could use to sell not just early investors and suppliers, but customers.

Musk was already talking about what would come after the Roadster. Eberhard's original business plan, two years earlier, had briefly touched on the company's future. However that plan had been long on hope and short on facts. Many of the plan's assumptions had quickly proven wrong, most notably its cost estimates. It had been based on the idea that Tesla would need to raise only $25 million to bring out a Roadster by 2006, after which profit was expected to roll in. By early 2006, the company had raised $20 million and that was far from what was needed. A new start of production was targeted for 2007. They were overbudget and behind schedule.

Board member Laurie Yoler was among those encouraging Musk to seek money outside his own pockets and his close network of wealthy friends. It was one thing for an eccentric rich guy to fund a vanity project. If some of the big names on Sand Hill Road bought in, it would give the company not just money but legitimacy—something it sorely needed, particularly when it came to recruiting employees and courting suppliers.

Investors would want to know what was next,

after the Roadster. Tesla needed to be more than just a one-off carmaker if it were to attract the kind of cash it sought. The original Tesla business plan anticipated revenue of $27 million in 2007. A car company with such little projected revenue wasn't going to be appealing to investors who had pumped tens of millions of dollars into it; it was a rate of return that didn't justify the risk. Tesla needed a re-formulated business plan, one that predicted much bigger growth—on the order of $1 billion of sales a year.

So Musk focused on a strategy that he'd later make public as "The Secret Tesla Motors Master Plan." The plan was almost laughably simple:

Step 1. Build an expensive sports car, starting at around $89,000, that could attract attention.

Step 2. Build a luxury sedan that could compete against the German luxury cars and sell for about half of the original sports car, at $45,000.

Step 3. Build a third-generation car that would be much more affordable and appeal to the masses.

The company's updated business plan projected Roadster revenue generating $141 million in 2008, with the company's total revenue eventually growing to almost $1 billion in 2011 through the introduction of the sedan. With back-of-the-napkin simplicity, it set a blueprint for the next decade of the carmaker's life.

More than just changing the kind of cars the auto industry sold, Musk was thinking about **how** vehicles were sold. He thought the car-buying experience was ripe for change, and he wanted Tesla to be able to take control of the experience. Eberhard's research backed up those concerns. Months earlier, he had sat across from Bill Smythe, who had a lifetime of automobile experience as the owner of a successful Mercedes dealership. As Tesla began exploring the complexities of the dealership model, Smythe's name had come up as somebody to seek out for advice. Eberhard had sought out the longtime Silicon Valley car dealer to learn more about the retail side of the auto industry.

Eberhard's original business plan depended upon using select franchise dealerships in wealthy enclaves to sell their Roadster—places like Silicon Valley, Beverly Hills, probably New York City, maybe even Miami. They wanted to use exotic car dealerships that already had experience with ultra-expensive brands such as Bentley and Lotus. Those dealers had skilled mechanics on staff, accustomed to working on high-end vehicles. They expected dealerships to mark the Roadster up to $79,999, $15,000 above what Tesla charged the stores, giving the dealer more than their normal gross margin, and thus plenty of incentive to do business with the upstart company.

Automakers in the U.S. had long sold new cars through a network of third-party-owned stores. These stores operated under franchise agreements— contracts that spelled out in painful detail how each side would go about conducting its business. It was a system handed down from generation to generation, from the days of Henry Ford and the advent of mass manufacturing, and that largely benefited the manufacturer, who booked its sale of the car when it shipped to the dealer. The financial burden of selling that car to consumers was solely on the dealership.

It was a system born out of the notion that a factory was most profitable when it was churning out as many cars as possible, allowing it to gain benefits of scale. But Ford Motor Co. didn't have the funds or the organization to open stores in every city in America. So Ford grew his empire not just on the wheels of the Model T, an affordable sedan, but on the backs of small business owners across the country, who aimed to make their own fortunes selling the iPhone of its day. Dealerships at first flourished as the new industry exploded, but they ran into trouble during the Great Depression. Ford couldn't have his factory idling; it would starve him of cash. So he not only aggressively pushed cars down the throats of the company's dealers, but the company underwent a dramatic expansion of its dealer network, aiming to have stores on seemingly every street corner.

Ford had dealerships in a vice. When, a hundred years later, Starbucks found itself with too many stores, exceeding customer demand, the coffee maker could pull back with closures, bearing the financial brunt of it. But franchise car dealers were independently owned, leaving them with little recourse in a downturn. If they were to pull back on their orders, the manufacturer could simply opt not to renew their contract at year's end, leaving the store owner with an expensive investment and little way to salvage it. In the post–World War II era, Ford and GM got into a sales race. The push came for higher and higher volumes, forcing cars onto dealers, who were forced in turn to push cars on customers. To avoid losing money on these sales, dealers would offer heavy discounts hoping to make up for it elsewhere, either offering less money to customers who sought to trade in or charging more for financing—practices that, if done poorly, could leave a bad taste in customers' mouths.

After generations of perceived abuse, dealers began banding together in statehouses across the country to find protection. In a small town, the owner of the Chevy or Ford dealership might well be one of the most successful local businesspeople, providing jobs to many, paying for high-profile advertising, offering donations to charities and sports leagues. Sometimes these dealers were also members of their respective state legislatures. Consequently, laws and regulations began appearing across the country—some aimed to limit where manufacturers put stores while others aimed to ensure that a car manufacturer couldn't sell directly to customers.

By the turn of the century, an uneasy détente had been reached. The truth was that each side needed the other. But like any system built up over a hundred years, it was byzantine in its complexities. By the very nature of the relationship, there was tension. Many of the dealers viewed themselves as independent and self-made, feeling they should be able to run their stores as they saw fit. Carmakers saw it differently, wanting to impose control as if they owned each franchise location. GM wanted its Chevy customers to have a uniform experience across the brand—an image the automaker was spending billions to create through its products and marketing.

Lost in all of this was the customer, for most of whom the car-buying experience was only slightly more appealing than a trip to the dentist.

Sitting with Smythe, Eberhard listened as the longtime car dealer counseled him on why he needed to work with franchisees, who, he said, would serve as the face of the Tesla brand to buyers. Smythe had made a fortune selling cars for the likes of Mercedes, so he was admittedly biased. He did caution Eberhard that some dealers could be shady.

Ok, Eberhard asked, which dealers did he trust? Smythe got quiet, looked at the table. "None."

Eberhard reported back to the board and his meeting helped sway the board: selling their car direct was the way go to. But Tesla would again be entering uncharted territory. Building an electric car was one thing—it had been tried in some form or another for decades. Selling a car directly to a customer was simply unheard of.

One of the first signs of how serious Musk was about changing the car-buying experience occurred when he put Simon Rothman on the company's board. The Harvard grad and former McKinsey consultant had forged his Silicon Valley cred in the burgeoning online retail space, creating eBay's automotive marketplace. It was a website for buying and selling used cars that was generating a billion pageviews a month and processing $14 billion a year in sales, representing about one-third of the company's overall merchandise sales. He brought to the board a perspective on selling cars unlike that of many in the auto industry.

Musk and his fellow board members debated whether Tesla needed physical stores to sell the Roadster, or whether online stores would suffice. Musk pushed for selling online only, but Eberhard and others worried that the buyers of electric cars were going to need some handholding at first to help them grasp the new technology. They needed a sales team to walk customers through how charging would work and how the car would operate differently from what they were used to. A store would also convey a sense of legitimacy to the new brand, reassuring the buyer that there'd be somebody there down the road if problems arose.

Their lawyer advised them that Tesla could sell directly in California on a technicality: the carmaker had never had any franchise dealerships, and therefore it wouldn't be cutting into its franchisee sales. That, at least, was the argument they'd be building off of. Now they just needed to figure out the rules for forty-nine other states.

Finishing the Roadster and developing their high-end follow-up sedan, which they would eventually call Model S, and opening a network of company-owned stores—taken together they would require the kinds of money that even Musk's large fortune couldn't support. Musk and the team set out to see who they could raise money from in Silicon Valley.

There was no bigger name on Sand Hill Road than Kleiner Perkins, which had been an early backer of Google and Amazon. JB Straubel had a connection at the firm from his days prior to Tesla, when he did some consulting work to make ends meet. KP managing partner Ray Lane, the former no. 2 executive at software company Oracle, came to meet with Eberhard. Lane quickly connected with him, impressed with Eberhard's smarts and the velocity of ideas that poured out during their conversation. They continued their dialogue beyond the meeting, and eventually Lane decided he wanted to pull together a team to conduct due diligence on Tesla. The idea of investing in a car company was going to be foreign to Lane's fund. He knew that a car company was going to require more capital than they were used to investing, and that it would take years to see a return. Still, he was excited by what he saw. From his days at Oracle, Lane had spent time getting to know many senior automotive executives in Detroit, including Jacques Nasser, former Ford CEO, and Brian Kelley, who had been president of Ford's Lincoln Mercury division. He asked them to help evaluate Tesla.

Due diligence is by its very nature intended to find holes in a company, and Eberhard seemed put off by parts of the process, according to one participant. "Martin was getting antagonistic and fighting. . . . It became just a total shit mess." Nasser warned that working from the Lotus platform would prove problematic. Asked about Tesla's idea to shun franchise dealers, he cautioned that at Ford he had made the mistake of fighting franchise dealers, calling it one of his biggest mistakes.

Still, according to Lane, the negotiation was moving smoothly—until the end, when Musk got involved over the company's valuation. Musk told them they had an offer from a competing VC firm,

VantagePoint, that valued Tesla at $70 million. Lane valued Tesla at just $50 million.*

Among Lane's partners, there wasn't a unified opinion on whether to move forward with a formal offer to invest. Half were against it, he recalled. They questioned whether Eberhard was the right guy to be CEO. "He seems like a crazy scientist," one told him. Some didn't like that in their meetings with the Tesla team, Eberhard flat out rejected the notion that experienced automotive industry executives from Detroit might be helpful.

The naysaying didn't dampen Kleiner Perkins head John Doerr's excitement. But ultimately the partners left it up to Lane—he could pursue the investment if he wanted to. That night he slept on it, and in the morning, he called Musk and Eberhard with the news: he wasn't going to go further.

"I was really excited. I wanted do to it," Lane recalled. "But I was feeling like I don't want to make an investment that the partnership didn't really want to make."

So Musk turned to their other offer, VantagePoint Capital Partners. Tesla appealed to them for a simple reason: the company didn't call for anything dramatically new. The batteries were a proven technology, and the demand for cars was well

* Musk would later say he told KP's chairman John Doerr that he was willing to take a lower valuation if Doerr, not Lane, joined the board of Tesla. But Doerr deferred to his colleague.

understood. The innovation was simply deploying these elements in a different way. They joined Musk in leading the $40 million fundraising round. It granted VantagePoint a seat on the Tesla board—something Musk would live to regret.

CHAPTER 5

MR. TESLA

Elon is the perfect investor," Martin Eberhard had confided to a colleague during the honeymoon period with Elon Musk. In the early days, Musk seemed to think highly of Eberhard too, showering him with praise. "The number of great product people in the world is tiny and I think you are one of them," Musk wrote Eberhard in a note as they worked through a problem one night. They shared a similar career path, though the scale of their success had differed. They both forged startups that took on the world of media—Musk going after phone books; Eberhard book publishers. They thought about customer desires and engineering solutions. They both believed electric cars were the future. Both could be charming, funny—demanding. They also shared a stubborn side and suffered no fools.

Employees didn't see Musk often, maybe for an occasional board meeting; but he was interested in the details of their engineering, and he often shared ideas with Eberhard through his late-night emails. He pushed for the Roadster's body to be made of carbon fiber, a lighter-weight material found in super cars, instead of using cheaper fiberglass. "Dude, you could make the body panels for at least 500 cars worth per year if you bought the soft oven we have at SpaceX!" Musk wrote to Eberhard. "The oven only cost us about $50k. The vacuum pump, storage freezer and assorted equipment cost us another $50k. If someone tells you this is hard, they are full of shit. You can make high quality composites in the oven in your home. Once you've made a few of these things, you realize that there is nothing magical about glue and string."

It was the kind of fanciful thinking that excited them both as they made their first steps into the automotive business, the kinds of decisions that would set the Roadster apart—and doom their working relationship. The two dined at each other's homes, celebrated personal milestones, and encouraged each other during down times. Musk's life was largely in LA at SpaceX, which was struggling to build a rocket that didn't explode on takeoff. Tesla was Eberhard's life, more than 350 miles north in Silicon Valley.

While Eberhard wasn't a traditional car guy, those around him generally marveled at his engineering skills. He showed an uncanny ability to drop into a problem and plumb a solution. Sometimes he could be overly curt when he disagreed with a colleague. When somebody during a meeting suggested they install solar panels on their new office, since by then they had moved to nearby San Carlos, Eberhard snapped, "Why the fuck would we do that?" But the flare-ups were often extinguished by his enthusiasm and his corny jokes. **What would Nikola Tesla say today if he was alive?** Eberhard asked. **Why am I in this coffin?**

As is often the case with startup founders, Eberhard's ego became intertwined in Tesla. He delighted in shuffling friends into the company's garage on the weekends to show off his baby, the first engineering prototype of the Roadster, taking shape in early 2006. Stephen Casner, Eberhard's former colleague at Packet who had introduced him to AC Propulsion, was one of the guests who got a sneak peek at what Tesla was working on as it got closer to completing its prototype. The team was pulling long hours and growing closer together in the process. Office assistant Colette Bridgman showed up for work one Monday to find that JB Straubel, Gene Berdichevsky, and the other Stanford guys who had joined the team had sprayed the ceiling with Nerf darts in a game of tag, as they blew off steam during a long weekend in the office.

Eberhard relied on Bridgman to be the bond that held the company of engineers together; she arranged weekly show-and-tells for employees to share about their personal lives. He even consulted with Bridgman when planning his second wedding—to the girlfriend who had encouraged him to move forward with the idea of Tesla, several years earlier. Straubel was in attendance, on the lawn of Eberhard's beloved home in the hills above Silicon Valley, to witness the nuptials. (Musk was invited but didn't make it.)

But as Tesla prepared to publicly reveal the Roadster for the first time in the summer of 2006, its inexperience was becoming an issue. The car's transmission was both a focal point and an emblem of the challenges the company faced.

A car's transmission converts the power generated by the motor into rotational force that drives the axles, which in turn spin the wheels. While the mule car from a year earlier used a one-speed transmission, one that had been hacked together from a Honda, Tesla's team in the UK, made up of former Lotus engineers, was charged with designing from scratch a two-speed transmission. This was a controversial choice. A key part of the promise of the Roadster was that it would have a quick off-the-line acceleration as well as a high top speed. They knew they didn't need as many gears as a traditional car. A gas-powered car has multiple gears in its transmission, around half a dozen, in order to channel the energy needed for both acceleration and speed—lower gears allow for the creation of torque to power acceleration, while higher gears allow for the car to continue going faster, even when the car's wheels are spinning faster than the engine is. The nature of an electric motor, unlike a gas-powered engine, is that it produces near-instant torque; it wouldn't need to shift through multiple gears in order to reach top speeds.

But the system would be challenged to have a 0-to-60 acceleration in 4.1 seconds **and** a high top speed without a two-speed transmission, the UK team argued. They had demonstrated their proposed design on computers, showing a simulation of a smooth transmission between first and second gear.

But when their transmission prototype arrived in San Carlos in May 2006, Dave Lyons, the former IDEO engineer hired to add seniority to the California team, realized there was a problem the moment they pulled it out of the create.

"Where's the actuator?" he asked. The transmission was there but not the part to connect it to the axle.

They called the UK, and a disagreement immediately erupted as it became clear that each side thought the other side was developing the actuator. Facing a deadline, they'd have to hack together a solution in order to demonstrate the cars at a reveal event that had already been scheduled for that summer, then circle back later with a more lasting solution to the transmission problem.

As Tesla prepared to reveal the Roadster for the first time, Musk checked in with Eberhard on the proposed pricing, telling him that he'd been telling friends the car would start at $85,000. Eberhard cautioned Musk against being too definitive; issues with finding a battery supplier and uncertainty around the transmission had caused their former estimates of $80,000 to feel understated. He suggested being vague, for instance suggesting a price tag between $85,000 and $120,000.

"I am nervous about $85k," Eberhard told Musk, emphasizing the point.

For the past three years, Tesla had been in "stealth mode," a rite of passage for Silicon Valley startups as founders try to get their feet on the ground—raising initial money and hoping to avoid the harsh spotlight that magnifies the mistakes inevitably made during the early days of being a company. Coming out of stealth mode tends to have a familiar playbook, aimed at maximizing exposure, whether the end goal is raising more money or gaining customers. In Tesla's case, their aim was simple: to pre-sell Roadsters. A thick order book would show parts suppliers that Tesla was the real deal.

It wasn't without risk. Unlike a piece of software, Tesla still had many significant steps ahead of it before the Roadster would really be ready. Even as Eberhard and Musk hoped for a quick turnaround in bringing the Roadster to customers, a road map for the typical Tesla release took shape:

Demo a prototype of what their car
    looked like.
Finish the engineering behind the actual
    car, to make it roadworthy.
Take it into production.
Launch it into the world, hoping to win
    positive reviews from jaded reviewers
    and picky customers, who may have
    already put down money but still haven't
    bought the thing.
Then repeat with the next car.

Anywhere in that complicated dance routine could trip them up. And though the reveal was undeniably important, some in the company were surprised to see just how nervous Eberhard was about making sure Musk was happy with it.

Musk wanted to reveal the Roadster with a **party.** No detail seemed too small to run past Musk's personal assistant—from the layout to the food. The guest list came from Musk, naturally enough. The plan was to hold the event at Santa Monica airport, where they could rent out a hangar and give rides around the party in the two prototype Roadsters that had been completed, a red and a black version. Guests were told to bring their checkbooks, and that a $100,000 deposit would get them on the list for a car once it was ready in 2007. Their goal was to pre-sell 100 cars within two or three months of the event.

Ahead of the launch, Jessica Switzer, who ran marketing for Tesla, hired a public relations agency in Detroit to help generate publicity from the automotive press. Musk fired them as soon as he learned about it. He didn't want to spend money on marketing before the car was done; he figured his involvement—and the Roadster itself—would create enough attention. Musk had already been unhappy with Switzer's decision to spend money—with Eberhard's approval—on focus groups to test the car and the brand. Musk ordered Eberhard to fire her. Eberhard was shocked, but he followed through, weeping as he did.

So Colette Bridgman, the onetime office assistant who had joined Tesla with hopes of more, found herself in a new role: marketing. She would now be tasked with overseeing the event. Another PR firm was brought on but it too found itself in Musk's crosshairs in the days ahead of the event, as major newspapers began rolling out stories about Tesla. **The New York Times** wrote a piece that referred to Eberhard as chairman, not president, and that didn't mention Musk at all. Musk fumed in an email to the PR firm. "I was incredibly insulted and embarrassed by the NY Times article," Musk wrote. "If anything like this happens again, please consider [your] relationship with Tesla to end immediately upon publication of such a piece."

When it came time for the event, Musk turned on his awkward charm. He held court among 350 guests, noting that his wife was due to give birth to triplets at any time (this after their twins). Guests included Michael Eisner, the CEO of Disney; Jeff Skoll of eBay; **The West Wing** actor Bradley Whitford; producer Richard Donner; and actor Ed Begley Jr. Then-California governor Arnold Schwarzenegger even attended. The real star of the night, though, was Eberhard, who introduced the car and company to the world, turning on his passion to explain why it was time for a purely electric car. "An electric sports car was the way to fundamentally change the way we drive in the USA," Eberhard told the crowd.

Inside the hangar, Bridgman had set up stations where engineers explained different parts of the vehicle; the technology was so new that they thought customers needed to be educated on it to feel comfortable making such a hefty investment. Every employee was tasked with trying to gently close a deal. They had a screen that tracked sales at the event.

But what Eberhard, Musk, or the engineers had to say, ultimately, mattered very little to the guests. What mattered was the Roadster. The car looked nothing like the amateurish tzero that had garnered attention in electric car circles. It looked like a real sports car, something a wealthy owner could imagine driving down Rodeo Drive to impress onlookers. JB Straubel, dressed in a slightly baggy black button-down shirt with Tesla's "T" insignia on it, was tasked with taking the governor along for a test ride. A crowd lined up to watch as Straubel quietly crept out of the hangar with the hulking Schwarzenegger stuffed into the tight cockpit, his knees jutting up against the passenger-side dashboard. "Step on it," somebody in the crowd shouted. Straubel didn't at first. He knew he needed to straighten out the car onto the airport runway to give it space to go. As he angled into position and pressed the accelerator, the car picked up some speed, its motor whirring like a spaceship. Then **zoom.** A small dust cloud was left behind as the car shot out of sight, the only sound the tire traction and the crowd in unison: "Whoa!"

This was no golf cart. What Eberhard, Musk, and Straubel had dreamed of had come true. They had a real (and really expensive) electric sports car, precisely the kind that Eberhard had sought out four years earlier, but that hadn't yet existed. And it was like riding lightning. As Straubel pulled in from the test drive, the governor could be seen wearing a giant grin.

For the rest of the night, the car remained the selling point. Blistering acceleration around the makeshift track. That instant torque. It did what no amount of salesmanship could. By the middle of the party, Tesla already had twenty reservations, buyers placing their $100,000 checks in a little cash box.

That night, more than just a sales pitch for the Roadster, Musk laid out his grander plan for Tesla. The cost of the car not only bought attendees an exceptional sports car, it would help generate money to invest in developing other green vehicles. A few days later, Musk would go further, posting on Tesla's website his vision for the company, an elaboration on his simple three-step premise. Or, as he called it, "The Secret Tesla Motors Master Plan."

"In keeping with a fast growing technology company, all free cash flow is plowed back into R&D to drive down the costs and bring the follow on products to market as fast as possible," he wrote. "When someone buys the Tesla Roadster sports car, they are actually helping pay for development of the low cost family car." He added another goal: He wanted to provide "zero emissions" electric power generation. The blog referenced his recent investment in a solar panel company, called SolarCity Corp. It was a venture with his two cousins (making him chairman in a third company, after Tesla and SpaceX) aimed at putting solar panels on homes, which, he wrote, could generate about fifty miles of driving per day worth of electricity. Musk was out selling not only a cool sports car but the notion that one could power it without fossil fuels.

It was an idea that appealed to many in California,

and even after the event, orders for the so-called Signature 100 edition kept rolling in. Joe Francis, creator of **Girls Gone Wild,** sent an armored truck to the San Carlos office to drop off his $100,000 deposit in cash. (Alarmed by having so much money in hand, co-founder Marc Tarpenning rushed to the bank.) Schwarzenegger touted it at the San Francisco auto show and put down his own money for one. So did actor George Clooney. Within three weeks, Tesla was sold out of its initial offering of 100 cars.

The success of the Roadster introduction also propelled Eberhard into the limelight. He was featured in a BlackBerry ad campaign, appeared on the **Today** show, and became a regular on the conference-speaker circuit, where he was sought out for his thoughts on the future of the car. Eberhard became the face of Tesla. As a gift for their wedding, his wife bought him a vanity license plate that read: "Mr. Tesla."

Now, he just needed to get the Roadster into production.

Along with the orders and attention came new scrutiny for Tesla. One potential customer emailed VantagePoint: "I hate asking you this, but want to get your helpful insight . . . if I were to put a reservation in for one of the Tesla electric Roadsters, if you were me, would you be concerned about the company becoming insolvent and losing my large deposit? . . . Part of me says just get the reservation and you wait to see it if it's worthwhile."

The message found its way to Musk, Eberhard, and VantagePoint's newly appointed representative on the Tesla board, Jim Marver. "Not sure how to field requests like this," an analyst at VantagePoint said as he forwarded the concern. "I need to tell him that his money is at risk and he should wait, or we need to provide some protection."

Musk responded forcefully. "I've always been clear that, although I think Tesla is very likely to be successful and that we will deliver a great car, the money is not escrowed or otherwise secured. My recommendation is that people buy the Sig 100 collector's edition cars, as these will probably have the greatest value over time."

Behind the scenes, officials at VantagePoint were alarmed that the money wasn't being placed in escrow, that it was commingling with the company's operating cash. The firm's CEO, Alan Salzman, worried that it opened them up to unnecessary risk, and he wasn't shy in expressing his displeasure. It was the first sign of the friction to come between VantagePoint and Tesla.

With the prototype in hand, however, Musk had more pressing issues on his mind. For the first time, he could see and feel what the car was going to be like. He wasn't satisfied with what he found. He began making more suggestions for changes. But unlike in 2005, when he was playful and excited about what was possible, he was growing frustrated that Eberhard didn't seem to share his sense of urgency. Musk thought the car was hard to climb into, the seats were uncomfortable, and the interior lacked polish compared to other high-priced cars. In the fall of 2006, Musk's temper flared with Eberhard over the quality of the dashboard. "This is a major issue and I'm deeply concerned that you do not recognize it as such," Musk told him in an email.

Eberhard deflected, saying that it was something that needed to be fixed only after other higher priorities were addressed, especially if Tesla was going to start production in the summer of 2007, less than a year away. "I just don't see a path—any path at all—to fixing it prior to start of production without a significant cost and schedule hit," he explained. "We have a tremendous number of difficult problems to solve just to get the car into production— everything from serious cost problems to supplier problems (transmissions, air conditioning, etc.) to our own design immaturity to Lotus' stability. I stay up at night worrying about simply getting the car into production sometime in 2007."

Eberhard continued, pleading: "For my own sanity's sake and for the sanity of my team, I am not spending a lot of cycles thinking about the dashboard and other items that I want to fix after [start of production]. We have a long list of things to think about once we are shipping, and I will think about them when I have free cycles to do so."

He may have been hoping to delay or pacify Musk, but he only enraged him more. "What I want to hear from you is that will be addressed after [start of production] and that customers will be informed that there will be an upgrade coming *before they receive the car*. I have never demanded that it be done prior to production, so I don't know why you would even raise that as a straw man."

At the end of 2006, the ever-expanding Tesla workforce could see the burden of stress that Eberhard was carrying. He sat in his office with his head in his hands, or else stared off into space twirling his beard. Late at night, Eberhard called board member Laurie Yoler in despair, recounting all of the changes Musk was demanding and the pressure he was putting on him. After a weekend of test-driving the Roadster prototype, Musk thought the Lotus seats were uncomfortable. But customized seats would add $1 million to the cost of development, money they were hardly flush with. Getting into the Roadster was hard, especially for Musk's wife, Justine. The seat was only slightly above the ground, which meant riders' knees didn't bend very much when they sat. It was more like sitting in a high-powered sled than a typical sports car. The original Elise design had a high lip on the doorframe, which you had to climb over. Musk wanted the lip lowered two inches. This would add $2 million to the cost of the project. Musk wanted special headlights, approving $500,000 for that work, and special electronic latches instead of mechanical push buttons for the doors, adding another $1 million. The change from fiberglass to carbon fiber added $3,000 to the cost of each car.

Musk was exerting control over the car itself. And it wasn't that Musk was wrong, but it was all too much. "Elon has all these ideas and I can't move fast enough," Eberhard confided in Yoler.

In late November of 2006, he delivered a presentation to the board in which the estimated cost of making each Roadster had swelled to $83,000 from its original target of $49,000. That number was based on production beginning the following fall—rather than the summer of 2007 as planned—with an expectation of turning out thirty cars per week by the end of December. Even that forecast was littered with caveats—they still hadn't finalized many of the production parts or picked suppliers. Eberhard hoped to whittle the cost down by $6,000 within the next year.

One of the challenges that remained was the electric powertrain—the batteries and the motor—which represented the car's most expensive system. The cost of battery cells was twice what they had forecasted. While Tesla was founded on the idea that commodity batteries could be found cheaply, in fact the company's engineers came to realize that not all batteries were the same, even ones sold under the same uniform ID (called 18650 for their dimensions, 18 mm by 65 mm). Each company had a different way of making the cells, and that had ramifications for their use in a car. After testing the industry's offerings, Straubel's team had found just a few companies' cells that worked best, including Sanyo's. If the battery companies discovered this limitation, Tesla would lose any negotiating power it had. As it stood, the suppliers thought that Tesla had an entire world of financial opportunity in their hands. In private, some employees, including Gene Berdichevsky, worried about being beholden to so few battery suppliers; they pushed for money to develop other options. Eberhard told them no—there just wasn't the funding. Eberhard's team was struggling, too, to find a supplier who could make a transmission to Tesla's specifications and price. The powertrain was the car's linchpin, and it was insecure.

Eberhard's performance at the November 2006 board meeting left Musk doing damage control with VantagePoint's Jim Marver, specifically over Eberhard's inability to address questions about costs. "It is not that Martin had no idea what the car costs when he was asked at the last board meeting but rather that he did not feel comfortable stating precise numbers off the cuff for a specific calendar quarter of production," Musk told Marver, offering a backhanded reassurance. But what Musk wasn't saying was that he, too, was beginning to wonder what was really going on within the company. A few weeks later, Musk flew to Lotus—a meeting that didn't include Eberhard. Musk wanted Lotus's view on how Tesla was doing on its timing of the project.

"I am sure you can imagine I find this a rather awkward situation where Elon has asked for Lotus' own view of the production timing," Simon Wood, a Lotus director, wrote Eberhard before the meeting, in which he offered his dim prognosis. "I recognize this conflicts with the plans that your team are developing."

PowerPoint slides warned Musk from the outset that Wood's opinions "could be viewed as pessimistic, but reflect the general feeling of key Lotus personnel involved in the project." He laid out that Lotus had a list of almost 850 concerns, ranging from items that would lead to a complete loss of the car's function, to safety and regulatory issues, to items that would affect customer satisfaction or require minor repairs. Wood calculated they could address 25 of the issues a week, which would add more than 30 weeks to the schedule before they could even begin production. All of that meant that Lotus thought it could make maybe 28 Roadsters by Christmas 2007. Tesla's plans had called for ramping up to 30 cars **per week** by year's end.

And yet, even with concerns mounting, the Tesla board began 2007 with optimism for taking the company public the following year. In January, Marver proposed the board begin meeting with bankers. The idea they discussed was to borrow money that would convert into an ownership stake at a certain price, a scheme that they hoped would allow them to avoid racking up costly debt payments while creating a bridge until they could take the company public. He suggested that Eberhard speak at an investment conference in New York City in the coming weeks. Musk disliked the idea of Eberhard spending his time that way, arguing he should focus attention on the Roadster and not on what he perceived as low-value PR and financing meetings. "There are several burning Roadster issues that need Martin's attention right now," Musk told the board. "We have slipped delivery significantly already and are at risk of slipping even more."

As they got further into the year, the board began to narrow its plan. Musk that spring told the board he believed Tesla needed to raise $70 million to $80 million to keep the company going until an IPO in March or April 2008. One deadline after the next seemed to be slipping. Marver cautioned against accessing the money they'd already collected from Roadster deposits. "It is the view of many that we should not spend these until we are shipping Roadsters, hopefully in October," he told the board. "For example, if we ship the 25 Roadsters in the business plan, we could spend 2.5m of the deposits this year. If we knew for a fact that private capital would be available to us in the fall at a good valuation, then of course we could do a more modest round now." But he worried about the potential for delays or a change in investors' willingness to put more cash into the company or lend it money.

Ultimately, the board decided to seek less than what Musk had suggested, deciding to raise $45 million, an amount that valued Tesla at $220 million. It was a decision that put faith in Tesla's ability to begin production that year that was still far from certain.

The impossibly tight schedules, blown deadlines, production setbacks, inadequate funds—it all pointed to a plain fact. Something was wrong at the core of Tesla. But what exactly was it? Fortunately for Musk, he knew just who to ask.

CHAPTER 6

THE MAN IN BLACK

Six years before Antonio Gracias became an investor in Tesla, he was on an airplane from Chicago to Switzerland for a trip that would reshape his life, and the electric-car maker, for years to come. In the spring of 1999, he arrived in the small town of Delémont in time for dinner. He had come to survey the latest addition to his burgeoning empire of small industrial factories, which he had begun buying up four years earlier, while still enrolled at the University of Chicago law school. The late 1990s had not been good to small manufacturers that supplied parts to electronics and automotive customers. Those larger companies were using their giant size to squeeze the little guys for cheaper and cheaper prices. In that environment, Gracias, who had grown up around factories in Grand Rapids,

Michigan, saw opportunity. These mom-and-pop operations might just need a little modernization, restructuring to lower costs, and some new focus to make them into better businesses. All of which brought him to Delémont to see a factory, one that had been part of a larger deal to buy a stamping plant near Chicago.

Gracias was met for dinner by the factory manager, who was joined by an engineering consultant named Tim Watkins. Watkins had been hired months earlier to help straighten out operations. Initially, Gracias didn't know what to make of Watkins, the UK-born engineer who kept his long hair in a pony tail, looking a bit like Sean Connery in the movie **Medicine Man,** except dressed all in black and wearing a fanny pack. Over dinner, though, it quickly became clear to Gracias that he and Watkins saw the world similarly. They read the same books, shared ideas about management and technology.

Gracias, twenty-eight, had taken an unusual path into the world of manufacturing. Born in Detroit, he grew up in western Michigan with his immigrant parents. His father was a neurosurgeon; his mother ran a lingerie shop, where Gracias helped out after school. As a teen, he bought stocks the way other boys collected baseball cards. His prized holding was Apple. In 1995, after working for two years at Goldman Sachs, he ended up in law school, not because he dreamed of becoming of a lawyer but because his mother had died, and it was his parents' dream to see all of their children become doctors and lawyers.

At school, he couldn't shake a desire to get into business, to **make** something. On the side, he created his own investment firm, MG Capital (named after his mother, Maria Gracias) and was joined in 1997 by a friend from Goldman. The firm raised $270,000, plus $130,000 from Gracias's own pocket, to get its start. Their first acquisition seemed too good to be true, an electroplating company in Gardena, California, being run by a manager with experience fixing troubled companies—meaning Gracias could continue at law school without trouble. The business was caught up in a bankruptcy case, a corporate orphan that nobody wanted. Taking on massive debt, the two bought it for the value of the company's assets (rather than on some multiple of its $10 million annual revenue, the way a company might typically be valued).

Very quickly, however, they found themselves in over their heads. Gracias stayed enrolled in school but all but abandoned the classroom to work the factory floor in California, relying on a close friend named David Sacks to keep him informed about what was going on back in Chicago and to help him stay prepared for the term-end finals that made up all of his grades. In Gardena, fueled on daily Starbucks coffee drinks, Gracias worked closely with the hourly workers to find ways to boost the company's production at a time when electroplating

was in ever greater demand, especially from electronics manufacturers. What had at first looked
like a foolhardy move turned into a cash-minting
machine in short order, generating annual sales of
$36 million—and fueling MG's buying boom. By
the time Gracias sat for dinner in Delémont, MG
Capital had acquired five companies.

The revenue had even allowed MG Capital to
invest in a startup called Cofinity, where his law
school friend, Sacks, had gone to work. That company merged with Musk's X.com, before they collectively became PayPal. It was how Gracias was
introduced to Musk.

As their dinner came to an end, the manager and
Watkins took Gracias to see his new factory. It was
late on a Saturday, presumably a good time to see the
facility in its dormant state—work at night and on
weekends was essentially prohibited through local
labor laws. As they got close to the door, Gracias
was surprised to hear the humming of stamping machines coming from the darkened factory.
Inside, as the lights came on and his eyes adjusted,
he couldn't believe what he was seeing: a fully automated factory. The machines were running; nobody
was standing around monitoring them.

Gracias had traveled the world looking at factories, studying the latest technology to improve line
speeds and process, but this was altogether new. In
an age when personal computers were still rare on
the factory floor, Watkins had figured out how to

automate, creating algorithms to anticipate when systems needed to shut down for maintenance. He timed the machines to run for four hours on their own. Then workers would return to the factory for eight-hour shifts, followed again by the machines operating independently.

With his innovation, Watkins had figured out how to operate a factory twenty-four hours a day, in a place where a maximum of sixteen hours of labor was normally permitted. In the process, he helped lower the costs of operation.

"I had never seen this anywhere on earth," Gracias recalled.

Gracias had stumbled onto one of the biggest finds of his career—not the factory but the inno-vator behind it. He spent the next several months trying to convince Watkins to join MG Capital, to improve its growing stable of factories across the U.S. He eventually succeeded. Their relationship would contribute to Gracias and his business part-ner selling MG Capital's portfolio companies for a ninefold return. The track record helped them raise a $120 million investment fund. MG Capital was renamed Valor, with Watkins on board as a partner. They were fast friends; even roommates for a time in Chicago. They shared a deep desire to evolve together. They wanted not to buy more distressed factories and run them; they wanted instead to in-vest in companies and lend their expertise to help improve them.

A key part of their plan was to remain in the background. They didn't want to cultivate a reputation that might scare off founders; they largely shunned publicity. Where they sought to pitch in was with their expertise in the mundane processes, like automation, that could make or break a company. In 2005, when Musk proposed that Gracias (along with some other friends) invest in Tesla, he jumped at it.

So it wasn't surprising when Musk called his friend Gracias in 2007 with a plea for help. Something wasn't right at Tesla. That year, Gracias had joined the board as well, and he already sensed the way Martin Eberhard had been struggling to manage the growing carmaker. It was increasingly clear that the founder was in over his head.

That's where Gracias and Watkins would come in. Musk needed them to dig into the company's books and figure out what was really wrong at Tesla.

It's not unusual in Silicon Valley for a startup to reach the size and complexity where a founder's skills are exceeded. In 2007, that time arrived for Eberhard. He knew it, and so did Musk. They began talking about bringing in a new CEO and hiring their first chief financial officer, as they'd intended to do since their formation. This would free up Eberhard to focus on the development of Tesla's next car after the Roadster—a step the company needed to nail if it truly was going to evolve from making what were essentially toys for rich people into an actual car company, as envisioned in Musk's not-so-secret master plan of the previous year. Musk was increasingly worried about how things were going with the Model S, still known internally by the code name WhiteStar. This vehicle was the crucial pivot point in Tesla's trajectory. Even in the best of circumstances, the Roadster would only ever be driven by a modest number of wealthy early adopters. The Model S was a mainstream car, with a mainstream audience in mind. It needed to represent to the public all of what Tesla stood for. There would no room for slipups.

Ron Lloyd, the project's leader, had come from the tech industry, but he built a team largely of car guys, going so far as to set up a large office in the Detroit suburbs. During Musk's first trip to the office, he was universally unimpressed. (Among their transgressions: One engineer made the mistake of having a typo in a presentation.) Built from former Detroit automakers and suppliers, the nascent team was developing a different culture from that of the engineers back in Silicon Valley working under JB Straubel, who were mostly in their first or second job out of Stanford. While Tesla had initially eschewed Detroit engineers, the challenge of building an all-new car from the ground up had led to an increasing reliance on their ranks. In 2007, the idea of joining a California startup making an electric car was pretty radical in Detroit, which had recently experienced several heady years, with sales reaching new records. Even though cracks were forming in the foundations of GM and Ford, the money was still flowing. Engineering and executive careers at General Motors, Ford, and DaimlerChrysler came with the expectation of lifetime employment and lucrative pensions. GM was often jokingly referred to as Generous Motors. The idea of giving that up for a startup that might not pan out was a hard sell in those days—to the bemusement of the team back in Silicon Valley, where the risk of losing a job for a potential payout was seen as the norm. It was an environment that rewarded risk and where failure was seen as part of the game—as long as you could rebound with another good idea.

"Fire all of them. Every one of them," Musk told Eberhard as they left the office. Like many of Musk's orders, Eberhard ignored this one. He still needed them.

To Musk, the Detroit team seemed overly focused on the cost of the vehicle, to the detriment of its quality. They were excited about a deal they'd arranged with Ford to use parts from its Fusion sedan program. The Roadster had taught Tesla that it couldn't simply buy a car body from another automaker, make a few tweaks to it, and expect everything to work out. Every change they made to make the body more Tesla-like added cost. So the Detroit team sought to do what big automakers did all the time: They would build a car from an existing catalog of parts, figuring it would be cheaper than sourcing unique ones themselves from suppliers, which were already reluctant to work with the small upstart. The Detroit team was, in essence, taking apart a Lego car and configuring the pieces to their own imagination of what it should look like. The problem was that Musk wasn't happy with the image they were conjuring.

Musk expressed his displeasure with Lloyd to Eberhard, who had the unenviable task of trying to reconcile their paltry budget with Musk's extravagant taste. "Many times, I have heard Ron imply that we can't make [the Model S] look like other $50k cars, because we have so little money left over after the battery cost," Musk wrote late one night in the spring of 2007. "If we lean on this crutch too hard, it will result in a crap car, which is why I've pushed back on our estimates of drivetrain cost in luxury cars. I am quite confident that the real difference in cost between our drivetrain and that of a lux gas car is not nearly as big as we think it is."

Elon summed up his concerns. "Main thing I wanted to be sure of was that our guys know that most American cars suck and how to change that. . . . do they have good product judgment?"

Despite his unhappiness with the Detroit team, Musk understood that automotive experience was a valuable asset for the company. He liked the idea of hiring a CEO who had proven experience bringing out a cool car with little money. The Roadster had shown him how quickly costs could spiral out of control. And even with the added expenses they'd incurred, Musk remained unhappy with elements of the Roadster, which he complained had a budget car interior and a luxury car price tag.

In weighing potential CEOs, Musk became curious about a product developer named Hau Thai-Tang, after seeing media coverage of his achievements at Ford. In 2005, Thai-Tang's profile had exploded for his role in overseeing the development of the new Ford Mustang, a project that had been under wraps and was now getting glowing press. Musk arranged for the executive to come to Silicon Valley for a weekend to see Tesla. They had him drive their car. Eberhard recalled that he had rarely seen somebody drive so aggressively. Thai-Tang finished and began rattling off all of the issues that needed to be addressed to improve the driving dynamics—the suspension was wrong, there was too much weight in the back, on and on. But he also was full of praise. "This is really amazing," he told them.

Ultimately, to their disappointment, Thai-Tang wasn't interested in leaving Ford.* Still, Musk and Eberhard came away bolstered: they were on

---

* In the spring of 2007, Ford dispatched him to oversee product development in South America, a stepping-stone to bigger assignments. By 2019, he was Ford's top product development executive.

the right track. Thai-Tang also helped in another way: He recommend a headhunter to help find a CEO.

Word eventually leaked out about Tesla's search to replace Eberhard. Two reporters reached out to the company's PR shop in June, and the leak frustrated Eberhard. He complained to the board in an email. "I do not know how this information has leaked out, but it can only be the board or our search firm," he wrote. "Needless to say, having this story out in public, and having the press calling to ask Tesla staff about my imminent ouster makes it especially difficult for me to do my job and is, I admit, very demoralizing for me."

Musk waved away his concerns, telling Eberhard this wasn't a "big surprise," given how publicly scrutinized Tesla had become. "Best strategy would be to get out in front of this and embrace it, just as Larry [Page] and Sergey [Brin] did at Google," Musk wrote, noting that Google's founders had handed over day-to-day control to Eric Schmidt as CEO. "Everyone knew they were looking for a CEO and the search took a long time, but Google operated just fine with Larry at the helm in the interim."

The two continued their back-and-forth in private. Eberhard told Musk that he believed that a member of the board had leaked the information. "A little sympathy would have gone a long way," he told Musk. "Tesla Motors has been my life for the last five years. It hurts to hear from the press that the board is firing me (that is the word that was used)."

Musk tried to comfort him, saying he would be happy to correct the perception of him being fired. "The objective fact is you brought up the CEO search yourself several months ago," Musk wrote. "By the way, I did encourage you [to] be open about it at one point, although I did not push the issue."

Eberhard knew he needed the help, but he wanted to move into a new role on his own terms. The comparison to the Google co-founders was disingenuous. It was true they moved over for a seasoned operator, but they remained firmly in control of the company after going public, thanks to a form of dual stock ownership that gave them a majority stake. Eberhard lacked such a deal. Tesla's biggest shareholder remained Musk, which left in his hands control over Eberhard's future with the company.

Embarrassed, Eberhard was determined to put on a brave face. On June 19, he held an all-hands meeting to rally the company. There was a long list of issues that needed to be addressed before production could begin. They needed to iron out kinks in the car's transmission. They had to build out a small network of stores. They needed to focus on getting the car reliable, root out engineering problems, deliver parts to the factory in the UK.

"Focus on Job No. 1," he told them. All of their jobs were at stake, not to mention "the viability of electric cars" and the "future of transportation."

If that message wasn't clear enough, he sent the team a follow-up email shortly afterward. "[GM CEO] Rick Wagoner said last November that now is a turning point for the automobile—that the change from internal combustion drive to electric drive is as significant as the change from horses to horsepower. If we succeed with the Roadster launch, history will remember you and Tesla Motors as the drivers of this change. If not . . . well, think about Tucker and DeLorean. We are better than they were, and the Roadster is vastly superior to their cars. Let's prove it!"

Tim Watkins could be incredibly polite but also shockingly blunt. He had the kind of precision and discipline that few can muster. Before it became fashionable, he came to believe that spikes in blood sugar could be harmful to your health, so he ate closely monitored amounts of food throughout the day. In his fanny pack, he kept oats from a store near his mother's home in the UK. His life had largely been lived on the road in recent years, dispatched to hot spots by Gracias. It could be draining work; often it involved delivering uncomfortable truths that led to people losing their jobs. Through the years, he had developed further habits. He'd arrive at an assignment and visit the local chain store to buy a pack of black T-shirts and black jeans. Once he was done, he'd discard the wardrobe for the next assignment, as if shedding his skin.

When he arrived at Tesla's San Carlos headquarters in July 2007, he found himself in yet another troubled spot. He quickly called Gracias with the news: Tesla had no bill of material for all of its parts—a simple accounting of every piece going into the car and the negotiated price the company agreed to pay. He'd have to construct one. There were other concerns: Lotus officials were warning that the Roadster would miss production in late August as had been planned. Tesla still hadn't given final design approval for some of the parts that suppliers were to make. And the team still hadn't found a working solution for a two-speed transmission.

As Watkins took the measure of the place, Eberhard's team was doing the same, creating a de facto race to figure out the company's finances. Tesla had begun adding new employees to its finance department, as they prepared for an IPO. Ryan Popple, a recent Harvard Business School grad, was hired to begin the work of preparing the company books for a public offering. In his first week on the job, however, Popple realized things weren't as rosy as they seemed from the outside. His first assignment was to build a financial model of the company—a state-of-the-business document. He asked to see the current financial model. "That model's bullshit," he was dismissively told. Popple began going around to each department, asking

Straubel and his like about their budgets. He heard similar answers from others: "I have no idea. Nobody has ever talked to me about that."

By the end of July, the finance department, with the help of an outside adviser, had a new estimate for how things stood. They sent the report to members of the board. It said that the cost of materials in the car was totaling $110,000 for each of the first fifty vehicles they expected to build. Dave Lyons and JB Straubel in engineering were pushing for reductions, and they expected the unit costs could come down as volume kicked in. "They will be working on this and hopefully we will have an update in a couple of weeks so we can get a better feel for the 'real' cost of the car once it is at volume production levels," the report said.

It included a projection of cash burn for the rest of the year. After raising $45 million in May, the company was on pace to be out of cash by September of that year—not including the $35 million in customer deposits that the sales team had collected. By year's end, however, that money would be gone, too, if something drastic didn't occur.

In other words, Tesla was in money trouble— again. Rather than raise a full $80 million in May, as Musk had pushed for, the board had bet that the company could make it another year with a smaller round. But that had been based on what was, they could now see, an unsteady estimate of actual costs. They were learning one of the hard truths of the auto industry: industrialization binges on cash.

The revelation only served to emphasize another point: that they still hadn't hired an experienced CFO to anticipate these concerns. Beyond that, they hadn't fully implemented an accounting system to properly track costs. In a staff meeting, Eberhard got visibly agitated over the dire projections.

"If this is true," Eberhard told his head of manufacturing, "you and I are both fired."

In July, Eberhard faced an unhappy board. They balked at the suggestion that the Roadster could still sell for $65,000, when the battery pack was still costing more than $20,000 in itself. A month later, it got worse for Eberhard. Watkins's initial findings came in to the board, painting a far bleaker picture than the internal Tesla estimates. He calculated that the cost per vehicle after the hundred were made would be $120,000. That was without overhead. Costs would presumably fall as production increased but never enough to be profitable. But at the company's projected delivery volumes, each early vehicle, with overhead factored in, would cost a staggering $150,000. Adding insult to injury, Watkins didn't see any way for production to begin that fall.

The board was shocked. Eberhard fought the findings, but in the end, his fate had been sealed. On Aug. 7, Musk called Eberhard, who was in LA at the time on the way to speak to a journalist group.

The news wasn't good. Musk told him that he was being replaced. Michael Marks, who had invested in a recent round, would be taking over his role as interim CEO. Marks, as CEO of Flextronics, had turned the company into a global electronics manufacturing powerhouse before retiring. Tesla had lured him out of retirement.

Eberhard was surprised by the news. He called other board members and learned that they hadn't been briefed about it in advance. Musk agreed to hold a board meeting on August 12 to approve Eberhard's resignation as CEO, and to announce his new position as president of technology. But in Musk's mind, the decision was already made. He counseled some of his fellow board members on the eve of the meeting. "Martin seems to be focused on his public image and position within Tesla, rather than solving these critical problems. If you should speak with Martin, please urge him to spend all his energy on making sure the Roadster works and arrives on time. He doesn't seem to understand that the best way to maximize his reputation and position within the company is to help get this right."

Even before it was official, though, Marks was at Tesla's San Carlos offices, wading into the mess, writing Musk late August 8 to say that they needed to talk soon. "Lots of issues at this company, as you know, but some are a lot scarier and pressing than I thought."

CHAPTER 7

WHITE WHALE

JB Straubel sat on Tesla CEO Michael Marks's chartered jet to Detroit with the newly demoted Martin Eberhard. With his oval face, receding hairline, and the weariness that comes from running a global company, Marks looked the part of the adult in the room at Tesla—especially so working with dozens of young engineers who had only recently graduated from Stanford, and for many of whom this was their first real job. He'd spent more than a decade running Flextronics, which built Xbox game consoles for Microsoft, printers for Hewlett-Packard, and cell phones for Motorola. It was the kind of third-party manufacturing company Martin Eberhard and Marc Tarpenning thought would exist in the automotive world for their Roadster, but were surprised to find so few of.

The new boss wasted no time jumping into the fray. He chastised the staff at a company meeting for his perception that they lacked hustle. "I've noticed a few things about this company—how promising it is, but this company doesn't work very hard," he told them. "I'm going to have official office hours in place and I expect people to be at their desks."

The Roadster clearly wasn't launching that August, so he delayed it six months to give the team time to work through the issues and find ways to cut costs from the program, creating "Marks's List" of the immediate issues that needed solving. Among the top items was the transmission, which continued to plague engineering. Marks was familiar with manufacturing, but he was new to autos. Fortunately, he knew who to turn to for advice: Rick Wagoner, the CEO of GM. The two had been at Harvard Business School at the same time and Marks called on Wagoner while working at Flextronics. That's how Straubel, the battery whiz, found himself on a private jet bound for Detroit.

The team was greeted at the airport by black cars and whisked to GM's downtown headquarters, the Renaissance Center, looming over the largely abandoned city center. Once the Paris of the West, the city had watched as many of its grand buildings went vacant. After years of neglect, some had trees growing out of their roofs, dozens of stories above the ground.

Their caravan pulled into a garage for senior GM executives, and the men boarded a private elevator that took them to the building's top floors, where the CEO's suite of offices was located. The entrance had views of the Detroit River, with small models of the company's cars lining the windows. Wagoner, who played freshman basketball at Duke University, greeted the trio. He'd spent his adult life at GM, rising up through the finance department, groomed as CEO with roles around the world.

When the team from Tesla visited in 2007, GM's business was on a knife's edge. It had endured years and years of escalating debt, labor costs, and swelling pension obligations, creating doubts about the company's future. Sales were falling. Still, Wagoner projected confidence that the automaker was on track to fix things—again—even if pesky stories kept predicting their eventual bankruptcy.

Straubel had never seen anything like it—neither the executive bubble that Wagoner seemed to live in nor the extravagant conference room he found himself seated at, complete with a catered lunch. He'd barely been able to scrape together a jacket for the occasion. It was his first look at what Tesla was up against, a massive corporation that seemed utterly foreign compared to the work he'd been doing not so long ago in his home garage with his friends. As the conversation progressed, Marks told Wagoner about Tesla's problem with its transmission, hoping his friend might be able to help.

"Yeah," Wagoner said, "we've been having problems with transmission for the past 80 years."

It wasn't entirely clear to Straubel what Marks was hoping to get from the trip. Still, he could sense tensions escalating between Marks and Eberhard. The two spent much of the trip arguing. Eberhard's demotion from CEO to president hadn't been smooth, leaving the staff feeling torn between the two sides. Many had long been friends with Eberhard and were loyal to him; others believed it was time for new leadership.

Their conflict pointed at yet another rift taking shape. When Eberhard had brought his idea for an electric sports car to Elon Musk, they had seemingly shared a vision for what the company might become. But with each hard-won milestone, the company's—and Musk's—ambitions grew ever larger. Those ambitions were running headlong into the realities of the moment: the Roadster was a mess, threatening to undo all of their future plans. Marks inherited the unenviable position of trying to answer what Tesla was **in the moment.** He didn't have the luxury of considering what it might become; he needed to save it today. He would suggest a different road from the one Musk had envisioned, and in doing so, he would quickly seal his own fate.

Wagoner's interest in Tesla's business was about more than a desire to catch up with an old friend.

The debut of the Roadster a little more than a year earlier had caught GM's attention in a big way. In 2001, Wagoner had brought in Bob Lutz, who had helped usher in the Viper during his days at Chrysler, to become vice chairman and help reinvigorate the automaker. Lutz's first trick was to tap a young designer from the company's California office, a man named Franz von Holzhausen, to create a two-door roadster called the Pontiac Solstice. (Von Holzhausen would very quickly thereafter go on to become director of Mazda's North American design operations.) Lutz's hope had been to impress the industry with the Solstice at the 2002 Detroit auto show—to demonstrate that the behemoth carmaker could move quickly, that it had some spark left in it.

As Lutz watched Tesla's all-electric Roadster come out—from an unknown startup in California, no less—he fumed about why his team couldn't do the same. "That tore it for me," Lutz, then seventy-five, recalled. "If some Silicon Valley start-up can solve this equation, no one is going to tell me anymore that it's unfeasible."

Lutz, a former Marine fighter pilot, was not somebody you'd mistake for a treehugger. He famously called the idea of global warming a "crock of shit," and as if to emphasize the point, he kept a massive V16 engine in his office. But he did understand marketing like few others of his generation. He understood that GM had ceded ground—not to that little Tesla, a company that few in Detroit thought had a shot at success, but to a stronger enemy: Toyota. In 2006, Toyota dethroned GM from its perch of seventy-six years as the world's best-selling carmaker. Toyota's Prius hybrid sedan had helped create an image for the automaker as a company on the cutting edge, whereas GM was seen as a dinosaur. The Detroit company had even starred as the villain in that year's documentary **Who Killed the Electric Car?,** which cast the auto-maker in an unflattering light for pulling the plug on its EV1.

Months before Straubel's visit to Detroit, Lutz, dressed in an impeccable gray suit, heavily starched white shirt, and purple tie, stood onstage at one of the 2007 Detroit auto show's most anticipated press conferences to reveal GM's idea of what an electric car could be: the Chevrolet Volt. The sedan prom-ised to run forty miles on an electric charge, then use an onboard gasoline engine to generate electric-ity to allow the vehicle to go further. The mashup was seen as the solution to combat a problem that was still plaguing Straubel: high battery costs.

The Volt and the Roadster were in different weight classes. Where the Roadster aspired to sexi-ness and prestige, the Volt aimed for accessibility. And while Tesla operated on a shoestring, GM had vast resources and decades of car-making history to draw from. Still, for anyone watching Tesla, the message was clear. The goliaths had awoken from their slumber.

Tesla hadn't yet taken its Roadster into production, but it was already refocusing more of its energy on the next step. They hired an outside design firm called Fisker Coachbuild to help come up with what the Model S sedan should look like. The boutique had been founded two years earlier by Henrik Fisker, a Danish design expert. He'd overseen Aston Martin's design department, designing the V8 Vantage concept car and bringing to market the DB9 coupe, both of which had drawn a lot of attention. In picking Fisker, Tesla was following its established path: going with designs that evoked not so much cutting-edge technology as adrenaline-soaked thrills.

They'd be building a car from the ground up this time, and the costs of that quickly became apparent. The board wanted to spend around $120 million on the Model S, and Ron Lloyd busied himself trying to find ways to do so. Tesla had learned that many of the Roadster's supposed shortcuts—working with outside vendors, buying at scale—didn't work the way they'd envisioned. So they began looking for new shortcuts. Was there a way they could build their own car without having to establish a costly factory of their own? Could they draw from an established car company's parts catalogue and create their own composite car, the way the Detroit team had been envisioning in their proposed partnership with Ford? These were out-of-the-box ideas—and they came with out-of-the-box problems.

While Tesla had been founded on the notion that fully electric cars were the future, the Model S team was beginning to think that trailblazing that future was going to be impossible—at least for the startup's next act. The balance between range and battery cost was still too difficult to strike. GM's argument for a hybrid electric vehicle had a kind of inevitable logic. Lloyd shared with Fisker details of a proposed plug-in hybrid—a plan that Tesla closely guarded, but that any design team would need to take into account so that their design might accommodate a traditional gasoline tank and engine, in addition to the batteries and electric motor of an EV.

But as they pushed forward with the Model S, the Tesla team puzzled over the designs they received from Fisker. Lloyd expressed disappointment with them to his colleagues, pointing to the proposed vehicle's front end, which had a rounded grille. It was far from emulating the smooth, inviting lines of the Aston Martins that Fisker had previously done. "I just don't understand what they're doing with it," Lloyd told one colleague. "How is it so ugly?"

During a design review at the Fisker studio, Musk echoed Lloyd's unhappiness. At one point, Musk ran a photo of a low-slung two-door McLaren F1 sports car through Photoshop, stretching the image in such a way as to make it look like a four-door sedan. Presented with Musk's idea, Henrik Fisker marched over to a whiteboard, where he drew the silhouette of a conventionally beautiful woman. "This is the form that the designers all design their fashion clothing for," Fisker said, according to a person in the meeting. Then, he drew the outline of a pear-shaped woman. "Eventually, they'll make the dresses to try to fit this woman as well, so they can sell them. But it's not the same thing." Musk turned beet red in anger.

Fisker defended his work, saying the problem was rooted in the assignment. Tesla wanted the vehicle to be midsized, similar to a BMW 5 Series, and wanted to place the batteries beneath the car. That inevitably raised the roof line in an unflattering way. Some Tesla managers began calling it the white whale for its bulbous appearance.

The looks of the car wouldn't matter much if the company ran out of money to pay for engineering and building it. As Marks looked at their finances, he quickly concluded Tesla couldn't fund the development of the Model S on its own. He directed Lloyd to find a partner to foot some of the bill—a path that took Tesla back to Detroit. The team began courting Chrysler, hoping a deal could be reached there. Tesla laid out their plan, taking Chrysler executives deep into the technology. The two sides discussed co-developing a vehicle platform: Tesla could get a fastback version of their prototype, while Chrysler could get its own version of the sedan.

But in 2007, Chrysler was in trouble of its own; its corporate owner, DaimlerChrysler AG in Germany, was spinning it off to a private equity buyer. By fall, under new ownership, Chrysler had pulled the plug on the idea of working with Tesla. (However serious they had been about teaming up, it was never conveyed to the level of company presidents, senior executives later said.)

The rejection was a gut punch to Lloyd and the team. They felt like Chrysler had been playing them to ferret out ideas for their own electric car program.* And just as Lloyd's team learned about Chrysler's decision, they received similarly disturbing news from Fisker. The boutique that had been working since February 2007 to develop the look of the Model S disclosed that it, too, was developing its own hybrid electric car—a direct competitor.

Tesla shuddered at the development. Fisker held in its hands Tesla's road map for bringing out the Model S in 2010. For months, the partnering companies had shared engineering and design ideas, discussed the constraints involved in engineering a hybrid vehicle, and considered how best to create

---

* In the fall of 2008, the Chrysler team would announce plans for electric cars, including one based on the Lotus Elise that would also run on lithium-ion batteries. That vehicle never went into production as planned, though.

a vehicle with the performance characteristics of a sports car. Not once had Henrik Fisker or his team told them that they were thinking about developing a similar car. The Tesla team looked over its contract with Fisker. It included a non-exclusivity agreement that let Fisker design for other clients, including potential competitors. It didn't say anything about Fisker creating its own car. And why would they suspect that Fisker would? Fisker had no experience in developing an electric car. They were a design firm. The two teams had envisioned a badge on the Model S that would read "Designed by Fisker Coachbuild"—the relationship would be part of the boutique's marketing.

Fisker appeared to be copying Tesla, but in many ways its strategy was the opposite of Musk's, who sought more control of Tesla's product than he had with the Roadster. Fisker intended to focus on the looks of his car and outsource much of the business of engineering to suppliers. It was a strategy more in line with what Martin Eberhard had first envisioned for Tesla than the company's current, hands-on trajectory. There was an added, even more painful element to the perceived betrayal. Fisker was being funded with an investment from Kleiner Perkins, the venture capital firm that Musk had rebuffed a year earlier.

Straubel found himself at the center of the debate about what Tesla wanted to be when it grew up.

His team's battery pack and its ability to manage thousands of cells without setting off a fireworks show was the most unique thing the company had. Some, such as the investors at VantagePoint, wondered if that was really their near-term future—selling the battery pack to other car companies. Months before his demotion, Martin Eberhard had crafted an updated business plan, projecting that revenue from the sale of battery packs could rise to $800 million each year. The team pitching the idea to other companies was having early successes. Think, a Norwegian electric vehicle startup, signed a $43 million deal to get packs from Tesla for its own small electric car. And GM, with its Volt project, was interested, too. Straubel had pitched their team and was preparing to do it again.

But Musk was skeptical. He thought it was a bad use of Straubel's time. Marks, too, felt the company need to focus on getting the Roadster out—the company was on the hook for it already, with hundreds of orders. Worried about the reputational risk for the nascent Tesla if a client had problems, Musk said to Eberhard: "If their powertrain has problems and they blame our battery (which will be their first instinct), how will we deal with that?"

Antonio Gracias and Tim Watkins had more bad news on the money front. As their team continued to dig into Tesla's cost structure, it became clear there was a bigger problem than not fully understanding how much it cost to build the Roadster. It turned out the company's entire financial house was essentially built on quicksand, threatening to bankrupt them the moment they began making cars at Lotus's factory.

Here was the problem: Tesla would be buying batteries from Japan that were to be shipped to Thailand to be assembled into a battery pack, then shipped to the UK to be put into the Roadster, then placed on a boat for California. It was a trip that would take months. Meanwhile, Tesla would owe money to those parts suppliers—before they could generate cash from those sales. Watkins calculated the cycle would require hundreds of millions of dollars on hand to sustain. Tesla didn't even have tens of millions of dollars. Their issue wasn't simply cost but cash flow.

The team debated their options. Musk wanted to shutter the Thai operation in favor of making battery packs in California. Tesla could then fly the cars from the UK to the San Francisco airport—something that plane regulations wouldn't permit if batteries were pre-installed. The time saved by air-freighting would allow Tesla to turn them around more quickly and require less cash. He urged Straubel's team to set up shop in Silicon Valley, making the all-important battery packs themselves. Marks, meanwhile, argued for relocating **more** work to Asia. He wanted to take advantage of low-cost labor. Straubel and his battery-packs sat at the center of a tug of war.

The discussion only served to highlight what was already becoming clear to both men: Marks wasn't a fit for the top role at Tesla. His short tenure as CEO would ultimately be a footnote in the larger Tesla history, although he stepped in at a time when things could have gotten much worse for the company. Soon, he would be missed.

For Tesla's third CEO, Musk tapped a friend from Beverly Hills, Ze'ev Drori, a figure in the tech world from another era. He had founded Monolithic Memories, a semiconductor firm that had done pioneering work and was acquired in 1987 by Advanced Micro Devices. Drori later gained a controlling stake in Clifford Electronics, which he developed into a leading car alarm company that was sold in turn to Allstate Insurance. He fancied himself a car guy and dabbled in Formula One racing.

With new leadership, there was a house cleaning of the Eberhard people, a move designed to help the company cut costs. Eberhard himself wasn't spared. In the late fall of 2007, he was shown the door of the company he founded.

Eberhard had been noticeably unhappy for months; the move shouldn't have come as a surprise. But it was painful nevertheless. A month earlier, Marks, in one of his last tasks, told Eberhard that his position with the company had become untenable given Musk's persistent calls for the co-founder's ousting. He offered him a severance package to quit, but Eberhard hung on.

Weeks later, Musk was more direct. According to Eberhard, Musk, who already directly controlled four of Tesla's eight board seats (including his own as chairman), threatened to convert enough of his preferred stock options into common stock that he'd be entitled to pick three more of the company's board members. That would give him seven of the eight seats, and total say over the company's decisions—full power to get rid of Eberhard, in other words.

In creating Tesla, Eberhard had gotten the broad strokes right: about the potential for lithium-ion batteries, and the untapped possibilities for a high-end electric sports car. But he made painful, naive mistakes, from minimizing the complexity of making a car to losing track of the growing organization's finances. His biggest trip-up, though, the one that would gnaw at him for years to come: He lost control of the board. Each time Musk raised more money for the company, his grasp on the company grew tighter. Tesla was a game of control, and Eberhard had lost.

The tables had turned from just three and a half years earlier, when Eberhard celebrated the arrival of Musk's first check with a bottle of champagne with Bernie Tse and Laurie Yoler—friends he had hand-picked to sit on the board with him. While it may have seemed for a time like Tesla was

Eberhard's company, Musk's influence had been growing. From how to design the Roadster to what lay ahead for the company—its evolution from a sports carmaker to a maker of low-cost electric vehicles for the world—Tesla was now, for good or for ill, Musk's.

And now Musk, according to Eberhard, offered him an ultimatum: six months of salary worth $100,000 and an option to purchase 250,000 shares in return for his exit. But if he didn't sign the agreement that day, Musk would exercise his stock options and Eberhard would get nothing. He signed it, drove home in his Mazda 3—still with the "Mr. Tesla" vanity plate his wife had given him—and sunk into a deep hole. His calls and emails to certain fellow board members went unanswered. On an internet chatroom popular among Tesla fans, he found some solace, responding a few days later to rumors about his departure: "Yes it is true—I am no longer with Tesla Motors—neither on its Board of Directors nor an employee of any sort," Eberhard wrote. "I have also signed a non-disparagement agreement with Tesla, so I must, by contract, be a bit careful about how I word things.

"But I am also not going to lie about it. I am not at all happy with the way I was treated, and I do not think this was the very best way to handle a transition—not the best for Tesla Motors, not the best for Tesla's customers (to whom I still feel a strong sense of responsibility), and not for Tesla's investors."

A few weeks later, Tarpenning, the co-founder who'd been a loyal friend and confidant of Eberhard's, decided to leave as well. He said he felt like he'd accomplished what he wanted to, with the Roadster ready for production (though it would still require a lot more work if Tesla was to produce it without going bankrupt). And for many of Eberhard's allies, Tesla just wasn't as much fun without him.

Drori may have been CEO, but Musk was clearly vying for the reins. The day Musk announced Drori's hiring, the two men, along with Straubel and Dave Lyons, took a trip to Detroit aboard Musk's private jet, where they had meetings planned to discuss the troublesome transmission. During the overnight flight, Musk appeared shaken.

"He was really obsessed that this whole situation was out of control," Lyons recalled. "And that he had to get it back into control. He didn't understand what was going on and that he had basically bet the farm and sold all of his friends on these cars and he had to deliver on his promises. He was incredibly personally invested at that point."

Tesla's success was looking less and less probable to Straubel. He was growing tired—in part by his frequent travel to Asia, where he was tasked with developing a supplier network. Eberhard's early expectation that Tesla would be welcomed by suppliers had proven wrong. Battery makers didn't want to go anywhere near electric car startups, or touch the potential legal and reputational liability that might come with it.

One such skeptical battery maker was Panasonic. In their Silicon Valley office, Kurt Kelty, who sought out new business for the company's lithium-ion batteries, was known for rejecting requests from startups like Tesla. But in early 2006, he was approached by a Tesla engineer with a different kind of proposal. The acquaintance, whom he knew from the conference circuit, had been hired by the then-unknown carmaker. After he was shown a picture of the yet-to-be-revealed Roadster, Kelty became intrigued. This wasn't like all of the other EV startups he'd rejected. This looked like a legitimately cool car.

And so, in what would become a defining feature of the Tesla narrative, the right person came along, with just the personal history and skills required, at just the right moment, to push the odds in the company's favor. To the surprise of his family and employer, Kelty quit Panasonic to work for Tesla. He would turn out to be the secret weapon Straubel needed.

Straubel had never been to Asia before joining Tesla, and the visits to China and Japan to find suppliers showed him a world far beyond Wisconsin and the Stanford University campus. Kelty, on the other hand, was deeply enmeshed in Japanese culture. He had spent his teenage years in Palo Alto as a young entrepreneur, going on to graduate from Swarthmore College with a biology degree. His first car was a 1967 Ford Mustang that he had fully restored. During a year studying abroad in Japan, he met a young woman. Their blossoming relationship was disrupted when her parents found out about it; they quickly put an end to the idea of their daughter dating a foreigner.

Two years later, Kelty was living in San Francisco and running a fish export business, which took him to Japan from time to time, and where he had occasion to meet his old flame for coffee. Things quickly reignited between the two, and they eloped in the U.S. over her parents' objections. During their first year together in San Francisco, Kelty realized that despite the depth of their feelings for each other, he needed to win over her parents for things to truly last. So, against her wishes, he went alone to Japan in search of a local job. Speaking just enough of the language to order a beer, he began taking beginner Japanese classes. He set his sights on a job at a large manufacturer, betting it would be the kind of status symbol that would impress his new in-laws. Ultimately, he landed one at Panasonic and she returned to Japan to be with him. As a **gaijin** who made himself fluent in not only the language but the culture of Japan, he stood out. Kelty developed a fifteen-year career with the technology behemoth, his wife at his side all the while. More importantly, he was able to win over his in-laws. With two young children, they eventually moved back to the States, and to Palo Alto, where he founded Panasonic's Silicon Valley R&D lab.

At Tesla, Kelty, then forty-one years old, would become Straubel's guide to a new world. On paper, they were an odd couple: the worldly family man and the sheltered bachelor, whose yard in Menlo Park was still piled with EV1 motors. But the two connected over a curiosity for the world and a common interest in energy products. Together they made an appealing sales team: With his industry connections, Kelty could land meetings. He would open them by introducing himself and Straubel in Japanese, followed by Straubel's presentation of Tesla's technology, Kelty translating all the while. Straubel demonstrated an impressive, convincing understanding of the technology, while Kelty grasped the intricacies of Japanese business culture.

Kelty weighed the options and felt his old employer, Panasonic, would provide Tesla with the best cells. Sanyo would be next best.

To Straubel, it sometimes felt like they were getting nowhere on **either** front—meeting after meeting with low-level employees who often seemed to have neither experience nor expertise in battery technology. All of it increasingly felt like a waste of time. But Kelty assured him things were looking up. He understood that breaking through with a Japanese business was often about forging long-term relationships and developing consensus over the best business idea or technology. Kelty was playing a game that the engineering culture of Tesla didn't understand. Every two months, he'd fly to Asia for a round of visits, using his connections at his former employer to find ways to meet with people. The meetings were always polite but noncommittal—fuzzy almost. Finally, the president of the battery division of Panasonic sent a letter to Eberhard that said the company would never sell Tesla batteries and instructed them to stop asking.

Such a letter was unusual, even by Japanese business standards, but Kelty appeared unfazed. Instead, he urged patience, at a time when it was in short supply. After a meeting with his second choice, Sanyo, Kelty and Straubel were called back a few months later. This time their reception was markedly different. They were ushered into a large conference room on the top floor of the company's Osaka headquarters. As is the tradition, one side of the table was set up for Sanyo, one side for Kelty and Straubel. But this time, instead of meeting with a few low-level salarymen, they looked across the table at as many as perhaps thirty Sanyo managers and executives; folding chairs had been set up behind the front row of seats to accommodate them all.

As Kelty and Straubel began their routine, the

questions they fielded were focused on the usual concern: thermal runaway. How could Tesla ensure that the occasional defective battery didn't lead to devastating explosions within battery packs? As in previous meetings, Straubel had answers. This time, however, one of the midlevel executives in the back of the room answered his colleague before Straubel had a chance. At first it was surprising, but then it became clear: the Sanyo side was **getting it.** What Straubel was proposing wasn't difficult to understand; he was simply the first to have figured out such an elegant solution to the problem. The idea that surrounding batteries could propagate heat away from a runaway was unprecedented, and it was stunning. By 2007, a deal had been brokered for Sanyo to provide Tesla the batteries it needed.

But was it too late? For three years, Tesla had been Straubel's family. He'd developed friendships with Kelty, Gene Berdichevsky, and others on the team. Whatever money he got, he plowed it into Tesla shares. The Roadster represented a dream that he'd long nurtured. But all the effort was taking its toll. No matter how hard he and his colleagues worked, how many sleepless nights they pulled, how many exhausting trips they took, it didn't seem to be enough to pull Tesla out of its never-ending stall. Increasingly to Straubel, Tesla was looking like it might go the way of Rosen Motors, his first job in the auto business, which ended after its founders realized they were just burning money.

After yet another long flight home from Japan, Straubel pulled into his Menlo Park home. He found the house dark, only to realize that during all of his long hours and travel, he had forgotten to pay his power bill. He opened the refrigerator to the smell of rotting food. He stumbled in the dark until he found a can of tuna fish and sank to the floor to eat his dinner.

Was Tesla going to make it?

Money continued to be the company's foremost concern. They needed to raise more of it to fix their defective supply chain before Roadster production could kick into high gear, now scheduled for late 2008. Instead of taking money from investors, the board decided to raise debt that could convert into shares down the road, when Tesla became more valuable. Musk again dug deep into his ever-shrinking fortune. By 2008 early, he had pumped in $55 million of the $145 million that Tesla had so far raised, all while SpaceX continued to struggle to build its rockets.

With a bit more money in the bank now and plans to start seeking deposits from European buyers for the Roadster, as well as some improvements to the car to justify an eventual price increase, Musk and the board saw a way out of their financial woes. The success of the Roadster would, they hoped, help Tesla raise one last round of investment money in late 2008, before taking the company public in 2009 on the excitement and promise of the forthcoming Model S. Fisker had seemingly screwed them, but it might have been for the best. Increasingly, Musk believed Tesla needed to control its own destiny, not depend on others.

With Eberhard's team mostly culled, new CEO Drori and Musk began rebuilding their leadership ranks, but with a different focus. Whereas Eberhard seemed to delight in hiring managers from tech rather than autos, the two quickly zeroed in on seasoned car executives. Within Ford Motor Co.'s finance department, they found Deepak Ahuja, whom they tapped to become CFO, a role that had been left essentially vacant since the company's inception. The former GM designer turned Mazda manager Franz von Holzhausen would take over where Fisker left off, overseeing design. And they looked for a seasoned product manager to help resume work on the Model S and take the Roadster over the finish line.

In March, Drori opened the pages of **The Wall Street Journal** to see a story from Detroit about a personnel change at Chrysler. Mike Donoughe, whom the paper described as one of the company's best engineers, had quit suddenly after twenty-four years, following the arrival of a new owner and CEO. Anonymous sources said he'd left after a disagreement over the direction and pace of the company's crucial development of a new midsized vehicle, dubbed Project D, to compete against the Toyota Camry.

Drori made quick work of tracking him down. Despite the team's previous entanglements with the automaker, by June they had reached an agreement for Donoughe, then forty-nine, to join Tesla as executive vice president of vehicle engineering and manufacturing. His offer gave a peek into the challenges a startup faced in hiring experienced executives. Donoughe's annual salary of $325,000 would be greater than what the former CEO, Martin Eberhard, had earned. Donoughe was granted the option to buy 500,000 shares in the private company at an estimated price of 90 cents apiece. He would partially vest those options over the course of four years. But more than that, Tesla agreed to cover some of his severance pay from Chrysler if it determined his new employer constituted a competitor, and thus a breach of his agreement. By Detroit's standards, the compensation plan wasn't so generous, especially considering the added cost of living in Silicon Valley.* For Tesla, he represented a significant line item.

Donoughe was to be responsible for a wide swath of the company, including development of the

* For contrast, the **Detroit Free Press** reported, Chrysler in 2008 promised about fifty executives retention bonuses, as it faced possible defections ahead of a 2009 bankruptcy reorganization. Donoughe's former colleagues were expected to get bonuses on top of their normal pay of between $200,000 and almost $2 million.

Model S. Very quickly, however, it became clear that he needed to focus his attention on righting the Roadster program.

It was time for a fresh accounting of the Roadster and its costs. Straubel took the car apart, laying out every piece. The team attached sticky notes to each, scribbling the price they were currently paying and what it needed to be. They reported each week to Donoughe. Then they set about trying to figure out ways to cut costs, either through engineering a cheaper solution or finding a less expensive vendor.

Donoughe held a morning meeting each day to plot the team's most important tasks. He wanted to convene at 6 a.m. but compromised at 7. In other ways, Donoughe wasn't so accommodating. Getting the Roadster to a build rate of more than twenty vehicles a month, from five when he joined, was like upending a game of whack-a-mole. The Tesla team's strategy had essentially been to bury problems when they arose, but in a way that didn't prevent them from rearing their heads again. Donoughe had a different strategy in mind. He wanted to cut off their heads so they could never return.

He had spent years climbing the ranks of Chrysler, with an early part of his career overseeing a shift at the Sterling Heights Assembly Plant's body line, where parts were welded together. A grueling environment, he was expected to deliver sixty-eight jobs an hour—and paid holy hell if he missed just one. He expected the same accountability at Tesla, where the team wasn't accustomed to such rigor. In one particularly tense meeting, an engineer reported his plan to address a problem with a supplier. Donoughe listened. He remained silent. No one talked, waiting for him to respond. Seconds ticked by like an eternity. Finally, Donoughe asked the man: What did the supplier say? The engineer said he hadn't called yet. Again, Donoughe asked: What did the supplier say? Again, the engineer said he was planning to call. But Donoughe didn't want such an excuse. If this supplier was causing a problem that was keeping the Roadster from ramping up production, the call couldn't wait. **Go now and call.**

Part of the trouble for Tesla was that the parts they required either came in late or with design problems that needed fixes. Tesla's original plan to use parts developed for Lotus had long ago been scrapped. Fewer than 10 percent of parts were now shared between the Roadster and the Elise. The structure of the car had had to be redone to handle a thousand-pound battery pack in the midsection along with a watermelon-sized motor in the rear. (They left a small space for a trunk, supposedly large enough for the stereotypical sports car accessory: the golf bag.) The overall vehicle was about six inches longer than the Elise. Basically, the only parts carried over were the windshield, dashboard, and front wishbone, as well as the removable soft top and side mirrors, which would have been costly to develop and test for safety.

The transmission continued to be a snag. The company had gone through two versions and still hadn't found a solution. It was in litigation with the large parts supplier that it had commissioned, Magna, over the issue, Tesla arguing that it wasn't getting the company's best engineers on its program. News of Tesla's delayed Roadster reached Detroit quickly; engineers in the hallways of auto component maker BorgWarner were talking about how the startup had stumbled with its transmission. One Tesla fan suggested to Bill Kelley, a long-time executive at BorgWarner, that his drivetrain R&D team might be able to help Tesla out. He'd been pushing for the company to prepare for the eventual arrival of electric vehicles, but early failures had left the company's board reluctant to make investments in a new area of business. He saw a deal with Tesla as a way to bolster his argument to the board.

Kelley sent an email to the address listed on Tesla's website offering to help. He quickly got a call back and, eventually, an invitation to present to the team in California. Kelley came ready with his pitch; he was surprised at its chilly reception. Musk sat quietly for about thirty minutes at the other end of a conference table, head down, looking at his phone, until he at last piped up: "Why do I need BorgWarner?"

Kelley was taken aback. BorgWarner was one of the world's best drivetrain suppliers, so established

that its name is on the trophy handed out for winning the Indianapolis 500. Kelley answered that
BorgWarner specialized in engineering challenges,
much like the ones the Roadster was currently facing. "And we're pretty good at it," he said.

Musk took the opportunity to reveal that he'd already signed a deal with another supplier, Ricardo,
to make the transmission. Kelley asked how much
Tesla was paying them.

"$5 million."

"I'll do it for $500,000," Kelley said. He laid out
a proposal: Let both companies compete to see who
can come up with a transmission that fits Tesla's
needs. Winner gets the business.

It was the kind of deal Musk had been pushing
the team to get with battery suppliers; he hadn't
liked the idea of being overly dependent on just
one. So, alongside Ricardo, BorgWarner was engaged to come up with a transmission. They did,
and it ultimately won out over their competition.

With that hurdle cleared, and as workers prepared to increase their modest production numbers
into somewhat **less** modest numbers, Musk called
Tim Watkins in Chicago. The UK supplier that
Tesla had engaged to make body panels had walked
away from the job after only making a few. Without
them, Tesla would give up its slots in Lotus's production schedule—slots it had to pay for whether
they were used or not. It was a disaster in the making. But Musk seemed to have it under control; he

even joked with Watkins about how it gave him an opportunity to drink wine in France, where they had found another supplier to do the work. He hopped on a jet and swung through Chicago to pick up Watkins, then flew to the original supplier's factory. There they personally yanked out the tooling, salvaging it for their new supplier, whose workers began cranking out panels by hand while Watkins figured out a more sustainable approach.

With so many challenges to deal with on the Roadster, traditional car guys like Donoughe thought it a distraction to work on a future car like the Model S, when failure with the first car would doom the second—and potentially end the company. Detroit automakers were naturally disinclined to talk about what was next in line for production, fearful that it would cannibalize sales of their current offerings.

But Musk didn't have that luxury. Produced or not, the Roadster had already served its purpose. Much in the way that the proof-of-concept tzero had sold Musk on the idea of Tesla, the Roadster had allowed Musk to sell other investors. Now, he needed the Model S—not just to drum up sales revenue, but to reach more people about his company and its mission.

Debate continued over the Model S prototype, specifically what size it ought to be. Donoughe had inherited the Detroit office, which, despite Musk's earlier protestation to Eberhard that they should all be fired, plowed ahead with its work. They continued to frustrate some back in California, too, Straubel among them. They seemed secretive and overly confident—maybe even dismissive of what Straubel's team had already accomplished. In his mind, they were indecisive. They kept arguing about vehicle size, when Straubel wanted to get back to the roots of what Tesla was about, the excitement of creating. "I just want to build the damn car!" Straubel thought. So with members of his team in San Carlos, he quietly began working on his **own** Model S prototype—an all-electric vehicle that would use the same kind of battery technology he had developed for the Roadster.

The Mercedes CLS large sedan seemed about the right point of reference. Straubel got one, hacked out the engine and gas tank, and began converting it into an all-electric electric prototype, just as he had done several times before. But this time was different. This was a real luxury car. His team preserved all the Mercedes refinements, working carefully to keep the interior intact. When they were done, the drive surprised even Straubel. The Roadster was still raw; their new electric prototype was magic. It was a massive sedan yet had the punch of a sports car. And unlike the Roadster, which rode rough, the electric Mercedes, with its finely tuned suspension system, fairly sailed over the road.

Musk was no less excited than Straubel about it—giddy, in fact. He drove it several times. **This was what the Model S could be.** Tesla may have been a total mess in its books, but on the road, behind the wheel of their prototype, they had new hope.

"It felt like it was going to change the world if we could have people experience this," Straubel said, "if we could actually productize it and make it work."

CHAPTER 8

EATING GLASS

When he was a kid growing up in South Africa, Elon Musk's mother called him Encyclopedia after his reading habits and ability to absorb information. "We could ask him anything," she'd later write. "Remember, this was before the internet. I guess now we would call him the internet." By his own account, Musk's childhood was troubled; he has alluded in several interviews over the years to issues with his father and to bullying he received from classmates during his school years. His parents' divorce in 1979 was followed by years of custody disputes. At age ten, Musk told his mother, who was struggling to pay the bills, that he was going to live with his father. "His father had the Encyclopaedia Britannica, which I couldn't afford," she later told a reporter. "He also had a computer, which was very rare at the time. That's why Elon loved it."

It was a period that clearly shaped him as an adult. While as a kid he may have doubted the wild ideas that made him question his own sanity, he since learned a certain kind of defiance and self-assurance to pursue those ideas, even when told they were insane. In many ways, he was devoting his life and fortune to the ultimate preparation for a coming disaster. SpaceX was about creating a way for humans to live on other planets in case Earth became unviable. Tesla was about developing technology to save the planet from climate collapse.

But what had started as a hobby with Tesla was becoming a second full-time job. Years later, Tesla executives would privately joke that Musk's first love was SpaceX; his relationship to that company was like a marriage. Tesla was the spicy mistress, offering him drama and passion. But instead of throwing it over the side in 2008, like the brothers behind Rosen Motors had had to do when the financial stakes became untenable, he became even more committed to seeing it through. Thorny as it was, he couldn't quit Tesla.

By the summer of 2008, it seemed that maybe the worst was over for Tesla and Musk. His four-year involvement in the company had certainly taken its toll. His marriage with Justine had frayed beyond repair; he quietly filed for divorce that spring. His former business partner, Martin Eberhard, bitter and hurt from how he was fired, was lambasting Musk on a blog that tracked the company's struggles, feeding news coverage in Silicon Valley and painting Musk as a villain.

There were still concerns from customers about the potential for lost deposits. Musk told them he would personally guarantee them. "Unequivocally, I will support the company to whatever extent is needed. I have a long way to go before [money's] a problem," he said. His assurances did what they were supposed to; only thirty of the now thousand-odd customers who had put down deposits wanted refunds. The car was too enticing a proposition for buyers to give up. Even Eberhard's buddy, Stephen Casner, who had helped connect him with AC Propulsion years earlier, remained excited to get his hands on a Roadster. "I really wanted the car," Casner said. "I suppose if I had really strong principles, and because of my friendship with Martin, I could have or perhaps should have . . . canceled my order, because of the way they treated Martin, but those principles were not as strong as my desire to have the car."

The arrival of the first true Roadster months earlier had also helped. In February, Musk greeted the delivery of the first production car, or P1 as the executives had taken to calling it, along with the rest of the Tesla team. The body had shipped from the UK and engineers rushed to install the battery pack. "I want to be very clear: We're going to put thousands of vehicles out there," Musk told a crowd of workers and gathered reporters. Next on the agenda would be the Model S. "Beyond that is Model 3," he told onlookers, "and we're going to do parallel investing, so we're not going to wait for Model 2 to work on Model 3."

Tesla wasn't going to stop until every car on the road was an electric, he continued. "This is the beginning of the beginning."

Early reviews of the vehicle were impressive—if not marked with asterisks. There were concerns over whether the company could stay afloat, and about its transmission. An editor from **Motor Trend** who took a test drive likened the experience to being "teleported down the barrel of a rail gun, head pulled back by a hard, steady acceleration." Michael Balzary, better known as Flea of the Red Hot Chili Peppers, wrote a blog post about his experience driving a prototype: "man it was unbelievable. it drove like nothing i have ever been in before, made my porsche feel like a golf car!" Musk would take his car to **Tonight Show** host Jay Leno, a well-known car enthusiast, for a test ride. Leno marveled: "You've managed to make essentially a true sports car."

It had been a rocky road, but the end was in sight. Musk had arranged a deal with Goldman Sachs to raise $100 million, mostly from Chinese investors. It was the kind of cash infusion that would relieve some of the financial stress on Tesla and put it on a path to going public, when it could then raise the kind of major money it would need to make the Model S. Musk even found new love that summer at a London nightclub, where English actress Talulah Riley caught his eye. Things were looking good for a change.

But just as Tesla's fortunes were improving, the bottom fell out.

It started when Lehman Brothers collapsed one weekend in early September 2008, leading to one of the largest bankruptcies in U.S. history and sending the global financial system into chaos. Credit markets froze. General Motors, Ford, and Chrysler began talking about the need for a government bailout of the auto industry.

If GM was in trouble, Tesla looked like toast. As companies and investors reined in spending, Musk's deal with the Chinese appeared to be in jeopardy. Musk complained to his colleagues that his bankers at Goldman weren't calling him back. Toward the end of September, Goldman Sachs announced that it had turned to Warren Buffett's Berkshire Hathaway for a $5 billion infusion to stabilize its business. By the time Musk reached his contacts at Goldman, the economic outlook was bleak.

Miraculously, Goldman, at Musk's prodding,

offered to put in its own money. But the terms valued Tesla at a level so low that Musk couldn't stomach accepting it.

Musk gathered the senior executives in a conference room in San Carlos to deliver the news. It was clear he was going to have to put more of his own money in. And with that decision came another one, about who ought to be CEO of the company. He decided to oust Ze'ev Drori—and appoint himself (making Musk the fourth CEO to hold the position in about a year's time). Musk told the senior leaders they needed to prepare for massive layoffs to preserve cash. The Model S would be the key to their survival, a risky gambit that would depend on every part playing out seamlessly.

The mechanics of the plan, reductively speaking, were this: Tesla would cut to the bone to save cash, and hope Roadster reservation holders didn't freak out and seek their deposits returned. They would quickly reveal a design for the Model S to stoke further interest in the company, then again begin taking deposits. This would give them enough runway to coast until further investment could be generated. If the plan succeeded, they could eke their way through to Model S production. If it failed, they would stiff an ever-growing base of customers, all but spelling their demise.

Darryl Siry, head of sales and marketing, pushed back against the plan, telling Musk he thought it was unethical to take deposits for the Model S when the company had no pending plans to actually build the car.

"We either do this or we die," Musk told him.

Tesla stormed into action. It began by cutting around 25 percent of its workforce. Inevitably, word leaked. Valleywag, a website that trafficked in Silicon Valley gossip, ran a story in October 2008 that said the company was cutting 100 workers, and that Drori was out. Reclaiming the narrative, Musk posted a message to the company's blog, announcing that the automaker was restructuring to focus its effort on getting the Roadster out and providing its powertrains to other companies.

"These are extraordinary times," Musk wrote. "The global financial system has gone through the worst crisis since the Great Depression, and the effects are only beginning to wind their way through every facet of the economy. It's not an understatement to say that nearly every business will be impacted by what has unfolded in the past weeks, and this is true for Silicon Valley as well." He added that there would be a "modest reduction" in headcount that he described as "raising the performance bar at Tesla to a very high level."

"To be clear, this doesn't mean that the people that depart Tesla for this reason wouldn't be considered good performers at most companies— almost all would," Musk wrote. "However, I believe Tesla must adhere more closely to a special forces philosophy at this stage of its life if we aspire to become one of the great car companies of the 21st century."

To many, Musk had stabbed his employees in the back, and now, by denigrating their performance, he was twisting the knife. Musk gathered those left behind for a blunt conversation about the state of the business. While it was obvious that times were tough, the depth of the trouble wasn't clear to everyone present. He revealed that the company had just $9 million of cash on hand—it had already blown through millions of dollars in deposits for the Roadster.

This revelation didn't sit well with everyone; word of the meeting again quickly spread to Valleywag, which published an email from a company insider that sounded the alarm, both about the distressingly low cash balance and concerns about spending through deposits.

"I actually talked a close friend of mine into putting down $60,000 for a Tesla Roadster," the email said. "I cannot conscientiously be a bystander anymore and allow my company to deceive the public and defraud our dear customers. Our customers and the general public are the reason Tesla is so loved. The fact that they are being lied to is just wrong."

The revelation wasn't just embarrassing to Musk, it torpedoed his plans to raise cash through orders of the Model S. How would new potential buyers feel if they knew how incautious Tesla was about holding its deposits? Furious, Musk wanted to know who betrayed him. He called in a private detective to fingerprint workers. A few days later, Musk sent out a company-wide email with a message from Peng Zhou, director of R&D, apologizing for revealing the company's finances. "The past month has been very difficult, sitting through planning meetings and watch employees make in or out of the layoff list. It is so sad to lose 87 employees in a week," Zhou wrote. "I became very upset and did the very foolish thing of writing a letter to Valleywag. I have never thought this letter would create such an upsetting situation for Tesla Motors and I should have never sent that letter."

His repentance didn't save him. Zhou was ousted.

On Nov. 3, Musk released a statement that the company had received a "$40 million financing commitment." Beyond saying that the company's board had approved a new debt financing plan, the details were sparse. The statement said the financing was based on commitments "from almost all current major investors," while saying the round would also be open to smaller current investors. "Forty million is significantly more than we need," Musk said. "However, the board, investors and I felt it was important to have significant cash reserves."

In reality, things weren't so clear-cut. Yes, Musk was asking his investors to put more money in; but behind the scenes, he faced opposition. The head of

Tesla's chief venture capital investor, VantagePoint's Alan Salzman, had been unhappy with Musk for months. He was furious that Musk had used his power to name himself CEO, and he worried Musk was overcommitted, with SpaceX and his cousins' budding solar company, SolarCity. Salzman threatened to withhold further funding; it seemed to some Tesla executives he wanted to become CEO and chairman himself.

Tension between the two men had been mounting for some time. Earlier that year, Salzman began playing a bigger role in Tesla after VantagePoint's representative on the board, Jim Marver—who had questioned Eberhard's grasp of the finances and worried about the company tapping Roadster deposits—was involved in a horrific bike crash that left him in the hospital for several days. By the time he'd recovered, VantagePoint decided it was time to step back from the board out of frustration with how things were going. "We just weren't in sync with a bunch of the thinking in terms of balancing risks and opportunities," Salzman said.

Still, Salzman stayed close to the investment. Employees had overheard he and Musk in a screaming match at the automaker's offices. At issue was disagreement about nothing less than the future of the company. Musk wanted to turn Tesla into a global automaker, plain and simple, a company that could compete with the titans of Detroit and force the industry into the electric car business.

Some at Tesla felt VantagePoint wanted to make a safer bet, either becoming a supplier to other car companies or else getting acquired by one of them. Tesla "was building a car, not a car company" was a phrase the VantagePoint team often used, according to Tesla executives, highlighting their thesis that the success of the Roadster would demonstrate to other car companies the might of its electric powertrain. The sports car would be a rolling billboard aimed not at consumers but at other carmakers.

As insiders saw it: VantagePoint thought Tesla could be the next BorgWarner while Musk thought it could be the next GM. Salzman later disputed the idea that he wasn't supportive of the broader vision of becoming a car company but noted that it was a tough sell in 2008 with a car that wasn't yet profitable. "The first rule of business is to stay in business," he said. And, he noted, the idea of selling the guts of the electric car to automakers had been a part of the 2006 business plan created by Martin Eberhard. "It seemed like an idea that could bridge the gap and raise access to important funds."

Few outside Musk's inner circle knew how much personal risk Musk was taking on. One evening as Musk and others pored over the latest financial projections, the phone rang. It was Musk's personal money manager. "Yeah, I know no one is selling anything right now," Musk said into the phone. "But Tesla has got to make payroll. Find something that you can turn into cash." He was writing personal checks to cover employee salaries and putting their work expenses on his personal credit cards.

Back in LA, Musk had dinner at a Beverly Hills steakhouse with a friend and early Tesla investor, Jason Calacanis. Musk was in a dark place. His third rocket had just exploded on liftoff, and SpaceX would go under if the fourth did. Calacanis had read that Tesla only had four weeks of money left; he asked Musk if that was true.

No, Musk said. Three weeks.

Musk confided that a friend had loaned him money so he could cover his personal expenses. There were other benefactors: Bill Lee, Al Gore's son-in-law, invested $2 million, and Sergey Brin put in $500,000. Some employees were even writing checks, not sure they'd ever see the money again. Things looked bleak. Still, Musk said he wanted to show Calacanis something. He pulled out his BlackBerry and revealed a picture of a clay mockup of the Model S.

"That's gorgeous," Calacanis said. "How much can you make it for?"

"Well, it's going to go 200 miles," Musk said. "I think we can make it for $50,000 or $60,000."

That night Calacanis returned home and wrote out two checks for $50,000 each and a note to Musk: "Elon, looks like an incredible car . . . I'll take two!"

—

With just a few weeks left of funds to cover payroll, Musk was close to finalizing the paperwork for
the fundraising round that would save the company
when he discovered that VantagePoint hadn't signed
all of the documents. He called Salzman about it,
and, according to Musk, Salzman told him they
had a problem with the valuations that were proposed in his terms sheet. Salzman suggested Musk
give the firm a presentation the following week to
sort it out.

Given Tesla's precarious position, Musk took
the request as a threat to the very existence of the
company, and his vision for what it could become.
"Based on the cash we have in the bank right now,
we will bounce payroll next week," Musk told
him. He offered to come by the next day, but, as
Musk told it, Salzman balked. It was a battle that
had been brewing between the two men almost
from the beginning—a battle of giant, forceful
personalities. Musk suspected the delay was part of
a strategy to send Tesla into bankruptcy, and for
Salzman and VantagePoint to seize control of
Musk's fledgling business.

It was bare-knuckle brinksmanship. Without
Salzman's money, Musk would need to find another
way to raise cash. And as an investor, VantagePoint
could block Musk from trying to raise money from
outside investors. Musk decided to double down.
He would borrow the money himself, from SpaceX,
a move that could give Tesla the lifeblood it needed

but also deepen his personal losses if things went wrong down the road. He proposed to the other investors that they pony up money in the form of a loan to Tesla. To stoke their competitive juices, Musk told them if they demurred, he'd raise all of the $40 million without them.

It was a risky gambit, but it paid off. Rather than miss out on what Musk saw, the other investors opted to match his $20 million. Ultimately, Salzman backed down. He had no desire to send an investment into bankruptcy and he disputed the idea that he wanted to take over running Tesla. The deal closed on Christmas Eve.

Musk, who was at his brother's home in Boulder, Colorado, broke down in tears. He had barely averted a crisis that could just as easily have cratered his dream of an electric car. What had begun more than four years earlier as a side project to SpaceX had evolved into a major drain on his time, money, and love. All of his fortune was now on the line. From the depths of the Great Recession, he'd done something that other U.S. automakers were unable to do: avoid bankruptcy. Congress had rejected a bailout of GM and Chrysler that December. President George W. Bush had temporarily rescued them from bankruptcy with temporary loans, but both companies were headed there shortly.

Not Tesla. Not if Musk, newly empowered, could remake the company in his own image.

—

Tesla still had an unsavory bit of business to attend to: They had to effectively raise the price of the Roadster. It was a gamble, as many customers had grown impatient with delays, not to mention concerns about a foundering U.S. economy. Hundreds of reservation holders had by now canceled their orders, and the company's treasury was seeing a run on deposits. Now Musk wanted to raise the price on those 400 remaining, for a car these would-be buyers had not only committed to buying but placed deposits on to the tune of anywhere from $30,000 to $50,000 (the first 100 customers paid $100,000 for the privilege), putting them on the line for even more than they'd planned to spend (the 2008 Roadster was supposed to start at $92,000 for some). This could be, for many, the last straw.

In January, Musk sent out an email to customers explaining why the company had to take this dramatic step. Several hundred of them got personal calls from Tesla sales reps telling them that they needed to reconfigure their vehicle's options. Many of the features that had previously been standard were now add-ons; previous add-ons became even more expensive. Going forward, the Roadster would start at $109,000, with about $20,000 worth of additional options available. That was a dramatic increase from the $80,000 starting price that had been detailed when the Roadster was first revealed in 2006.

Responses were mixed. Billionaire Larry Ellison, the co-founder of Oracle, told the team he wanted to configure his car so that it was as expensive as possible to help them generate whatever revenue they needed. One customer posted the price-increase email on his personal blog along with his accommodating reply. "We complained a lot but, in the end picked a set of options and agreed to pay the price increase because we want Tesla to be successful and we want our car as soon as possible," wrote Tom Saxton, a vocal early-reservation holder, part of the grassroots community surrounding Tesla that had taken root on internet chatrooms and blogs. "It didn't seem worth it to spend a week complaining and arguing about it, not when our car was ready to go into production."

As negative reactions mounted, though, it became clear inside Tesla that Musk needed to hold a round of town halls with customers, to answer questions and allay concerns. He had done so in the past, after firing Martin Eberhard, and it had gone smoothly; Musk generally received a hero's welcome. This time, though, customers began to show their frustration.

They weren't the only ones put out the by the constant delays, Musk wanted them to know. "I cannot understate the degree of grief that I've

personally gone through and many of the other people at Tesla have gone through to make this work,"
Musk told a crowd in Los Angeles. "When I say it
was like eating glass, I mean glass sandwich every
bloody day."

With Tesla's focus finally shifting from sales of the
Roadster to increasing production, in early 2009,
JB Straubel's work on the Model S prototype needed
to accelerate too. Musk needed a shiny new car to
sell. They were well short of a production-ready vehicle, but Musk wanted something he could showcase, a car that at the very least looked and drove
like the Model S he'd dreamed up. They didn't have
time to squander: Musk scheduled a reveal party for
late March, just a few months away.

Franz von Holzhausen, the former GM designer,
set to work in a corner of SpaceX's rocket factory,
beneath a white tent to differentiate it as the Tesla
area. Tesla engineers began with another full-sized
Mercedes-Benz sedan and began cutting it apart.
They would use the chassis and wiring beneath the
body of the Mercedes as the foundation for their car,
then couple it with what had quietly been agreed
upon for the Model S's body, built from fiberglass.
For their part, Straubel's team would have to figure out a way to fit a preexisting Roadster battery
pack and motor into their improvised vehicle. Von
Holzhausen worked on the design during the day

while at night the engineers worked to figure out how to secure the body of the Model S to the chassis of the Mercedes and make it run.

They operated at a grueling pace until the last minute when, on the night of the reveal, Musk gathered Roadster customers and other distinguished guests to see his latest creation. A party was set up at SpaceX, orange trees brought in to decorate the space. The centerpiece of the night was when Musk rolled out behind the wheel of their Frankensteined creation.

The Model S was stunning, a sleek sedan evoking the contours of an Aston Martin but with interior space rivaling an SUV. They claimed that a mountain bike, surfboard, and fifty-inch TV could lie flat inside—at the same time. Instead of storing the batteries in a giant box in the trunk, as they'd done with the Roadster, Straubel's team imagined them in a shallow rectangular box beneath the floor. A motor, much smaller than the typical gas-powered engine, would be fitted between the back wheels. With the bulk of the drivetrain beneath the car instead of under the hood, it opened up a ton of interior room.

Musk exited the car to woo-hoos, cheers, and "wows," audible over the thumping bass that had ushered in the reveal. "I hope you like what you see," Musk told the crowd, Straubel behind him nervously fidgeting, his hands in his pockets.

"You're looking at what will be the world's first mass manufactured electric car," Musk continued. "This is, I think, really going to show what's possible with electric vehicles." He went on to promise that five adults could sit comfortably inside, plus two children's seats facing rearward in the back. In the front seat, a giant video screen replaced the center console. Instead of a radio with its usual nobs, there was a touch screen, with the same functionality as the Apple iPhone that had been released less than two years earlier. (Tesla's screen was unveiled a year before Apple would release its iPad.) While the car looked comparable to a Mercedes E-Class or BMW 5 Series, Musk rattled off performance promises that would, if true, far eclipse those cars. The Model S would be able to go from 0 to 60 in less than six seconds. It would have a range of 300 miles on a single charge. It would start at $57,400, which meant that—with a newly instated federal tax credit giving buyers $7,500 off the purchase of battery-powered cars—customers would be effectively paying just under $50,000. The Model S would begin production in 2011, he said.

"Would you rather have this or a Ford Taurus?" he asked to laughs.

Musk had laid the foundation for his dream of an electric car—if not exactly one for the masses just yet, then at least one for the comfortable few. Now, he just needed to figure out how make it real.

To automotive insiders, his vision seemed improbable at best, a joke at worst. Detroit had tried to make a consumer-level electric vehicle; the world had seen how badly that went.

They wouldn't think it was funny for long.

PART II

THE BEST CAR

CHAPTER 9

SPECIAL FORCES

P eter Rawlinson went straight from Los Angeles International Airport to Santa Monica for dinner. Fresh off a flight from London, he wasn't hungry. His internal body clock said it was the middle of the night. But he was excited to hear what Elon Musk had to say. Just two days earlier, Musk had called Rawlinson for the first time. Rawlinson was at his farmhouse in Warwickshire, a roughly two-hour drive northwest of London, where the Imperial College grad had carved out a consulting career working with car companies looking to try new things. Even before Musk called, Rawlinson had followed the news of Tesla's struggles. The idea of creating a car company had long been a fantasy of his; he had even designed and hand-built his own roadsters years earlier.

In mid-January 2009, Musk had staved off

bankruptcy, but the executioner remained in waiting. After three months as CEO, he faced balancing three tall tasks: continuing to deliver Roadsters to customers, to keep cash coming in; building a team that could execute his vision of what the Model S could be; and finding the money to do it all. Sitting with Musk and his newly hired car designer, Franz von Holzhausen, Rawlinson wasn't sure how he fit into those demands. Musk, he assumed, was like all of his clients—looking for advice on new ways to engineer a car using computerized tools, say, or how to make one without the standard materials.

Of course, that was in normal times, and there was nothing normal about the moment they were living in. The automotive industry was undergoing painful changes following the financial markets' collapse the previous fall. General Motors was heading toward a U.S.-government-backed restructuring—a move that would shed billions of dollars of debt but require cutting thousands of jobs, and that would spell the end for hundreds of franchise dealerships. The newly elected Obama administration was eager to frame their support of the auto industry as a means toward building more fuel-efficient vehicles, including loans from the Department of Energy to help retool factories for electric vehicles. For several years, Tesla had been trying to secure money through the government. A Model S costing around $50,000, aimed at a broader swath of the public than the Roadster, might be the way to sell the Department of Energy on a bit of financial support.

Musk's glimpse into the abyss a few months earlier had also made him more pragmatic about generating revenue beyond selling his high-end car. While in 2006 he had wanted to focus solely on getting the Roadster done—and not becoming a parts supplier—he had since grown more open to partnerships. That may have coincided with big automakers suddenly awakening to the need for their own EVs, amid a spike in oil prices that made it tougher to sell gas-guzzling cars. Musk wanted to be picky about who he partnered with. He thought there could be a benefit from being connected to a luxury brand, such as Daimler AG's Mercedes-Benz. And after months of talking, just days before his dinner with Rawlinson, he was able to announce a deal that would generate millions of dollars for Tesla, supplying a thousand battery packs for Daimler's small car brand, Smart.

All this was simmering in the background on the night they met, but it wasn't Musk's focus. He was thinking about the team he would need to build the Model S. With funding on the horizon and deliveries starting in earnest, Tesla needed a reboot if it was going to eventually compete against the likes of Daimler and evolve into what Musk envisioned: a car company providing affordable electric cars. From the beginning, the Roadster was about proving that an electric car could be cool, but it made a lot of necessary compromises—from comfort to function. The next act couldn't compromise, not if it was going to beat big automakers. Musk wanted the Model S to be the best car out there, one that just happened to be electric. That was the way to win: to show that there wasn't a trade-off in owning a car that was better for the environment. In fact, he wanted to argue, it was a better overall experience than a gas-powered vehicle.

To do that, though, he needed a team that wasn't going to be swayed by past practices, how things used to be done. To build the Model S, he needed to create innovative organizations to design and engineer the car, then to build and sell it. The company needed to go from making 20 cars a month to 2,000 a month.

Next to Musk sat von Holzhausen, who had clearly developed a rapport with his new boss. What Rawlinson didn't know was that Musk had become skeptical of another recent hire, Mike Donoughe, the former Chrysler executive who had helped save the Roadster, and who had been charged with ushering the Model S into production. The alpha males were already clashing. Musk wanted a chief engineer, someone to put into practice von Holzhausen's designs.

The roles of chief engineer and car designer can be fraught with tension, as they weigh trade-offs between what's cool—what the designer wants—and what's buildable—what the engineer can reasonably make. If all goes well, the two jobs work in conjunction like a pair of smooth-fitting gears. If not, they can (and often do) misfire, leading to stalls—or worse, breakdowns.

Rawlinson's name had come up from a member of von Holzhausen's team who'd worked with him years ago on a consulting project, and who'd offered an endorsement that Rawlinson could be trusted to put into reality a designer's vision. What Musk wanted would require no average engineer. It was one thing for von Holzhausen to sketch out, for instance, door handles that recede into the door when they're unused; building such a part was another thing altogether.

Sitting at the table picking at his dinner, Rawlinson at first blush had little in common with the coarse-spoken rocket maker before him. Musk wore T-shirts; Rawlinson preferred a sport coat. Rawlinson had polite English manners, enjoyed skiing, and stood about a foot shorter than Musk. But as the two chatted, it became clear that they held the automotive industry in similar disregard. Rawlinson talked of his frustrations with the inefficiencies he'd observed over twenty-five years working in the business, how he had spent a career trying to forge improvements, emphasizing the use of computers to speed along design and engineering. He'd also experimented with using smaller teams, to cut through corporate bureaucracy and shave months of work off the cumbersome process of creating a car.

Rawlinson had begun his career at Rover Group and quickly discovered that large corporations could be dull, slow, and averse to work that's on the cutting edge. He spent his time trying to figure out how to use computers to aid the design of his engineering, his eyes growing weary from staring at monochrome green screens for hours at a time. He eventually went to work at Jaguar, which at the time was a separate company, with teams that had begun using computer-aided vehicle development, a rarity in the 1980s. He found himself doing challenging work, and with a group still small enough that he could gain a wide variety of experience. He was especially attracted to the engineering work involved in a car's body. It touched nearly every other function of the vehicle, giving him a broad view into how a car was developed. He learned about suspensions, transmissions, drivetrains, engine manufacturing—and how all of these pieces fit into a car like a giant jigsaw puzzle. Jaguar afforded him a rare opportunity: As modern automotive companies evolved, engineers had often spent their careers hyperfocused on one area. One might aspire to be the foremost in-house expert on door latches, for instance, and never get a chance to learn up-close how the rest of a car works.

But when Ford Motor Co. acquired Jaguar in

1989, he began to see the U.S. automaker's bureaucracy creep into the company's operations. He left
to work on developing his own car instead. In his
home garage in Warwickshire, he designed a two-
seat roadster that was featured in **Road & Track**
magazine, including pictures of the frame. A year
later, he got a call from Lotus, which was low on
cash and looking for ways to quickly and efficiently
develop new vehicles. He showed them pictures of
his car, only to get odd looks from executives in return. Rawlinson would later realize that his design
looked similar to a secret project Lotus was working
on, the Elise sports car.

He wound up getting hired by Lotus as a chief
engineer. He finally had the clout and experience to
put into practice his ideas, effectively cutting vehicle
development time to just months rather than years,
and with a fraction of the staff. When Rawlinson's
boss left Lotus in favor of consulting, Rawlinson
followed, working with automakers around the
world, before finally hanging out his own shingle.

At their first meeting that night in Santa Monica,
Musk quizzed Rawlinson on different parts of the
typical car, asking, for example, about what kind of
suspension he might use. Rawlinson, who recalled
the night years later, enthusiastically picked up
empty plates as props to demonstrate how the parts
work. Then Musk turned to materials, then welding techniques. In Rawlinson, Musk found an engineer who enjoyed drilling down into the very basics

of why and how a car worked, and what could be tweaked to make it better. Rawlinson saw in Musk the potential for an enthusiastic supporter.

Dinner progressed, but Rawlinson had been too busy talking to think about his food. It was Musk's turn to talk. In following conversations, he confided to Rawlinson that his team of engineers in Detroit had drafted a plan to hire 1,000 engineers by Christmas that year. They claimed to need an army to develop the car that had been dubbed the Model S. A back-of-the-envelope estimate suggested they'd spend more than $100 million per year on engineers if they staffed a team of the kind traditional Detroit auto executives were accustomed to for a new car project. "I don't have the budget for that and I can't even recruit the **recruiters** to do that," Musk said. "How many would you have?"

"Let me think," Rawlinson said as he did the mental math, remembering how many engineers he'd used for projects at Lotus. "By June, I'd have about 20," Rawlinson said. "By July-August about 25 . . . I think about 40, 45 by Christmas."

"That's one-twentieth the size!" Musk said. "What's wrong with the automotive industry? . . . Why do they need so many people?"

"Let me tell you how the motor industry is run," Rawlinson said, winding up like a professor beginning a lecture. "It is run like a World War I battlefield." As Rawlinson saw it, car companies, like armies, hired phalanxes of ill-prepared, ill-trained troops who marched forward as cannon fodder, while generals led from miles behind the front lines, unaware of the conditions on the ground.

Musk wanted to know what Rawlinson would do differently. "Elite fighting forces," Rawlinson said. "Take the paratroop regiment. The big difference in a paratroop regiment is the leader is on the ground . . . You have direct leadership which adjusts to battlefield conditions."

Musk's eyes widened. "Paratroopers! You mean **special forces**?"

"Oh—" Rawlinson paused, realizing he had struck a vein. "Yes!"

At his SpaceX cubicle, Musk greeted Rawlinson, who was just a week into his new job at Tesla and already back from suburban Detroit, where he had gone to inspect the team working on the Model S (even after Musk had unsuccessfully given orders to fire them since Martin Eberhard's days). Rawlinson told him afterward that he thought the two should sit down so Musk could tell him everything he was envisioning for the Model S. "I want you to give me an absolute brain dump," Rawlinson said.

At his desk, Musk turned from his screen toward his new executive. "Beat 5 Series," he said, then turned back to his screen.

For Musk, there was no simpler aim than dethroning BMW's popular midsized sedan, which sat in size between the 3 Series compact and 7 Series large sedan. If the 3 Series was a luxury version of the Toyota Corolla, then the 5 Series was the luxury Camry.

Rawlinson hesitated, mulling over what he had just learned about the Model S's progress in Tesla's Detroit office. He had spent a few days there, meeting with soon-to-be-former team members, delving into how they planned to engineer the Model S. At that point, the engineers had spent about a year on the work, and the project had so far cost Tesla $60 million. Rawlinson quickly became concerned with what he saw, which seemed to favor cost-cutting over performance. For example, the team was excited that it had negotiated a deal with Ford to acquire the front suspension for the vehicle—such a good deal, in fact, that they planned to repurpose the front suspension as the back suspension as well. Rawlinson thought it would result in a poor ride. It was the kind of sacrifice he knew Musk wouldn't abide.

Rawlinson drew Musk's attention away from his computer. He'd seen the Model S project as it was progressing, and it wasn't looking good, he said. "I'm sorry but it's going to have to stop—we're going to have to stop the program."

Musk turned to face him again. "All of it?"

"Yes, all of it," Rawlinson said with confidence. They needed to start over, begin the Model S from scratch. He paused to watch his new boss's reaction.

Musk remained quiet for a moment, his head cocked slightly upward as he stared into the distance. His thumbs twiddled. Beneath Rawlinson's assuredness was a hint of doubt. Was his new boss going to fire him for his insubordination?

Musk shifted back to Rawlinson, his eyes piercing into Rawlinson's own. "I thought so."

And in that moment, Rawlinson began to reevaluate his role. This was more than just a typical assignment—one he'd carry out for six months before moving on to the next adventure. Musk was an anomaly in the auto industry, someone who didn't care how things had been done before (especially if their end product was a car that he thought sucked). Musk seemed to care only about making the best, coolest car they could.

Game on, he thought to himself. **This is a bloody big chance.**

Rawlinson's arrival surprised no one more than Mike Donoughe, who'd ostensibly been hired to usher in the Model S. And now in came Rawlinson, who had been hired—personally by Musk—for a vague product development role. He was spending his time reviewing Donoughe's engineers and development plans. It didn't bode well for the former Detroit executive.

It was all the more surprising because, as the new year began, Donoughe's approach to whack-a-mole was showing results. Issues with the carbon fiber panels were being worked through, and other supplier issues were getting resolved. Production of the Roadster could now increase from 5 a month, when he joined Tesla the previous summer, to around 20 to 25 a month in the first quarter and 35 in the second quarter. It was a run rate that paled in comparison to his days at the Sterling Heights Assembly Plant back in suburban Detroit, when the factory at its height churned out considerably more than those numbers in a single day. But for Tesla it was a milestone worth celebrating. One afternoon, he dragged some kegs into the shop for the team and toasted their victory.

The revelry would be short-lived for Donoughe, of course. He could see the writing on the wall. By summer he'd orchestrated a graceful exit. Unlike other recent departures, he wouldn't bad-mouth Musk on his way out the door; he had seen enough to think Tesla might have a chance.

He wasn't the only one. Daimler, which Musk was trying to entice as an investor, was taking a greater than expected interest in the Model S. The sprawling German automaker had a division called MBtech that worked as a consulting agency on tough automotive projects for automotive clients. With the corporate parent's interest in Tesla growing, MBtech's Detroit office had been given clearance to lobby Musk to turn over engineering development to them. They argued that Tesla lacked the time, money, and expertise for such an ambitious car. During a day-long meeting at SpaceX, the Daimler team recommended using the vehicle platform that was used to make the Mercedes-Benz E-Class, the midsized sedan comparable to BMW's 5 Series—the very same platform that van Holzhausen and Straubel had hacked for their Model S show car.

It made a lot of sense. Such a deal would conform to the original Tesla business plan, the one co-founder Martin Eberhard had crafted years earlier. Back then he was left struggling to negotiate with smaller carmakers, such as Lotus. Now, Tesla had the ear of the second-largest luxury carmaker on the planet. Tesla could do what they did with the Elise, but this time with a more extravagant car as a foundation.

Rawlinson sat through the Daimler team's presentation. He had arrived only weeks earlier and now the chance to design a car from the ground up seemed to be in jeopardy. After the team finished its presentation, Musk turned to Rawlinson: "What would you do?"

Rawlinson didn't like it, and he wasn't shy about it. He began laying out an alternative plan that seemed rather unbelievable to the German team, one that called for an entirely new vehicle platform that incorporated the battery pack into the vehicle's structural design—as Musk had said publicly it would. Rawlinson went further, suggesting the battery pack could theoretically help the vehicle bear the force of a crash in an innovative way. It was so radical an idea it seemed unhinged. The team at MBtech was beside itself, promising them their company would fail if they pursued Rawlinson's plan.

As Musk weighed the two options in the days that followed, others inside Tesla evaluated Rawlinson's approach. It depended heavily on skipping steps that had long been held sacred by big automakers, such as researching the market and developing several rounds of prototypes. He wanted to run as much testing as possible through computer simulations, which he thought would save not only time but manpower.

Deepak Ahuja, the new CFO formerly of Ford, went over the numbers and came away impressed. If they were able to design a vehicle with so few people, it could give Tesla a competitive edge on cost against the bigger players. "This is revolutionary," he told Rawlinson. "I've never seen anything like this."

A lesson Tesla had learned from the Roadster informed their decision. They ultimately had needed to replace just about all of the Elise's parts to get the performance and appearance they wanted. To take advantage of its new battery pack technology, Tesla needed to build a car around it, not shoehorn it into something that already existed.

Later, Rawlinson flew to Detroit and decided to call on the head of MBtech's office, who laid out his plans for an electric car based on the E-Class. He told the fledgling Tesla executive to show him why their plan wouldn't work. Rawlinson proceeded to pore over a list of about three hundred parts, one by one. He sat on the floor, explaining why each part wasn't suitable until, after several hours and having reached the sixty-fifth part, Rawlinson was cut off.

"I've seen enough," the German said. "You're right. There's no way this is ever going to make sense." He phoned Musk to withdraw their proposal. It was Rawlinson's show now. He had either been handed the reins—or given enough rope to hang himself.

CHAPTER 10

NEW FRIENDS & OLD ENEMIES

On January 27, 2009, Justine Musk's lawyers filed a motion with the family court in Los Angeles handling her divorce from Elon. Her legal team wanted to add Tesla and Musk's other companies into any divorce fight. Until that point, Musk had largely expected their separation to be straightforward. She'd signed a financial agreement, drawn up ahead of their wedding in 2000, that protected his then comparatively small fortune.* It provided Justine with their home in Bel Air if the two had a child and then divorced, as well as child support payments—an arrangement worth a combined

---

* While the original agreement was drawn up before the wedding, they didn't sign the final document until after they officially tied the knot.

$20 million. Nine years later, though, she felt entitled to much more than just the house.

The end of a marriage is never pretty. From her side of things, their relationship changed as their lives took on the trappings of wealth, following the sale of PayPal. They went from a tiny apartment in Mountain View to a mansion in Beverly Hills. They struggled with the death of their first child to sudden infant death syndrome in 2002, followed by parenting a set of twins, then triplets. Justine began to feel secondary to Musk's other endeavors. He often criticized her, she claimed, saying, "If you were my employee, I would fire you."

Musk's divorce lawyer, Todd Maron, told the court that including Tesla in the divorce case threatened the very survival of the company; it was simply a brazen attempt by Justine to leverage a settlement. The practical effect, Musk worried, was that if she was successful she could have required participation and permission in every significant corporate decision. "If Justine succeeds in embroiling Tesla in this matter, through which it essentially would be cast into quasi-receivership, Elon and the 324 other shareholders may lose their investments," Maron told the court. At that point, Musk had sunk what fortune he'd possessed into Tesla, SpaceX, and SolarCity. Maron warned of the ramifications for Tesla of a costly and public divorce battle.

Luckily for Musk, the filings went unnoticed at the time. It was the kind of publicity that could have hindered Tesla as he was trying to pull off a lifesaving funding round from skittish partners already worried about Tesla's ability to survive.

While Musk simultaneously fought to keep his company and stoke interest in the Model S that spring, newly elected president Barack Obama was making further moves to save General Motors. The effort included ousting CEO Rick Wagoner and announcing that the administration was considering a plan to send the automaker through a government-backed bankruptcy reorganization, one that envisioned a nimbler GM—fewer brands, dealers, and workers.

For months, the nation had been consumed with debate over what role the U.S. government should play in saving both GM and Chrysler. After Congress's failed attempt to agree on a bailout in late 2008 and President Bush's interim loans, Obama had followed up with further loans. Looking for a way out, the companies presented restructuring plans to the federal government.

In the upheaval, Musk saw opportunity for Tesla. For months, one of his key deputies, Diarmuid O'Connell, had been working the halls of Congress, lobbying to include Tesla in a Department of Energy (DOE) loan program to help jump-start America's green technology companies. A former chief of staff for political military affairs at the U.S.

State Department, O'Connell had joined Tesla shortly before the Roadster was revealed in 2006, motivated by nothing less than a desire to slow global warming. He brought needed experience in Washington to the California startup.

Then-CEO Martin Eberhard endorsed O'Connell's idea: to lobby for legislation aimed at enticing electric car buyers with a tax credit on their purchase of a zero emission vehicle. Its successful passage added a powerful tool in helping market the Model S at a lower price. (It was this legislation that Musk had alluded to when he announced, at the car's reveal, that it would run for just under $50,000 after a government tax credit.)

President George W. Bush's administration had implemented a government loan program that the DOE, in late 2008, was already trying to get going when GM began foundering, amid the collapse of the greater global economy. That winter, Musk and CFO Deepak Ahuja had submitted a loan proposal to the DOE, hoping to raise more than $400 million to support the development of the Model S.

March's reveal of the show car, built by hand at SpaceX, gave O'Connell the perfect prop to take to Washington. The same vehicle that had been shown to customers and the media in Los Angeles in late March was then quickly shipped across the country for an East Coast tour, including a stop at David Letterman's Manhattan studio, to appear on his popular CBS **Late Show** on TV. A **New Yorker** writer tagged along, penning a lengthy piece that ran months later, featuring pictures of Musk and his young sons. The media attention gave his project a renewed sense of credibility.

Perhaps most importantly, O'Connell arranged for a tour of the car around Washington. The Model S was taken to the Department of Energy's headquarters to give a ride to the small team of managers tasked with doling out the loan program's funds, including Yanev Suissa, a recent graduate from Harvard University's law school. As he walked out to the car, he noticed colleagues were looking down at him from the office complex, taking in the unusual sight. These bureaucrats weren't accustomed to such attention. Sitting in the Model S, Suissa was impressed with the airy cabin and large display screen in the dash.

Suissa's team's goal was to issue loans to companies that had a chance of paying the government back. With Tesla, they weren't so sure. Tesla wasn't among the popular projects being considered, Suissa recalled. "Early on it wasn't clear that Tesla was going to get it done," he said. "It was incredibly risky. They're not just producing a new version of something already proven, they were creating an entirely new industry."

The government didn't want to be the only ones putting money into Tesla. The company was told it needed to find additional backers. Frustratingly for Tesla, the DOE team wasn't the only one showing reluctance.

Herbert Kohler, head of Daimler's advance engineering team, had met with Musk early in the Tesla days, and he was eager to have Daimler invest in the startup. Such corporate investments, however, are frowned upon by many startups, fearful they'll gain a reputation for being merely an ancillary project for a big company or, worse, for prioritizing the patron's business needs over its own. Musk wasn't interested.

By 2008, however, Musk's opinion had changed. As he was looking for ways to raise cash, he visited Germany and met with Daimler executives, where he learned they were looking for a supplier to provide battery packs to create an electric version of their Smart car. A few months later, Kohler emailed to say he would be in the Silicon Valley area in six weeks and was open to seeing Tesla's technology. Musk turned to JB Straubel for what was now one of his specialties: to convert Daimler's tiny, two-seat Smart car into an electric vehicle. Only this time in just a matter of weeks.

Just months earlier, Musk had eschewed the suggestion of working with GM and others to provide battery packs, but now he had little choice. Plus, there could be an upside in being associated with Mercedes. The first challenge was purely logistical: Daimler didn't yet sell any Smart cars in the U.S. The nearest one Tesla could find was in Mexico. They went to the corporate treasurer's office and asked for $20,000 in cash. They would send a Spanish-speaking friend to Mexico to buy a used Smart car, then drive it back to Silicon Valley. Once the car was in hand, the team would race to get it converted as quickly as possible, paying especially close attention to making sure the interior of the car was unaffected by the teardown.

When the day came to meet with Daimler, Musk sensed that the Germans were unimpressed with his PowerPoint presentation. He cut off the discussion and asked if they'd like to see a demonstration, shuffling them out to the company's parking lot where the hacked Smart car awaited. The German team proceeded to take gleeful rides in Straubel's invention, which, converted with the instant torque of an electric motor, suddenly became a hellcat. Daimler was impressed.

The companies had worked out a supplier relationship by January 2009, and they were thinking about sinking an investment of cash into Tesla as well. Some in Stuttgart, however, were reluctant. Like the DOE, they were fearful for Tesla's financial future.

Musk faced a conundrum: He had Daimler interested in putting in money. He had the U.S.

government willing to lend him money. But neither side wanted to go it alone.

Timing, however, was finally on Musk's side. As GM and Chrysler struggled, the Obama administration was putting pressure on the DOE loan office to begin announcing projects—even if the deals weren't yet ready for final approval. So the office announced that a Silicon Valley solar company called Solyndra would get money—followed shortly after with news that Tesla would get a loan, too.

Daimler followed suit. In May they announced a $50 million investment, giving the company a 10 percent stake in Tesla.

As it happened, the DOE announcement reflected not a true deal but a highly contingent term sheet, Suissa said. "It became about the press release," he said. "So when people thought the deal was done, it was nowhere near done." The details would still need to be ironed out—a process that would take months. But it was a public relations win for both sides. The government appeared to be pumping money into the economy while Tesla won a government's endorsement. For a moment anyway, the perennially cash-poor startup would have some extra scratch.

While Musk had been able to avoid publicity around his messy divorce from Justine, another breakup from his past was about to become public.

Martin Eberhard stewed for almost a year about his ousting from the organization he had founded. It was Eberhard who had built the initial team, who had worked to bring out a Roadster prototype in 2006 to sell to buyers. It was him, and not Musk, who had come to embody the role of Mr. Tesla, the name burnished on his license plates.

In the weeks and months that followed his departure, however, he watched as many of the friends he hired were fired or else departed on their own. Eberhard still loved Tesla, but he despised Musk. He continued venting his frustrations and detailing the company's changes on his blog—until Tesla board member Laurie Yoler asked him to tone it down. His anger was hurting the company, she said. Tesla's lawyer was more direct than Yoler, informing him that the company believed he'd violated his non-disparagement agreement. In response, they withdrew his ability to vest in the 250,000 stock options.

A string of media hits in mid-2008 only enraged Eberhard further. In newspaper and magazine stories, Musk blamed Eberhard for everything that had gone wrong with Tesla. But the final straw came late that summer. In the company's early days, Musk and Eberhard had playfully argued over who would get the first Roadster off the assembly line—something they imagined would one day be worth many times its sales price as a collector's item. A compromise was reached. Musk would get

Roadster No.1 and Eberhard would get No. 2. But after months of runaround about getting his car, once production began, Eberhard got a call from Tesla in July 2008: His Roadster had been driven into the back of a truck during "endurance testing" and was almost completely totaled. Also, he ultimately learned that he wasn't getting Roadster No. 2 as promised; instead that had gone to Antonio Gracias, the board member whose due diligence led to Eberhard's ousting in 2007.

By the spring of 2009, Eberhard struck back with a lawsuit claiming libel, slander, breach of contract, and more. It was a full-throated attack pointed straight at Musk's own vanity and insecurities. Eberhard questioned Musk's claim to being called a founder of Tesla. He raised doubt about Musk's often-recounted story of having moved to California to pursue a Ph.D. from Stanford University before dropping out after two days to launch a software company. "Musk has set out to re-write history," the suit began.

It was a perfectly aimed jab at his former partner. Musk was notoriously thin-skinned about his place in Silicon Valley history. When Valleywag suggested he didn't deserve credit as a PayPal founder, Musk responded with a more-than-2,000-word rebuttal, complete with footnotes. Musk didn't wait for his day in court with Eberhard; he responded on the company's website with a massive recounting of Tesla's history as seen through his eyes, noting that when he first met with Eberhard about Tesla, Eberhard "had no technology of his own, he did not have a prototype car and he owned no intellectual property relating to electric cars. All he had was a business plan to commercialize the AC Propulsion tzero electric sports car concept."*

A war of paperwork began. Musk's assistant, Mary Beth Brown, worked to track down documentation that he had, in fact, been admitted to Stanford for that short stint. And while such antics made for high drama in the tech world, it proved a distraction to Musk's aides, who were then trying to raise necessary funds. When a court struck down a claim by Eberhard that Musk couldn't be called a founder of Tesla, Musk issued a statement declaring victory. "We look forward to proving the facts in court as soon as possible and setting the historical record straight." The company added in its own statement that the ruling was "consistent with Tesla's belief in a team of founders, including the company's current CEO and Product Architect Elon Musk, [and] Chief Technology Officer JB Straubel, who were both fundamental to the creation of Tesla from inception."

In private, some cautioned circumspection. Michael Marks, who had served briefly as interim

* Although Eberhard had filed paperwork to acquire the rights to teslamotors.com in 2003, Musk would later say he ended up paying a man in Sacramento $75,000 for the rights to the name. The backup option: Faraday.

CEO after Eberhard was demoted, wrote Musk and the board that summer urging them to cool down the rhetoric. He called the statements about Eberhard "terribly unfortunate" and "hurtful." "Probably the most difficult of those are the comments about [Eberhard] lying to the board. You can all imagine what that does to his employment opportunities," Marks wrote. "I also don't believe it's true." The challenge for Eberhard, he continued, was that Tesla had lacked an experienced CFO, and the management team was unrealistic in their expectations of cost and timing. "I am not saying that Martin should not have been removed for his lack of competence to manage these issues, for which he was responsible as CEO. I supported that idea. And I was not privy to whatever happened before I got to the company. I would add, also, that in the first months that I was there and while [he] was still working and reporting to me, he did everything I asked him to do, and enthusiastically. That should also be part of his record."

Whatever effect Marks's entreaty may have had, by September Eberhard and Musk had settled their dispute, though the terms were confidential. It included a non-disparagement agreement between the two, according to a person familiar with the details; Eberhard got his shares and, more importantly, he got a Roadster. Tesla issued a statement referring to both men as "two of the co-founders of Tesla," the others being Marc Tarpenning, Ian

Wright, and JB Straubel. Both parties issued statements, their effusiveness a perfect contrast to the vitriol of just a few months prior. Eberhard's statement read: "Elon's contributions to Tesla have been extraordinary"; Musk said, "Without Martin's indispensable efforts, Tesla Motors would not be here today." The two had retreated from their war of words. But the hard feelings would linger for years to come.

Taken together, the legal battles with his ex–business partner and his ex-wife illustrated the wake that Musk's hard-charging personality could leave. His single-minded focus in 2008 probably saved Tesla from bankruptcy, but the collateral damage to people he had been close to created new threats to Musk down the road. As Tesla continued to grow in its attempts to bring a mainstream car to market, such fights could have even greater consequences— the stakes growing ever larger, the margin for error smaller and smaller.

CHAPTER 11

ROAD SHOW

CFO Deepak Ahuja had been given a crash course in startup life when he joined Tesla in 2008, after a career spent in Ford Motor Co.'s vaunted finance department. He had grown up in Mumbai with parents who had started several businesses in the garment industry, making jeans and lingerie. A gifted student, he attended the Banaras Hindu University, earning a degree in ceramic engineering before heading to the U.S. to get a Ph.D. in material science. His plan was to return to India to partner with his father in a business manufacturing ceramic insulators for the electric grid. He landed at Northwestern University outside Chicago, where the bitter cold winters were as much a shock as the academic resources now at his disposal. While he had programmed on mainframe computers back in India, he'd never worked on a personal computer before; he struggled even to find the power button. His requests for help were greeted with quizzical looks; his accent was so pronounced that he struggled to be understood.

But he quickly settled in, meeting his future wife and ultimately deciding against a Ph.D. He headed to Pittsburgh for an engineering job at Kennametal, developing ceramic composites for the automotive industry. On the side, he studied for an MBA at Carnegie Mellon University. In 1993, he landed a job at Ford, which was known for its intensive training of finance-side workers. He worked at one of the automaker's stamping plants, learning how U.S. carmakers do business before working his way up the corporate ranks for the next decade and a half, ascending in 2000 to the rank of CFO of Ford's joint venture with Mazda, and later the same role with the U.S. automaker's South African operations. In 2008, he'd only recently returned to Michigan for a new assignment when a Tesla recruiter called. He ended up joining Tesla just ahead of the economic collapse, spending his first months on the job fearing his position would be one of the many to be cut as Musk looked for ways to reduce cash burn.

As Ahuja watched Musk dip into his own pocket to pay Tesla's bills, he didn't tell his wife and daughters how bad the situation was. He worked to cut costs on the Roadster and to prepare Tesla's financial books for what was ahead, assuming they could

keep the lights on until then. Namely: until they
became a public company. It would be a learning
experience for him and for Tesla. The company
would be forced to take on all of the pressures that
come with being publicly traded—quarterly reports, Wall Street expectations. But it could also
unlock millions—if not billions—of dollars of capital for the often cash-strapped carmaker.

Wall Street bankers had been working for a while
to develop relationships with Tesla.

By the middle of 2009, however, the market for
an IPO seemed dubious at best to many members
of the Tesla team. GM was winding its way through
bankruptcy at the time, shedding thousands of
dealers and tens of thousands of employees, and the
auto industry as a whole was taking a hit on
the market amid depressed sales.* That fall seemingly presented a glimmer of hope for a fledgling
IPO like Tesla: a small automotive battery supplier
named A123, which had delayed its 2008 offering, returned to Wall Street with an IPO in late
September 2009. It jolted the market. Shares soared
50 percent on opening day. It was enough to give
some within Tesla's finance department confidence
that Musk's plan might work after all.

Despite all of his experience at Ford, Ahuja had

---

* U.S. auto sales hit a twenty-seven-year low in 2009 of 10.4 million
deliveries, 38 percent fewer than in 2003, when Tesla was formed,
according to researcher Autodata Corp.

limited experience in dealing directly with Wall Street. He would learn from the bankers that a Tesla IPO faced several unrelated hurdles. The effect of Musk's ongoing divorce with Justine and the fact that Tesla lacked a factory to make the Model S were only two among them.

The bankers had held fast to that second concern, that before it could make its IPO, Tesla needed to lock down a factory to build its cars. The entire point of Tesla's public offering was to raise money to bring out the Model S; revenue forecasts were mostly based on the sales of that vehicle alone. How could you go to an investor, they argued to Musk, and say "We're going to make these numbers," when you don't even have a factory in which to build them?

It wasn't as though the company hadn't considered it. The effort to find a place to build the Model S had bedeviled Tesla for years, largely because Tesla simply couldn't afford one. They aborted an effort to build a facility in New Mexico, which left local developers unhappy. Another effort, in San Jose, had died just as Musk took over as CEO. He began taking weekend excursions around Los Angeles with a commercial real estate broker in attempts to find a location. About thirteen miles from SpaceX's new headquarters in Hawthorne, in a city called Downey, Musk found an old industrial site, where a former sprawling NASA facility sat after shutting down in 1999. The space history was appealing to Musk, as was the fact that the facility wasn't far from his home.

But not everyone was as enthusiastic. The facility would need a lot of work to be turned into a car factory. Inside Tesla, some worried about the time it would take to get the proper permits, in particular the lengthy and complicated process involved in establishing a paint shop. While state government officials promised to expedite matters, some executives worried it could take years to complete, all while Tesla bled cash.

There was an alternative: a former GM-Toyota factory across the bay from San Francisco. It was properly permitted to make cars, and it was in the process of being mothballed by Toyota after its partner in the facility, GM, filed for bankruptcy. Gilbert Passin, recently recruited from Toyota Motor Corp., where he had run a Lexus assembly plant in Canada, thought his former company might have an incentive to let Tesla take the factory off its hands. Toyota had been building up its manufacturing presence in Texas and was eager to consolidate its U.S. operations there. Passin and others surmised that the Japanese automaker might be looking for a graceful way to off-load their factory—Toyota's culture would be aghast at the idea of a closing, a sign of perceived bad planning and inconsistent with its belief in lifelong employment.

For all of the mutual benefit, Tesla managers hadn't been getting much interest from Toyota.

Musk even tried turning to his doctor in Beverly Hills, who had a connection to Toyota's CEO, Akio Toyoda. In early 2010, Tesla heard back, and to everyone's surprise, Toyoda was excited for a meeting, which was quickly arranged at Musk's home in Bel Air. A formal entourage arrived at the mansion, where Musk put the esteemed visitor, the grandson of Toyota's founder, in the passenger seat of the Roadster for a quick spin around the block. Toyoda, who raced cars, was impressed. He also liked the entrepreneurial spirit of Tesla. When Musk asked about the possibility of buying the factory, Toyoda quickly agreed. More than that, Toyoda wanted to invest.

The money would help, of course. But Musk liked the idea that Tesla would be seen as having received an endorsement from one of the world's best carmakers. Getting the seal of approval from Toyota **and** Mercedes-Benz, on the eve of going public—it was an enviable spot for a young carmaker to be in.

The factory deal caught city officials in Downey off guard. They thought Tesla was coming to their community; they fumed that Musk had betrayed them.

JB Straubel was likewise surprised by the deal. Tiny Tesla had looked for whatever negotiating leverage it might gain against the bigger rivals. When it sought a relationship with Daimler, managers in Europe visited with BMW AG as well, making sure to park their Roadster in a conspicuous spot in hopes that word might travel within the gossipy industry. Similarly, as Musk sought a relationship with Toyota, he had Straubel working to make in-roads with another mass-market automaker, Volkswagen AG. Musk had sent him to Germany to try to sell the German company on a supplier deal similar to the one Daimler had agreed to. Straubel's team had equipped a Volkswagen Golf with an electric powertrain and taken the vehicle to Germany to demonstrate how they could work together, walking through the benefits of lithium-ion cells. Volkswagen CEO Martin Winterkorn took drives around a test track at full speed in the Roadster that they had also brought along.

The Toyota deal was announced on a day of Volkswagen meetings. Winterkorn summoned Straubel to his office. "What the hell just happened?" he demanded. Straubel had no idea what to say. Any potential deal between the two companies instantly died—already Winterkorn's team had been privately objecting to a partnership, in part, over safety worries about lithium-ion cells but also because the idea hadn't been invented in house. Straubel and his team packed their bags for a frustrated return to San Carlos.

"I don't have time for this," Elon Musk bellowed. "I've got to launch the fucking rocket!" And with that he stormed out of the glass conference room at

SpaceX, abruptly ending a meeting to go over the finer points of the marketing material for Tesla's initial public offering.

While Musk had shown he could be a nano-manager, he also had little patience. A passionate writer, he argued with lawyers about the phrasing of the prospectus that was issued in early January 2010 after nine months of work by Goldman Sachs and Morgan Stanley bankers. One of those bankers was Mark Goldberg, just twenty-four years old. An interest in renewable energy drew him to work at Morgan Stanley, which was how he found himself the newbie involved in a very unusual IPO process.

On more than one occasion, Musk threatened to fire them all. "This needs to be more exciting," he'd tell them, pushing for such claims as that Tesla would take over the entire midsized premium sedan market. Or else he'd push back against slides in which Audi was cited as one of Tesla's competitors. He fumed about how the carmaker had used its marketing might to get Tesla kicked out of appearing in **Iron Man 2,** after Musk refused to pay to place his cars in the movie. (It didn't prevent him from making a cameo in the film, playing himself.) "Why is Audi even in there?" he demanded about the presentation. "They're not even a factor . . . We're going to crush Audi."

It was the kind of behavior expected of a startup founder but not the kind typically found in the c-suites of publicly traded companies, where attention was paid to shareholder return, communications were scripted, and unnecessary risk avoided. If Musk was chafing at the chance to become public, what would it be like when Tesla was owned by the thousands of investors and he was at the mercy of their whims? But Musk had little choice. Going public would allow the company to raise money at the cost of his being able to run it like his own personal fiefdom. There was other atypical behavior as well. While Morgan Stanley had been initially picked to hold the prestigious lead banker role, it was later bumped to the No. 2 spot behind Goldman Sachs. Records later filed with the IPO would show that Goldman had given Musk a personal loan—an early indication of how tenuous his own finances had become.

For help with the IPO, Ahuja hired Anna Yen, a former executive at Pixar, to oversee investor relations. The job involved a tedious amount of paperwork that had to be filed with the Securities and Exchange Commission. In the process of submitting forms, Ahuja's team realized that Tesla had failed to submit proper documentation to the EPA that year, a mistake that could lead to a fine of $37,500 per vehicle sold, or an almost $24 million weight to be carried by the company's already beleaguered balance sheet. In the final days of 2009, they rushed to get the paperwork in, settling with the DOE to pay just a flat $275,000 fine. More importantly, the EPA agreed to cover all of the vehicles sold in 2009 as if they had been properly certified to begin with. (Musk had personally called EPA administrator Lisa Jackson, appealing for help in expediting the matter amid concern it could drag on for months.)

It was during this period that Musk hired a general counsel and, just a few weeks later, watched him depart. It was a position that would prove notoriously, almost comically difficult for Musk to keep filled; he didn't seem to have much interest in the advice of his lawyers.

One thing Tesla elected not to do when preparing for public ownership, which would have ramifications years later, was introduce a dual-class stock system. This was what allowed Larry Page and Sergey Brin at Google (or Mark Zuckerberg at Facebook two years later) to keep control of their company, even as they held a small fraction of its total stock. It's unclear why Tesla's IPO paperwork, which it filed in January 2010, contained no such provision to ensure Musk's continued oversight of the company. Those who worked on it said the idea was omitted, in part, because selling Tesla as an investment was already going to be hard enough. Having an entrenched, unpredictable leader tethered to the company could make the IPO more challenging still. (Plus, Musk's brother, Kimbal, was on the board, already raising one corporate governance question around nepotism.)

The best that Musk got was a provision requiring that any measure by shareholders to force changes, such as an acquisition or sale, must pass with a two-thirds tally of outstanding shares—a supermajority provision that could effectively give Musk a veto over articles he didn't like. As long as he maintained his stake in the company, which in January 2010 stood at around 20 percent, his fellow shareholders would need to gain the approval of roughly 85 percent of the outstanding shares to get a measure past Musk—an exceedingly high bar for a company that was more and more becoming defined by its outspoken leader.

He had another form of protection too, at least for a couple of years. A provision of the deal cut with Daimler effectively ensured that Musk would remain CEO through 2012—a tacit acknowledgment by Daimler that it saw Musk as key to Tesla's future.

Weeks after filing the paperwork for its IPO, the company was racked with disaster. A team of engineers from San Carlos was taking a private airplane to Hawthorne for meetings when it crashed upon takeoff. All three of the Tesla engineers onboard died. Musk himself had been scheduled to go to Hawthorne that day, but he had had to cancel his trip at the last moment upon learning that Kimbal had broken his neck in a sledding accident in Colorado. Musk was rattled as his team briefed him on the incident.

Two days later, Tesla was rocked with a blow that was decidedly less tragic, but that cast a shadow over its IPO prospects. The Silicon Valley industry

online publication VentureBeat broke the news
that Musk's divorce proceedings had taken an unexpected turn. Musk, the billionaire, was pleading poverty. "About four months ago, I ran out of
cash," Musk had told the court in filings. He had
been living off personal loans from a friend since
October 2009, spending $200,000 a month.* In
The Times of London, Musk's wife, Justine, who'd
been blogging about the couple's divorce, ampli-
fied their spat. Musk found himself in a troubled
spot. Tesla had survived because of his fortune; just
a few years earlier he had promised customers that
he would personally refund their money if Tesla
went bankrupt.

The company's lawyers did damage control.
They drafted an update to Tesla's IPO paperwork
saying it was "no longer dependent on the financial
resources of Mr. Musk" and that "we do not believe
that Mr. Musk's personal financial situation has any
impact on us." Behind the scenes, however, Tesla's
bankers worried that the divorce threatened to disrupt the budding public offering. If Justine Musk
suddenly had a claim to Tesla shares and didn't
agree to the lock-up period, for instance—the time

---

* Musk later complained that the $200,000 monthly expenses in-
cluded about $170,000 for both of their legal fees tied to the divorce.
Much of the rest, he said, was for nanny salaries and supporting
Justine's household. He noted that they split custody of their five
children. "Almost all of my non-work waking hours are spent with
my boys, and they are the love of my life," he wrote in 2010.

during which insiders aren't allowed to sell their shares, which might devalue newly issued stock—the IPO could suffer. All of which was putting pressure on Musk to settle his divorce, and quickly.

When Justine lost an effort to throw out the couple's postnup, the two quietly settled. That threat, at least, was put to bed.

Musk proved he could be his own worst enemy. A fan of Stephen Colbert, he wanted to go on the comedian's Comedy Central TV show **The Colbert Report.** Tesla's bankers and lawyers urged him to hold off. Tesla was in a quiet period ahead of the IPO; anything he might say could result in new filings and further delays. Musk bristled at the attempts to contain him. He threatened to fire the team handling the IPO, going so far as to begin efforts to set up his own road show with would-be investors on the East Coast. When the bankers asked him to rehearse for them what he planned to say on this tour of investors, Musk refused. He didn't care about their opinions. Finally a compromise was reached. Musk was persuaded that the bankers' sales teams would need to hear his pitch if they were to help sell potential investors on Tesla.

Musk didn't want a typical road show: A staid pitch made by a private company that yearned to be public, held in investor conference rooms, punctuated with PowerPoints. Sure, he would do those things. But he also wanted investors to get a true feel for his cars of the future. The young Goldberg worked for weeks to get permission to have the glass doors of Morgan Stanley's Times Square office building removed so they could bring the Model S into the lobby. Important investors were given test drives at a store nearby. "I'd never seen investors with ear-to-ear grins on their faces before," he recalled.

The presentation attracted those curious to see Musk in person. He wasn't yet as famous as he would become but was already developing a reputation as a rebel in the tech world. That summer he had begun flexing his marketing muscle on a new social media platform called Twitter, which allowed him to send out 140-character dispatches from his smartphone. During his presentation Musk encouraged investors to look at Tesla differently. "Think closer to an Apple, or Google, than a GM or Ford," Musk said. To emphasize the point, he had a slide that showed Tesla's headquarters in Silicon Valley surrounded by tech giants. Despite the distance from Detroit, he did want to use Tesla's connection with Daimler and Toyota as endorsements. Daimler, he noted during one talk, invented the automobile and now was turning to Tesla for help. "It's like Gutenberg saying can you make a press for me?" he joked.

One of the topics on some investors' minds was an obvious one: How would he balance his time serving as CEO of two companies, Tesla and SpaceX (not to mention his role as chairman of SolarCity)? He'd grow angry at Ahuja and the team whenever he got questions from potential investors that he thought were dumb, telling them they should've done a better job educating them ahead of time.

For those who bought in to the pitch, Musk allowed them to reimagine what the auto industry might look like. Morgan Stanley analyst Adam Jonas had begun his career inspired by Steve Girsky, who gained recognition as a leading authority on the industry before eventually joining the General Motors board after its bankruptcy reorganization. Musk's vision excited Jonas enough to help inspire him to return to the U.S. from an assignment in London. His job was to monitor a pack of auto industry companies, providing unbiased advice to investors on their performance. He would study their financials and place in the overall industry, publishing regular notes on how he thought things were going and what he thought was a proper price for their shares. In the case of Tesla, he thought there was potential for the stock to rise in value over time to $70 a share. This constituted a rosy outlook, to say the least.

Opinions like Jonas's could help form a narrative around the company's performance. Was it meeting or failing to meet expectations? The answer to that question could impact the stock's price, pushing it down with disappointments or up with unexpected good news.

Jonas could see the potential for Tesla cars to move from a "rich man's toy to mass market," he told investors, while cautioning that the company's long-term independence could only be achieved through Musk's long-held goal of coming out with an electric car at a price of around $30,000. The risks were great, though. He warned that missteps and delays in bringing out the Model S could hinder that ambition, as could the entry into the electric market of more experienced automakers.

"As is not uncommon with start-ups, the biggest question is if Tesla can remain solvent long enough to capitalize on the forthcoming technology breakthroughs," Jonas wrote to investors in financese, when initiating his coverage of Tesla several months later. In a worst-case scenario, the company's shares might be worth nothing.

Would-be investors fell primarily into three camps: Those who questioned why they should buy Tesla shares before it had proved it could build the Model S. Those who felt that now was precisely the time to buy in, because once Tesla proved itself, it would be too late to get shares on the cheap. And a third, more unusual group. As part of the terms of the IPO, Musk had insisted the banks set aside a pile of shares to be sold to early customers of the Roadster, to give them a chance to buy a stake in the company they had already backed as consumers. It was an acknowledgment that without them—their patience and support—there wouldn't be a Tesla. They would serve as vocal advocates for years to come, a Greek chorus reflecting on Musk's every move.

With investors suitably wooed, Musk huddled with his bankers on a call to discuss what they'd price Tesla's stock at. The bankers recommend starting at $15 a share.

Said Musk: "No. Higher."

Goldberg hadn't been doing IPOs for very long but in his three years at it, he'd never seen any CEO push back on price like that. After all, these bankers from Goldman Sachs and Morgan Stanley were the experts. Now the experts were stunned. They muted their phones, filling their side of the conversation with profanity as they debated their next steps. **Who the fuck does he think he is? Who here can convince him otherwise? Is this whole thing going to fail? Is it too late to pull out?**

In the end, they had gone too far to back down. Musk had them over a barrel, and after watching him for months as he pushed back against custom, they knew it was well within his MO to walk away if he didn't get what he wanted on arguably the most important part of the IPO—the decision that would impact how much money Tesla took away from the arrangement. He had been right so far. They relented.

The stock was ultimately priced at $17 a share. At that price, it would raise a much needed $226 million for the burgeoning automaker. On the day

Tesla was set to go public, Musk and his team arrived at the NASDAQ stock exchange in lower Manhattan in Roadsters. He was accompanied by his girlfriend, soon-to-be second wife Talulah Riley, whom he'd met at the London nightclub two years prior. Musk rang the opening bell. As Tesla shares rose 41 percent that day, he stood outside for a TV interview with CNBC's longtime automotive correspondent Phil LeBeau, who didn't go easy on the newly crowned public company CEO, asking when the automaker would turn its first profit and noting that many in the industry doubted Tesla would be able to ramp up production of the Model S as promised. "People at this point ought to be a little bit more optimistic about the future of Tesla because we've confounded the critics at every turn so at a certain point people have to get tired of being wrong," Musk said with his typical bombast.

He and Ahuja then raced to Musk's jet for a trip back to Fremont and a party at the company's newly acquired factory. Ahuja watched as champagne was poured for the team. That day represented not only the culmination of months of preparation but years of building: The company had now successfully gone public. It was a huge milestone for any startup.

Musk raised a glass to the occasion. "Fuck oil," he toasted.

CHAPTER 12

JUST LIKE APPLE

The charts told the story. Tesla's weekly sales of Roadsters numbered in the handful. Zak Edson, director of product planning, was presenting the results at Tesla's headquarters as Elon Musk looked over the figures. "Sales suck," Musk said. "They don't just suck—sales suck monkey dick." The group tried to swallow their laughter.

It was a glib observation of a growing problem for the company. For the first seven years of Tesla's existence, the automaker's primary focus had been creating the Roadster: designing, engineering, sourcing parts, and finally building the car. Things needed to change in 2010. **Selling** the Roadster needed to become paramount—a skill the company had little experience with and quickly needed to learn.

The reveal of the car in 2006 had helped seed

interest in early customers, followed by the open-ing of several company-owned stores in 2008. They had stood as calling cards to those interested in the electric car startup, primarily in California and other wealthy enclaves outside the Golden State. The increase in pricing in 2009 helped the com-pany cover its costs, even if it had been a hiccup for some buyers. But it also hadn't mattered much. Those buyers were early adopters, less sensitive to price than typical shoppers. In many ways the car sold itself to them.

The problem, in 2010, was that Tesla had spent the previous year running through its list of people who had put down deposits on the car. They were finding it a tougher go to make new sales now. Of the 2,500 vehicles that Tesla had contracted Lotus to build, the company still needed to find a home for 1,500. Tesla's efforts to raise money to develop the Model S were built on the assumption they could sell their entire Roadster production run. Musk was counting on that money to defray the cost of operating the company as they raced to bring the new car to market. Revenue in the fourth quarter of 2009 plunged 60 percent from the third quarter. The investments and public offering softened some of the financial urgency, but it would be hard to make the case that Tesla's answer to the common car would be a huge hit if they couldn't sell nearly half of the initial vehicles they planned to build. Something needed to change.

But what? As a consumer of high-end cars himself, Musk thought he knew a lot about the sales end of the car business. He had little interest in the hard sell, disliked the idea of advertising. PayPal had been built on viral marketing that came from happy customers telling friends and family about the new service. That's what Musk envisioned for Tesla. The quality of product would sell the cars; if sales were slow then the product must be off. With all of those strongly held opinions, he tended to show little interest in the boring nuts and bolts of running a sales operation.

For a solution, Musk turned to board member Antonio Gracias and his business partner at Valor, Tim Watkins, the black-clad fixer with the ponytail and fanny pack. They had addressed the issue of the Roadster's runaway costs and straightened out the company's supply-chain problem. In the process, the three had developed a deep respect for one another.

While the issue at first may have seemed to be one of brand recognition, the data suggested something else. As Watkins investigated, he found that Tesla had plenty of interest—the company had 300,000 potential sales leads, thanks to all of the publicity surrounding the Tesla and Musk. But it was doing a horrible job at closing them.

Gracias and Watkins's theory was that Tesla needed to narrow in on what they dubbed the "sales event," or the moment when the customer decided to make the decision to buy. It was not, they argued, when the check was written but rather at the moment when the customer felt an emotional connection with the car. And it was clear to the two that the test drive was key. Going back to Martin Eberhard's ride in the tzero at AC Propulsion, and to the prototype Lotus used to sell investors on Sand Hill Road, the torque of the electric motor had provided an exhilaration that was hard to replicate in any amount of advertising copy or sales calls. Consumers had to get behind the wheel.

By 2010, Tesla had opened just more than ten stores, and they had plans to open fifty more. After digging into the sales organization, Watkins proposed an alternative, arguing that they should delay efforts to open stores for the moment and focus instead on hosting test-drive events. To Watkins and Gracias, it was simple math. If the odds of somebody buying the car increased after they drove it, then Tesla needed to focus on creating more opportunities to drive the Roadster. Retail stores, geographically confined by their nature and requiring real estate and long-term staff, couldn't connect broadly enough for their purposes.

Gracias and Watkins began hiring a group that could be deployed as Tesla's roving sales team, picking up leads for potential buyers at guerrilla marketing events and following up with individuals. The two interviewed potential candidates, often eschewing people with car sales experience in favor of youngsters fresh out of college, who were attracted by the company's environmental mission. Gracias, in particular, focused on college athletes, because they were accustomed to team environments. The new hires—more than thirty in the first twelve months—were put through a training program overseen by Watkins. In their first month on the job, the new recruits each called about three thousand sales leads, then were sent out to a different region of the country. Instead of focusing on the typical sales pitch—about financing or upselling, say—they were trained to talk extensively about the car's engineering capabilities.

The marketing team had already been holding events to publicize the car; Gracias and Watkins wanted these new sales recruits to be more deeply integrated into them. Events were staged every weekend across the country. Some weekends they might lure a thousand potential customers to test-drive. Each salesperson was given a script, and after each test drive by a would-be customer, they would rank the likelihood that the person would end up buying. That way the team could prioritize their time in closing a deal (especially given that Tesla was only able at that point to crank out about fifty Roadsters a month).

One of the early hires was a Stanford grad named Miki Sofer. On the day of her first sale to a female buyer that fall, she greeted a potential customer named Bonnie Norman, a fifty-seven-year-old Intel executive who had inquired about the Model S that was getting so much attention in the press. One of the sales team members had arranged for her to take a test drive of a Roadster near Sacramento on a Saturday. Norman had owned Porsches and BMWs before, but concerns about the environment led her to a Toyota Prius (Martin Eberhard's stereotypical buyer almost to a T).

The test drive was simple enough. Sofer handed Norman the keys and rode shotgun so she could answer questions. She said little, letting Norman hit the accelerator. Norman marveled at the fact that she could get a speeding ticket with such a car. Afterward, they got out. Instead of a sales pitch, Sofer offered details: that the car could get a range of more than two hundred miles, but given Norman's aggressive driving style she might get less. Norman asked how she could buy it.

The approach worked. Quarterly bookings tripled once Watkins and Gracias's system was put into place. It was a temporary solution, though. If Tesla planned to move beyond selling a few hundred cars a year, especially as it moved into its next phrase with the comparatively affordable Model S, it was growing increasingly clear that it needed a viable network of stores.

Musk had watched the explosion of Apple retail centers. He asked his assistant to find him the person responsible for them.

—

George Blankenship's career in retail began after he dropped out of the University of Delaware. He was working at the Gap in Newark, Delaware, as a store manager in training. He enjoyed customers, focusing time on them, trying to figure out what they wanted and how to reach them. His team won all sorts of company sales contests; one employee won a car during the back-to-school shopping season. He dreamed of becoming a store manager himself, then a regional manager.

But he wasn't good at attending to the details, for instance making sure the store shelves were stocked with the correct sizes and colors. When corporate asked if Blankenship was ready for the next step, his boss was pessimistic. "He'll never be ready to be a store manager," his boss apparently said. "He spends way too much time waiting on the customers and not keeping the store clean."

His sales numbers, however, told another story. He was soon given his own store, and he proved a success. Eventually, he got moved into real estate, where he'd spend the bulk of his career, tasked with a rapid expansion of the business. He became known for being able to make deals quickly with malls—understanding how to navigate the negotiation with the developer and being prepared for the internal politics of it all. As he moved up the corporate ranks, Blankenship oversaw store design and planning; real estate on the West Coast; and then overall corporate store strategy—a job that allowed him to focus on what the customer experience would be like. He kept a toe in day-to-day store life, too. Every Christmas holiday season, it was all hands on deck for the company; he'd pick a store that needed a little extra help and embed himself there, working until late on Christmas Eve.

As he approached his twentieth anniversary with the company, he'd opened 250 Gap stores. He and his wife were looking forward to an early retirement, enjoying some Florida sunshine. An unexpected call from Steve Jobs at Apple, however, changed the course of things. Blankenship joined the tech company ahead of the release of the first iPod, as Jobs was moving into a company-owned retail store strategy. He met with Jobs every Tuesday morning for three hours, plotting out what the Apple Store shopping experience would be like. He'd scout out locations across the country that were uniquely "Apple." He helped open more than 150 of the company's now-iconic stores before semi-retiring, settling this time into the world of consulting.

That was when he got an email from Musk's indefatigable assistant, Mary Beth Brown. He ignored it—having worked in the real estate business, he was accustomed to unsolicited offers, typically from local shopping centers. Brown persisted, though, and one of her emails caught his attention: "Elon Musk would like to speak to you about the things you did at Apple. Please give me a call."

Blankenship called, was greeted by Brown, and

then unexpectedly found himself connected with Musk himself. They spent an hour talking about the company before Musk said he wanted to meet in person.

"I'm sorry to be pushy but can we do it this afternoon?" Musk asked.

"Sure," Blankenship said. "But I'm in Florida."

Musk asked if he could meet the next day instead. "I have to meet at Cape Canaveral tomorrow with Obama at noon and we're doing a presentation," he explained.

Blankenship agreed. The conversation with Musk had piqued his interest, and the idea of doing something that might help the planet was appealing. Before they met, Blankenship glanced online at the handful of stores Tesla had already established—many were either former car showrooms or designed like them. A customer had to drive to get to one. He thought that would be fatal for Tesla. Car customers tend to be loyal; a majority of new vehicle buyers in 2010 returned to the brand that they bought the last time. Some brands had even better return rates, such as Ford Motor Co., which saw a 63 percent loyalty rate. Tesla needed not just to convince people of their brand, but to persuade them to leave their old one. It didn't help their cause that Tesla was asking buyers to take a chance on a new, unfamiliar technology.

If Blankenship had the helm, he'd want to ambush people when they **weren't** thinking about buying their next car. He'd want to educate them in a comfortable, non-confrontational environment, and get as many people in front of the brand as possible. He thought of the work he had done for Jobs. They opened Apple's network of stores ahead of the launch of the iPod music device in late 2001 as the company began evolving from desktop personal computers to mobile technology. The knowledge-able staff and Genius Bar became tools to educate customers that needed a nudge into the digital age.

His advice to Musk would be simple: "I'd do just like Apple."

That first meeting in Florida went well. It was followed by a trip to California to meet the team, and to offer Blankenship his first chance to drive a Roadster—a moment that, as with so many others, made up his mind about joining the company. In Tesla, he saw potential for change. He was re-minded of Apple in its early days; he thought Tesla was sitting on a product so amazing that all they needed to do was unleash it on the world. For his part, Musk saw in Blankenship the potential to rep-licate Apple's titanic success.

The stores that Tesla had launched to date typi-cally cost half a million dollars to build out. Musk wanted to double the number of stores by the end of the year, with the goal of having all fifty stores open for the rollout of the Model S, now planned for 2012. Because Tesla didn't intend to keep a lot of inventory onsite, the company figured it could save money on smaller spaces, no lot full of cars required.

And so Blankenship signed on. He would lead the automaker's efforts to open stores across the globe, looking for locations primarily in high-end shopping malls. Mall operators may have been initially skeptical of the notion of a car seller setting up shop within their facilities, but Blankenship's longtime relationships helped open doors.

He inherited a powerful weapon, one that Musk had been nurturing from the early days: a core group of customers who acted as evangelists for the brand. Bonnie Norman, who bought her Roadster in late 2010, was among their ranks. After her purchase, she was surprised to learn about the community of owners online. They congregated on a website called the Tesla Motors Club, offering each other tips and answering questions for newbies. There she found help pairing her cell phone to her Roadster's Bluetooth. She traded jokes with owners who fretted that women weren't interested in electric cars. (She facetiously suggested they paint them pink.) They made up a pillar of Musk's strategy to avoid spending money on TV ads to publicize the car. He wanted to rely on word of mouth. These would be Blankenship's foot soldiers, amplifying the campaign he led with his stores.

As Blankenship began working closely with Musk, he found some similarities to Steve Jobs, but also key differences. Jobs had likewise been super focused on many aspects of the business. With Jobs, he had spent their hours-long meetings delving into such details as the wood grains to be used for the legs of the tables that the stores would need to showcase their products, or else weighing the position of the holes that would be cut into those tables to accommodate cords—even discussing the size and shape of those holes.

While Musk could be super focused on engineering issues or car design, he had less interest in other parts of the business, such as how the stores should look. He wanted it to be like Apple—he wasn't up to picking wood grains. With Jobs, Blankenship had gone through several iterations of the store design in a physical warehouse; for Musk, all that was needed were some renderings.

"Is that what it should look like?" Musk would ask Blankenship, sincerely.

Blankenship explained there would be graphics on the walls and places for storage for apparel and brochures. It would be reasonably inexpensive to build—an open layout with the car at the center of attention.

"OK," Musk said, and left it at that.

Unlike his engineering colleagues, who might have to be prepared to debate with Musk the very basic elements of their decisions, Blankenship was being given wide latitude. Musk trusted him to do his job. If he failed, it was on him—and he knew there'd be hell to pay.

Almost a year after joining, Blankenship opened Tesla's first store of this new generation in a high-end shopping center in San Jose, California, called Santana Row. Blankenship situated a Roadster in the middle of the airy store, surrounded by displays about Tesla's technology and a large interactive monitor allowing shoppers to check out how different paint colors and leather interiors might look together. Gone were the salespeople typically found at car dealerships. Instead, he hired product specialists to help educate shoppers about the new technology. Much in the way that Apple had been starting from a place of small market share and unfamiliarity with its products when its stores opened, Blankenship was betting that buyers would need to first become comfortable with the Tesla brand, so that when they decided to buy their next car, Tesla would enter the equation.

"We are revolutionizing the auto purchase and ownership experience," Blankenship told the **San Jose Mercury News.** "At a typical car dealership, the goal of the dealer is to sell you a car that's on the lot. At Tesla, we're selling you a car that you design. The shift is people say: 'I want this car.' "

He quickly followed San Jose with a similar store just south of Denver, in the high-end Park Meadows mall. Foot traffic didn't disappoint. The San Jose store received about 5,000 to 6,000 visitors per week after the initial burst of interest, double what Blankenship's team had expected. The

Denver-area store was getting 10,000 to 12,000 visitors per week.

One of the next places Blankenship targeted wouldn't offer such easy entree. Texas was among a handful of states that banned carmakers from selling directly to customers. Musk wasn't about to lie down, though. Tesla simply needed to find a way around the law.

Shortly after Texas's regular legislative session came to an end in May 2011, Blankenship went to Austin to meet with state officials in hopes of circumventing dealership statutes. He brought along a Roadster to give regulators test rides; he knew that a recalcitrant bureaucrat became more pliant after a spin around the block.

Together with a Tesla lawyer, Blankenship sat in the state office going through the books, line by line, exploring the limits of what was possible. They presented an idea: Can we open a showroom that's purely educational—no sales, just staff members on hand to tell people about the electric car?

An official looked at the existing language. "Well, that's not in here."

"So that means we can do that?"

"Well, we'll have to . . ."

"No, no, no," Tesla's lawyer interjected. "It's not in the law, it cannot be in the law next week."

Tesla had found the loophole they needed, and

they sped right through it. The timing of their meeting, just after the legislative session had ended, meant they'd have fully two years on the ground before the statehouse had a chance to revise the laws on the books.

Blankenship wasted no time. He set about crafting a facility that he dubbed not a store but a gallery. He'd staff it like Tesla's other stores, include the same educational materials, but no pricing would be allowed on the premises. If somebody was interested in buying a car, they'd be directed to a computer where they'd enter their contact info. A call center in Colorado would follow up. The law dictated that their vehicle couldn't physically be located in the state prior to the sale—but that wasn't a problem for Tesla, as there wasn't exactly a ton of inventory lying around. Customers who ordered vehicles generally had to wait for them to be made. Only after a Texas customer's check had arrived at Tesla could the vehicle enter the state. After that, the car could drive wherever it pleased.

It was an ingenious workaround, and it opened up parts of the country beyond Texas, where similar statutes applied. With Blankenship's innovation, one of the early touchstones of the Tesla sales model—avoiding franchise dealerships in favor of direct sales—had been put into place.

Increasingly, Blankenship believed they were on the right path. Every Friday afternoon before heading to dinner with his wife at a steakhouse near the San Jose store, Blankenship went to the Tesla showroom to study the customers. He watched them linger as they took in the floor-to-ceiling display windows, as they took test drives, as they used the free charging station in the parking lot behind the store. He listened, too. He noted a small group of super enthusiasts—they may have already bought a Roadster and were excited for what was next.

Most visitors were just curious. "Why is there a car here?" "What is this?" "Is this a temporary location?" It was clear that few had any idea yet what an electric vehicle really was. The ones that did tended to have to be disabused of their preconceived notions, likely from what they'd read about the EV1. All of it reinforced Blankenship's belief that he needed to educate a new generation of buyers.

Tesla prepared to open its first gallery showroom in Houston, with a sign advertising its imminent arrival. Again he observed as shoppers stopped to take note, listened in on their conversations to get a sense of what they thought about the brand. He was excited to find growing enthusiasm.

Sometimes, though, the interest was misguided. He watched two ladies stop in front of the temporary display.

"Oh? Tesla, what's that?" one asked.

"That's that new Italian restaurant," the other replied.

CHAPTER 13

$50 A SHARE

Bitter, below-normal January temperatures greeted Peter Rawlinson as he arrived in Detroit for the 2011 auto show. It was a jolt compared to the pleasant weather in Hawthorne, California, where he had set up the automaker's vehicle engineering operations to be near Elon Musk's desk at SpaceX headquarters. Rawlinson and Musk faced twin challenges that called for close coordination. They were engineering a car from the ground up that would serve as the foundation of the company for years to come. Equally important, they were creating the culture of what it meant to be Tesla. Getting these right could help set the tone for a generation to come.

As Rawlinson approached his two-year anniversary with Tesla, all the effort and strain of overseeing the development of the Model S had mounted. He'd come down with a terrible flu, just his latest health complication as he worked from early morning until late at night, breaking for dinner at an upscale Manhattan Beach hotel, where he and some of the other senior executives were living before moving north to be closer to the Fremont factory. Ahead of leaving their desks on the second floor of SpaceX, they'd call the hotel bartender to put in a food order to catch the kitchen before it closed. There, they'd rehash the day. A year earlier, Rawlinson had injured his hip in a skiing accident—he had been racing Dag Reckhorn, the second in command of Tesla's vehicle assembly.

Now, Rawlinson just wanted to be in bed. But Ricardo Reyes, the head of public relations, needed him in Detroit to talk about Tesla's latest developments at a press conference for automotive journalists gathered from around the world. Between events, Rawlinson lay down for a nap in an exhibit hall coat closet.

He had become the face of the engineering effort. When Reyes was hired away from Google, Musk gave him a clear directive to move the narrative away from the drama surrounding Tesla's CEO and to put the attention instead on the cars. A key part of that, Musk said, was to understand that what Tesla was proposing to customers was still new to many; skepticism was only natural. "They're going to have tough questions and they're supposed to ask tough questions and we're supposed to answer every single question," Musk said.

It was that mentality that put Rawlinson on-stage in Detroit, to reveal what the Model S's vehicle platform looked like—the skeleton beneath the fancy sheet metal. He took great care to describe the efforts by his team to ensure that every inch of the car was used efficiently: it would have lots of interior space, taking advantage of the fact that the battery pack was located beneath the car; the front end wasn't bogged down with a typical engine; and the back end lacked a gas tank. He also stressed the care they took to ensure safety.

In press interviews, he emphasized what made Tesla different from the hometown rivals. "Culturally, we're so different from traditional automakers that have silos of expertise—body people, suspension people," he told one reporter. "We have an emphasis on process."

At the show that year, General Motors celebrated winning the prestigious North American Car of the Year award for its Chevrolet Volt plug-in hybrid sedan. The Volt, which had begun rolling out to showrooms just a few months earlier, had been inspired by the Tesla Roadster's 2006 reveal. GM had beaten the Silicon Valley automaker to market with its own sedans, but a sigh of relief came over Tesla's people as they looked at what the Volt offered. It felt like an econo-box compared to what they were imagining for the Model S.

If asked what the hardest part of the Model S had been, Rawlinson was quick to say building the team. It was taking more than just the dozens of staffers he had suggested to Musk it might during his job interview, but still vastly fewer than what his predecessors had wanted. In a two-year period, Rawlinson personally interviewed hundreds of applicants, and in the end hired more than a hundred for his team.

Rawlinson's approach fit well with Musk's. He wanted to find the very best in their respective fields—whether it was design or welding—and give them free rein. Initially, Musk interviewed most candidates, often asking a simple question: What have you done that's extraordinary? Engineers would marvel at his ability to dive into the finer points of their work. A wrong answer would end the interview quickly—and often draw Musk's wrath to the recruiter who had sent the person.

Tesla had raised enough money to keep the lights on and fund the development of the Model S, but for all that, cash wasn't exactly free-flowing. Tesla might be able to match salaries of engineers coming from Detroit, but the compensation looked weak when factoring in the cost of living in Los Angeles or Palo Alto. It was downright stingy compared to what the big Silicon Valley tech firms were offering.

It fell to recruiters such as Rik Avalos, who had spent years as a corporate headhunter—including at Google, where he recruited recruiters—to help find workers and sell them on the Tesla dream. For some it was the idea of helping the environment, for others the chance to build something. He told them Tesla was on the cusp of tremendous growth. "What if we hit 50 bucks a share?" he'd say. Such a notion seemed almost fanciful in January 2011 when the stock traded at around $25, only slightly better than where it had sat at the end of its first day as a public company a year earlier.

Musk's high bar for hiring proved challenging for the recruiting team, which quickly learned it needed to better prepare recruits for their interview. One engineer left Musk unhappy, for example, when she told him that using aluminum for the Model S body structure was a bad idea; it could prove tricky and expensive to weld. Musk and Rawlinson had already decided aluminum was the way to go, in an effort to reduce weight from the vehicle and boost range. About 97 percent of the car's structure would be aluminum, with a few specific areas made of higher-strength steels, such as the pillar that sits in the middle of the length of the vehicle, between the front and rear doors. The plan was to do their own aluminum stamping in-house. The interviewee was quickly rejected.

As he hired new managers and engineers, Musk was building a culture around how he wanted Tesla to operate. He subscribed to "first principles thinking," a way of problem solving that he attributed to physics but that was rooted in Aristotle's writings.

The notion was to drill down to the most basic ideas, ones that can't be deduced from any other assumptions. Put in terms related to Tesla: Just because another company did it one way, that doesn't mean it's the right way. (Or the way Musk **wanted** it done.)

But he also acknowledged that he needed to be quick to change course if an idea didn't work. "Rapid decision-making may appear as erratic, but it is not," according to Musk. "Most people do not appreciate that no decision is also a decision. It is better to make many decisions per unit time with a slightly higher error rate, than few with a slightly lower error rate, because obviously one of your future right decisions can be to reverse an earlier wrong one, provided the earlier one was not catastrophic, which they rarely are."

New hires were quickly learning what the mercurial Musk thought was important. He was paying close attention to spending now, really trying to understand what needed to be purchased and what didn't. Engineers would email requests for expenditures, explaining why they were required, and if lucky, Musk would quickly reply with an OK.

One mistake some made early on was justifying an expense by saying that it was within budget. An engineer would long remember the response from Musk: **Don't ever use the word budget with me again because it means you've turned off your brain.** Generally, Musk would agree to an expense if it was needed, but he wanted assurances that there was a real reason.

While Rawlinson's team sat in Hawthorne, the company headquarters remained up north. Musk flew from Los Angeles to Silicon Valley about every week on his private jet. (In 2009, alone, he made 189 trips in his jet in general and spent 518 hours in the air.) If Rawlinson's engineers needed to go north too, they would email Musk's assistant to inquire about open seats. The trick was to arrive ahead of Musk, because when he pulled up, the airplane door shut and they took off. Some days he boarded without saying a word, totally engrossed in his smartphone. Other days, he'd be charming, engaging in speculative conversations about life on Mars or entertaining the wacky idea of tying three Falcon 9 rockets together to create a super, heavy-lift rocket. (He called it the Falcon 27.)

One engineer recalled using the time on the flight to ask Musk his opinion on the suspension characteristics for the Model S, a subject they had been debating. Because Tesla was building their car from scratch, such questions were purely up to them. Was the car's handling going to be sporty, like a BMW, or more giving, like a Lexus? Musk paused, looking directly at his engineer. "I'm going to sell a fuck load of cars, so whatever suspension you need so I can sell a fuck load of cars—that's the suspension I want."

Maybe the engineer caught Musk on a bad day.

Or maybe it got to the heart of how Musk ran triage. Running two complex operations, there was only so much maniacal obsession he could offer. He would give flippant answers until precisely the moment when something became the object of his focus, at which point he gave **every** fuck. In such a world, you're delegated to and given full authority—until he turns his attention on you and your little fiefdom.

This engineer decided the best way to keep a career at Tesla was to avoid more travel with Musk—best not to fly too close to the sun.

As Rawlinson's team worked to develop the Model S, one of its senior leaders, whom Rawlinson had inherited from his predecessor, presented to Musk a plan for helping the team prioritize the vehicle's features. He said GM and Ford engaged in similar processes; if Ford was developing the Fusion, it would gather all of the data it could on competing cars, rank each function, then decide which attribute it wanted to exceed, making trade-offs when needed.

Musk listened to the senior leader for about twenty minutes before cutting him off. "This is the stupidest thing I've ever seen," he said, before walking out. "Don't ever show it to me again." Musk didn't want to prioritize one thing over another, he wanted to prioritize **everything.**

About a week later, the manager was gone. This wasn't a new occurrence; the team had seen a lot of turnover, much of it attributable to either the company's dire financial situation or the need to move engineering prowess from Detroit closer to headquarters. Musk gathered the departed leader's remaining colleagues together.

"Look they're good engineers, but just not good enough for the team," he said.

While Rawlinson set up shop in Hawthorne, JB Straubel and his battery team were among those still in Silicon Valley. Tesla had moved into its new headquarters in Palo Alto in 2010, ahead of the IPO. Far from Rawlinson and his staff, Straubel's team was developing its own culture. By comparison it was a bastion of stability. Many of the recruits he'd hired in the early days through his connections at Stanford had stayed around, growing into new roles or developing even greater expertise in the world of batteries.

The success of the Roadster coupled with Kurt Kelty's persistence began to open doors in Japan. Straubel had chastised Kelty for continuing to stop by Panasonic in Japan every couple of months, especially after they had found in Sanyo a willing partner for the Roadster (not to mention Panasonic's letter indicating that they had no intention of doing business with Tesla). But Kelty still believed Panasonic's batteries were superior enough to be worth the trouble, that each cell could hold more energy than what they were getting with Sanyo.

It was in pursuit of their business that, in 2009, Kelty had found himself in a small meeting room at his former employer, sitting with Naoto Noguchi, the president of the company's unit responsible for battery cells. The walls were yellowish from Noguchi's chain-smoking. Kelty sat on his heels at a traditional short-legged Japanese table, trying to balance his laptop computer as he presented. Data from continued testing and real-world results of the Roadster allowed Kelty to demonstrate how their battery pack system was working. In particular, he was able to say that no Roadster had caught fire from thermal runaway.

In a major breakthrough for Kelty, Noguchi agreed to provide sample cells for testing. He and Yoshi Yamada, who ran Panasonic's U.S. operations, visited Tesla's headquarters that year. Their interest came as Panasonic was beginning to acquire control of Sanyo (they would take a majority stake in the company in December of 2009).

And just like that, the tide turned for Tesla and Panasonic. The manufacturer became excited to be part of a high-profile Silicon Valley startup and was on board to provide cells for the Model S. More than that, Kelty and Straubel wanted Panasonic to invest in Tesla, which was still hungry for funds. The Japanese company agreed to kick in $30 million.

For its successes, Straubel's team still had some

maturing to do. Tesla was no longer a little startup converting Lotus cars into Roadsters, leaning on the experience of their UK counterparts. Tesla was planning to build thousands of Model S sedans a year—on its own. The team got a taste of what that would be like when they became responsible for building the battery packs for the Roadsters. Then more so when they began building components that Tesla was going to sell to Daimler and Toyota, as part of the lifesaving deals Musk had signed a year earlier. Those deals, and the way they would influence the company's culture, would play a larger role in the success of the company than even the money would.

When Akio Toyoda and Elon Musk celebrated an arrangement that included the two automakers working together on electric cars, the details remained far from complete. The idea was that the startup would do for Toyota's popular RAV4 compact sport utility vehicle what it had begun doing for Daimler's Smart cars: providing electric powertrains. But for the teams charged with implementing the agreement, things were less clear. To make the Smart car, Straubel had reconfigured the Roadster's powertrain to fit into the tiny two-seater. Now, Musk wanted him and his small team to do it for a new, larger vehicle, while also continuing to work on battery packs for Daimler, and also developing a new powertrain for the Model S.

Straubel's team assumed they would just be

handing over powertrains to their business partners. The Toyota team, which was inexperienced in dealing with lithium-ion battery technology, thought it was getting help designing an electric RAV4 from the ground up. Some on the Tesla team wondered whether Toyota was secretly trying to steal its technological know-how.

There were cultural differences, too, as the first meeting between the Tesla and Toyota teams made clear. Greg Bernas of Toyota, who had been the chief engineer on another Toyota vehicle, arrived with a newly purchased book on the basics of electric cars. One of Tesla's engineers, meanwhile, carried around a harmonica to play between meetings.

It took months of haggling before they were ready to get to work. Because the teams only had twenty months to complete the vehicle, Toyota agreed to use an old RAV4 platform, avoiding all of the testing and research that the larger automaker's internal rules would require for putting out a car on a new platform. Toyota wasn't accustomed to the way Tesla made on-the-fly changes—to its battery pack, for instance, and the computer software controlling it. As the teams managed cold-weather testing in Alaska, they struggled to diagnose why the prototype vehicle was vibrating on slick roads. Tesla engineers tapped on their laptops, adjusting the vehicle's traction-control algorithms.

They resolved the problem in a few hours of work rather than taking the data back to the lab for further review.

While that kind of speed impressed Toyota, the executives weren't happy with the quality of the product they were receiving. A show car delivered for the Los Angeles auto show in 2011 left executives fuming. A Toyota manager was at a University of Michigan football game tailgate party when he got word from his crew in LA about the poor quality of the SUV delivered by Tesla the weekend ahead of the show. The SUV looked sloppy; that kind of inattention to detail wasn't going to fly for Toyota, especially when they would be presenting it to the press and public. He dialed up a Tesla manager. "What the fuck is this?" he barked before demanding that the Tesla engineers meet him in LA the next day to fix the mess.

One big source of tension emerged over how Tesla validated its powertrains. To Toyota's dismay, Tesla engineers told them that they took their suppliers' word that parts were up to snuff, instead of conducting quality control tests to ensure durability for real-world use. That was a major no-no in the established world of car building.

For all the headaches, Tesla's team was getting a useful primer on how to develop a powertrain that could not only deliver muscle, but that could last in the real world. Toyota's handholding with the RAV4 was having unintended benefits—Straubel and his colleagues took what they learned and poured it straight into the Model S.

Tucked away in a complex of industrial buildings next to the SpaceX headquarters in Hawthorne, Elon Musk opened a new design studio for Tesla. He picked an old airplane hangar that had been converted into a basketball gym a few years earlier. The proximity meant he could easily sneak out of SpaceX to see what chief designer Franz von Holzhausen's team was working on.

In 2011, even as the Model S remained in the distance, they began to consider the **next** vehicle in Tesla's lineup. Musk had long touted his goal of a third-generation car aimed at the masses, but there were too many obstacles to bringing that to market yet. Selling just 20,000 Model S sedans a year wouldn't fuel enough growth—in revenue or brand exposure.

So the Tesla team considered alternatives. They could use the Model S's platform to make variation cars, such as a van or a sport utility vehicle. This would allow Tesla to get savings in parts and tooling, spreading the development costs over a greater number of vehicles sold. It was something that larger automakers had done for years—much in the way that the Roadster, eight years earlier, had been built on the platform of the Lotus Elise as a means to reduce Tesla's costs (while defraying some of Lotus's).

That memory should have been instructive. Musk had dictated so many changes to the Elise in his design of the Roadster that costs soared far beyond what had been expected. His demands for the vehicle that was to share the Model S's platform threatened to repeat his mistake.

The next car, which Musk had taken to calling the Model X, needed to be a family vehicle with three rows of seats. Musk's personal experience weighed heavily on the debate surrounding the new car. It wasn't lost on the designers and engineers that Musk's own five children were getting bigger; he had firsthand experience with sport utility vehicles. They knew this because he complained about it a lot.

He had a few clear ideas. For one, he wanted to make it easier to put kids in the second row. A minivan's sliding door might open further than an SUV's, but still it was a struggle to put a kid into a car seat without bumping your head on the roof, especially for Musk, at over six feet. For another, there was the third row. Musk recounted to the team how he used his SUV. He first placed his youngest kids in safety seats in the second row, so he could attend to them from the front. His older twins went in the third row. But he didn't want Tesla's vehicle to be like the Audi Q7, where you had to do gymnastics to get back there past the second seat.

The team imagined an SUV with the overall curves of the Model S, but with rear doors that opened upward like bird wings (a bit like the DeLorean in **Back to the Future**). These would allow for the largest opening. This way Musk could stand up straight while installing his kids. In theory, such doors could create a huge opening to enter the SUV's back two rows. They created mock-ups of the design so Musk could physically see and touch the rear doors. At first, the opening wasn't big enough, he declared. He wanted it to be as effortless as a magic carpet ride to get into the third row. The concept car they sketched out kept getting longer and longer, to accommodate a wider rear-door opening.

Development of a car can be influenced by many factors. At Tesla's major competitors, designers and engineers might watch in agony as various levels of bureaucracy, from marketing to finance, weighed in on a project, not to mention senior leaders and their last-minute suggestions. At Tesla, it was becoming clear that the decider was Musk and Musk alone. He had flexed his muscles, in an early instance, when Martin Eberhard didn't take him seriously about the quality of the dashboard. With the Model S, his personal preferences had likewise been written all over the car. Musk has a long torso, sitting higher in the seat than a typical driver. Consequently, he pushed the team to hang the sun visors in a way that, engineers worried, wouldn't be useful for most drivers. He rarely carried anything but a phone, his assistant trailing him with whatever he might need. So he had no interest in cluttering a center console area with cubbies. Instead, the team laid a strip of carpet between the car's two front seats, with little walls that formed a sort of gutter.

Even the placement of the external charge port was influenced by Musk, specifically the layout of his garage. Most American drivers park nose first; the design team figured that placing the charge port at the front of the car made intuitive sense. But Musk wanted it at the rear of the car: that's where it best aligned with his home charger.

The cars were being built in Musk's image. It remained to be seen whether car buyers would share that vision.

Even with the IPO the prior year and the partnerships with major automakers, it was painfully clear in 2011 that Tesla wouldn't have enough money to get the Model S out, not without yet another infusion of money. That year, Tesla had grown to more than 1,400 employees, mostly in northern California, and the company was racing to fill a factory with expensive tools to build the Model S, all while sales of the Roadster came to an end that year after completing their contracted run. That meant the only money coming in was from the deals with Daimler and Toyota.

Rawlinson's team was looking for ways to save pennies. They would run countless computer simulations to understand the car's aerodynamics and how it would affect performance. Then they'd rent Chrysler's wind tunnel, to the tune of $150,000 or $200,000, going in on a Saturday night and staying overnight until 6 a.m. on Monday to see how the car worked in the real world.

The team struggled with getting a sunroof that met Musk's standards. Suppliers were either asking for more than Tesla could afford or else offering up compromised versions. Musk got so angry that he ordered his team to hire sunroof designers away from the suppliers and make the part in-house, figuring it would be cheaper to do it themselves.

Still Rawlinson progressed with the vehicle to the point where they were ready for the all-important crash tests. Each car cost $2 million to make, and he sent those gems barreling into objects, destroying them. They discovered that the welding of the car's aluminum wasn't holding as intended; the structure was falling to pieces on impact. It needed a redesign, something that added time and money—and increased pressure on Rawlinson, with Musk hovering over him at each test.

Rawlinson wasn't the only one who felt it. A dark humor bonded Musk's senior executive team. Musk, who split his time between Hawthorne and Palo Alto, would hold an executive committee meeting at the start of the day every Tuesday that would often stretch into the lunch hour. Depending on Musk's mood, the running joke among the team was Musk's lunch plans: Who would he be devouring this week?

Many around the table could see that, more and more often, Rawlinson was on the menu; he drew Musk's increasing irritation. Those working for Rawlinson in Hawthorne couldn't miss it either, overhearing him on calls with the CEO, then feeling the fire of his outbursts afterward. Musk's anger with Rawlinson erupted at one point. Musk, with a rugby player frame, towered over Rawlinson during a disagreement. "I don't believe you!" Musk screamed as he jammed a pointed finger toward Rawlinson's chest. "I don't believe you!"

The relationship was continuing to take a toll on Rawlinson, who also privately worried about his ailing mother back in the UK. Her health had worsened, and with no one else to help, he was trying to make arrangements for her from the other side of the world.

Musk had his own home troubles. His relationship with Talulah Riley was on the rocks. After getting married in 2010, they had been living apart for the past several months. In his rare free time, Musk could be found in the basement of his 20,000-square-foot French Nouveau mansion in Bel Air playing BioShock, a dystopian video game based upon the ideology of Ayn Rand. At Tesla's Christmas party that year, Musk could be seen passed out on a pool table in a room whose entrance was blocked by his brother, Kimbal.

Rawlinson and Musk squabbled over a lot of issues, but none greater than one that dated back to the very beginning of the Model S, when Henrik Fisker, the designer turned Tesla rival, was still at the helm and had churned out what Tesla employees derided as the "White Whale." The problem was tied to the placement of the battery pack at the vehicle's floor, which added height to the vehicle. To address its bulbous appearance, Tesla designer Franz von Holzhausen had stretched the car out, allowing the battery cells to be better distributed beneath. This made for a lower roof line and proportions consistent with a sleek sedan, rather than when Fisker was trying to fit everything into a typical midsized car footprint. But Musk remained concerned about the roof line rising too high. He rode Rawlinson to make the battery pack as slim as possible. Rawlinson worried that going too thin would leave the pack vulnerable to road debris piercing the underside of the vehicle. They fought over millimeters. Eventually, Rawlinson relented. He told Musk he'd cut as much off the bottom as the CEO wanted. He lied.

Rawlinson's team was hard at work on the Model X, too, with an idea for how to make the miraculous rear doors a reality. They had studied the "gull-wing" doors of the Mercedes, a similar feature to what Musk had demanded, but they concluded they needed something stronger. Their doors would be much larger; they'd need to be dual-hinged, not only rising outward but folding in at their halfway point, like a falcon hovering in flight. They settled on a hydraulic system, which would automatically lift the door rather than requiring a passenger to do so. To test it out, they welded a sample door onto a car frame, behind the SpaceX building, and expectantly they pushed the button.

Psssh. It rose like a falcon.

"Bloody hell," Rawlinson proclaimed.

It was a small victory, but even after its completion, Rawlinson sensed his days working for Musk would be numbered. As the Thanksgiving holiday approached, he returned home to the UK; those close to him didn't expect him to return. They were surprised when he showed up in the office in December—determined to give Tesla one more go.

Specifically he wanted to show Musk the Model X concept car and demonstrate the doors that his team had engineered. "Elon, I've done this to show you I can," Rawlinson told him, but "you shouldn't do this." He thought the doors were too risky, adding unneeded complication to the car even as they wrestled to get the Model S into production-ready shape. The two got into another of their heated arguments. Musk dismissed the technical problems Rawlinson raised.

But it wasn't just his concerns about the rigidity of the vehicle, or what would happen when you deployed the doors when there was snow on the roof; Rawlinson didn't think the doors made commercial sense. He had taken the concept car for a drive, and the ride pained him. The Model X had been conceived as a simple top-hat to the Model S's platform, but it was ballooning to be so much more than that, which would mean added cost and complexity. He didn't want to be responsible for it. Rawlinson returned home again for Christmas. This time he didn't come back. He phoned Musk with the news: he was out.

The departure caught Musk by surprise, and he spent the early part of January 2012 pleading with Rawlinson to return. CFO Deepak Ahuja tried to broker a peace between the two men. Musk turned to Rawlinson's key deputy, Nick Sampson, to help as well. But in the end, Rawlinson was too exhausted for Tesla. In a rage, Musk ordered Sampson fired.

News of the departures leaked out late on a Friday, sending shares down 20 percent in after-hours trading as investors worried about a loss of engineering leadership just months away from the start of production of the Model S. It was Musk's first taste, now that Tesla was publicly traded, of how unforgiving the public markets could be. Investors saw two key deputies unexpectedly departed and their faith in Musk's venture was shaken. Could Tesla deliver the Model S as promised? If not, then his long-promised affordable car seemed like a fantasy.

Rawlinson's departure wasn't the first major exit

for Tesla, and it wouldn't be the last. In many ways, his fate was similar to Martin Eberhard's. Both men had at first won favor with Musk, providing him with something he needed. In Eberhard's case, it was a concept for how to build a business around an idea. But Musk ultimately lost faith in his executive's abilities, as the challenges grew exponentially harder. In Rawlinson, Musk had found sorely needed expertise, but ultimately he needed something else—an executive to oversee the start of production, not just the development of a new car.

The next Monday, Musk tried to contain the bad news, holding a conference call for reporters before the market opened in New York, framing Rawlinson's departure as marking a new phase of the company's development and downplaying Rawlinson's role as simply chief engineer for the car body.

"I'm highly confident of the following things: that we will meet or do better than the delivery date for the first Model S car in July." Tesla would deliver at least 20,000 cars in 2013, he promised.

The next day, Riley filed for divorce. Late that night, Musk posted a message on Twitter: "It was an amazing four years. I will love you forever. You will make someone very happy one day." He got on the phone with a **Forbes** reporter the next morning. "I still love her, but I'm not in love with her. And I can't really give her what she wants," he told the writer. "I think it would be extremely unwise for me to jump into a third marriage without spending considerable time figuring out if the third one will work—it was never my intention to have a short marriage. Essentially I'd want to be super sure before getting married again, but I certainly would love to be in a relationship. For sure."

CHAPTER 14

ULTRA HARDCORE

Deep within the cavernous factory that was to be Tesla's 5.5 million-square-foot manufacturing home in Fremont, California, sat a deep pit, a hole that stretched like a giant Olympic-sized swimming pool, the remaining trace of a giant machine that had been used to stamp metal into door panels. Toyota had yanked it out before departing the factory. Its gaping absence left a visible mark of how deep a challenge lay ahead. Peter Rawlinson's team had to begin from scratch engineering the Model S; now the factory team had to figure out how to build the car.

Starting up a new factory is a challenge for even experienced automakers, but it's made somewhat easier through generations of institutional experience. Lessons learned from making cars have been passed down, etched into a playbook of rules and procedures. Tesla had nothing—except a ticking clock. The company needed to figure out how to build the Model S by the summer of 2012 when, Musk had again promised, vehicles would start shipping to customers.

Musk hired manufacturing experts. The factory was divided into two kingdoms: battery pack assembly and vehicle assembly. JB Straubel oversaw the battery pack effort. Vehicle assembly was led by Gilbert Passin, the former Lexus factory manager who had served as a cultural bridge during the acquisition of the Fremont factory, and his deputy, Dag Reckhorn, whom Elon Musk hired for his experience working with aluminum.

They had spent months haggling over what pieces of machinery Tesla would buy from Toyota and what would go. They knew that in many cases it would be cheaper and easier for Toyota to sell the equipment to Tesla at a significant discount rather than to yank out and salvage decades-old machinery. This was another godsend for Tesla. Instead of having to piece together parts of a factory from around the world, they in effect purchased a starter kit. Musk ordered the factory to be painted bright white, the robots colored red instead of traditional yellow. He debated with Passin the placement of giant windows to add light to the dark workspace.

To fill the massive pit, Reckhorn found a giant stamping machine from a bankrupt manufacturer in Michigan that he could buy on the cheap. It ultimately cost more to ship it to California than it did to buy it, but it was a crucial tool, and worth the effort. It would be the first milestone in the factory for the creation of the Model S.

Building the cars began with the arrival of giant aluminum coils weighing as much as ten tons each, almost the weight of a bus. The aluminum would be unrolled, then cut into large, flat rectangles to be fed into Reckhorn's stamping machine, where forty-ton dies would stamp each piece into a three-dimensional part, such as a hood. The thundering sound of a die falling onto aluminum with more than one thousand tons of force was ear-splitting. It was the metronome of a well-tuned car factory. The power impressed into the metal the shape of a part at a top speed, theoretically, of one every six seconds.

But Passin's crew was learning life as a small-volume carmaker. A big car company might use a giant stamping machine to bash metal together at a consistent pace, with perhaps 2,000 parts emerging from a single die. Tesla didn't need nearly that many. Instead, it might bang out 100 parts before changing the die over to something else—a fender, say. That process could take an hour. Then it would spend another couple of hours stamping out a new part before repeating the process again. It was a crawl compared to what many of the factory workers had become accustomed to, banging away on the same part for days at a time, creating an army of parts that would be sent to assembly plants to make, ultimately, millions of vehicles.

From stamping, those individual pieces traveled to the body shop, where they were welded together to create the framework of the car, then joined with the outer panels to give the vehicle its skin. At each stage robotic arms moved in a coordinated dance—seemingly herky-jerky movements executed with lightning-fast speed and precision. The sound of warning buzzers. Clanging metal as the frames moved from station to station. The buzz of soldering, the attendant sparks flying. An acrid smell in the air. This was where, through bonding, riveting, and welding, the production car took physical shape.

Here, the team was also learning how to handle aluminum, still a rare material for cars at the time. Stray dust particles from cutting could lead to dents, and some gauges of aluminum were prone to cracking, which meant the crews couldn't hammer or drill beyond a certain point without risking their expensive materials. All of this required the teams to figure out just the right cadence.

Once completed, the body of the Model S traveled to the paint shop, where it would be submerged into a 75,000-gallon tank of electrocoating solution that would adhere to all of its exposed surfaces. An electrical current passed through the tank—a process that helps protect the material from corrosion. The body then emerged and traveled to an oven heated to 350 degrees to cure the solution. From there, a primer was applied, then a base coat of bright red, blue, or black, followed by a clear coat to give it that new-car shine.

A new paint shop could easily cost $500 million, an amount Tesla couldn't remotely afford. Reckhorn found a supplier to refurbish one of the factory's existing paint shops for around $25 million—about $10 million over what Musk had planned, but by then the team had learned that their CEO could be persuaded. They argued that the deal would leave them with better robots, giving them greater flexibility in the future. "He loves robots," an executive said.

As the factory teams expanded, they caught on to certain trigger words that were a ticket to unemployment. "We learned that quickly: 'no' is not the answer and we'd train our people that they would not say no," the manager said. Instead, managers were coached to tell Musk that they needed to check on whatever he was asking for, and keep playing that line until he eventually (hopefully) forgot—a dangerous game with a guy who could have a vault-like memory for certain things and totally forget others.

Leaving the paint shop, the newly coated body headed to final assembly, where it would take on critical final steps: adding the windshield and seats, and, crucially, marrying it with the battery pack. This stage added incredible complexity, often requiring a small army of workers to attach parts by hand. Passin needed five hundred workers for the factory; one thousand showed up for a single job fair alone, attracting workers who'd spent careers working under GM-Toyota and happy to be employed again.

It was a daunting job by any measure, made all the more so for Passin's team by the fact that engineering, down in Hawthorne, weren't anywhere near ready enough with the car for them to be able to lay out workstations for final assembly.

In February 2012, Passin and Jerome Guillen, who was leading the Model S program in Peter Rawlinson's absence, were delivered the bad news that a low-speed safety test of the Model S had turned up a potential problem. While the car had passed its crash tests, engineers noticed a deformation in a critical safety part, one that was designed to absorb the force applied to bumpers in the event of a crash. They knew the car wouldn't pass a more rigorous test scheduled for the following Monday, just four days away, in a facility outside Los Angeles.

The company had just eleven prototype vehicles available for crash tests, and given the seven-figure cost of each, they couldn't afford to waste a car on a drill they knew it would fail. Postponing wasn't an option either, with production scheduled to begin in four months.

From his experience at big automakers, Philippe Chain, who had joined months earlier as vice president for quality, knew that this was exactly the

kind of snafu that could delay a new car and result in six months of internal investigations. At a
company such as Renault or Audi, there would be
a painful autopsy to pinpoint where the error had
arisen and who was to blame. All the while the issue
would navigate internal politics and multiple levels
of bureaucracy, punctuated by finger-pointing and
time-sucking meetings.

Musk didn't have the bandwidth for such things.
His response: "Solve it, guys."

The team in Fremont huddled with the engineer
responsible for the design of the part in question, and
they began to brainstorm solutions. They arrived at
one quickly: They didn't need to redesign the part,
they simply needed to use a stronger grade of steel.
A purchasing manager got on the phone in a hall-
way and tracked down the material—a thousand-
pound coil was available in North Carolina. They
had it shipped in twenty-four hours to a special processing plant in the Midwest where it could be cut,
formed, and welded. Progress was delayed when a
snowstorm grounded a flight bound for California.
When the part arrived on Saturday night for final
processing, the company's oven, needed to harden
the material, malfunctioned. A repairman had to be
roused from bed. With the piece still warm Sunday
night, Chain loaded it into his car and drove all night
to make the Monday afternoon appointment.

The effort paid off: The part passed the test.
Eventually the Model S would be granted a five-star

rating by the U.S. National Highway Traffic Safety Administration.

It was the kind of story that would become legend, pointed to as an example of how Tesla was nimbler than its competitors. But it also illustrated that Tesla had yet to develop the kinds of systems to prevent such an error to begin with, as it prioritized the pace of development over process, sacrosanct at a traditional car company. Those painful, self-reflective investigations to reveal how an error was made were done not just to ferret out mistakes but to deter future ones.

Tesla didn't want to take chances with safety, of course, but Musk was willing to let some quality issues slide if addressing them meant slowing down their schedule, which could result in costly delays. German automakers, for example, drove test cars through 6 million miles and two winters to find any problems that might emerge. Tesla didn't have that time. Instead, Chain got approval to do the equivalent of one million miles over six months, to spot potential problems and fix them. Musk's approval for even this abbreviated run came on the condition that it didn't affect the start of production. That meant problems discovered during testing would come to light only after work had begun on production, requiring late changes that would add cost. Vehicles already sold would need to be recalled for fixes. In many ways, Tesla was building the airplane as Musk was heading down the runway for a takeoff.

And to complicate things even more, Musk was still making his own cosmetic changes to the car. A few weeks before production was to begin, he ordered larger tires to be installed; he thought they looked better. The direction came over the protests of his engineers, who worried about the complicated tweaks required to the vehicle's anti-lock braking system and the risk that they might shorten the vehicle's battery range.

At times, he seemed to acknowledge that what he was asking was hard. As they approached the start of production, late one night in June, he sent out one of his all-company emails, the kind that served to bind the growing company through a common sense of purpose. In talking with workers, he had been quick to share how the challenges were affecting him, showing a vulnerability that seemed authentic and motivating to some. For many, this would be their first vehicle launch; he wasn't going to sugarcoat how it hard it would be. The scars of the Roadster's near demise still ran deep with him. The subject line to his email read: "Ultra hardcore." It was followed by a warning: "Scaling Model S production without hurting quality over the next six months is going to require extreme effort. Please prepare yourself for a level of intensity that is greater than anything most of you have experienced before. Revolutionizing industries is not for the faint of heart, but nothing is more rewarding or exciting."

The vehicle kept changing. Passin couldn't very well install a costly assembly line if he didn't have anything but a best guess of what would be needed. So instead he and his team came up with the clever idea of using automated carts to move the vehicle frame from workstation to workstation. The carts would move around the factory floor guided by magnetic tape, a solution that proved prescient, as they ultimately needed more than twice as many workstations as had been originally estimated. When production began, the car assembly team had about five hundred staffers, with about as many working on the factory's second floor wiring battery packs together by hand.

After a dry run in the fall of 2011 (conveniently offering a show for the media and test rides for customers, including Bonnie Norman, who teared up as she pulled up to the factory and saw the company's large sign), the team ran ten cars through the assembly process in July 2012, building manually rather than using robotic arms. Each of those cars was to be turned over to an investor, such as Steve Jurvetson, an early board member who was close to Musk and had invested in other Musk ventures, including SolarCity. The crew worked almost

nonstop for a month, finishing up at 3 a.m. most
nights; every panel that came out of the stamping
machine had to be manipulated by workers wield-
ing hammers, who banged the panels into shape.
When cars went to general assembly, they were in
such bad shape that their trunk lids wouldn't shut.
But each day they aimed to get better and faster.
Internally, Musk was targeting five hundred a week
by year's end.

By August, they had made fifty cars. GM or
Toyota would've considered those early cars prototypes to test the line, spending months tweaking
the processes to make sure that when official production began, every car off the line was ready for
the showroom. Not Tesla. Those early ones were
treated as sellable, but only to people close to the
company, like Jurvetson, so engineers could learn
what was wrong and continue tweaking. The managers knew they were flawed; Jurvetson's vehicle
quickly broke down. (The company sent a flatbed
truck to retrieve it, covering the Model S with cloth
so as not to telegraph to the world that the first
Model S had already crapped out.) A team was created to drive every vehicle for one hundred miles,
looking for problems and seeing if the battery pre-
maturely died—an event they internally, morbidly,
dubbed infant mortality. Some of Straubel's battery packs were having problems involving cooling
liquid leaks, which rendered them useless.

The exterior of the car was also plagued with gaps

between body panels, the sorts of imperfections that stand out to car aficionados. Traditional car companies wouldn't allow those cars to be shipped from the factory. Tesla crews made adjustments with mallets and foam until the pieces fit.

Another perplexing problem was water leakage. The cars came off the end of the line and were tested for leaks, and several interiors were ruined in the process. The team rearranged the line so the car could be water-tested before seats were put in place, but the flaw was frustratingly inconsistent, suggesting it may have been a problem with how the car was being assembled.

Tim Watkins, who had helped diagnose the Roadster's cost problems and set up the new sales team in 2010, returned to Fremont to help. By this point the low-key and polite Watkins had developed a bit of a reputation as a harbinger of doom among those who had seen him around during bad times. Some referred to him as the Wolf; others, behind his back, called him the Hangman, because whenever the man dressed in black showed up, somebody was likely getting terminated.

As he reviewed the workflow, he realized there was no standardization of what was being done at workstations. This was because the design of the vehicle kept changing, and without standardized work, it was harder to know if the installation of parts was the problem or if there was a legitimate design flaw. Normally, the steps would have been plotted out years in advance; tested innumerable times; certified and documented in a picture book for workers to memorize. Tesla hadn't done that— each worker had come up with their own process.

Watkins thought he knew the answer. He ordered up GoPro cameras to attach to some workers to document their steps. They were going to reverse-engineer how they were building the car to figure out how to fix the problems. They also implemented a buddy system; subsequent stations down the line checked the work of preceding ones. As Chain, the quality executive, would later write about his experience: "What would have been deemed as unacceptable by any carmakers was seen as part of an ongoing process by Elon Musk who believed, rightly so, that the user experience of driving a truly innovative automobile would outweigh minor defects that will be eventually corrected."

As the year progressed, Tesla increasingly found itself entangled in U.S. politics. The 2012 presidential elections grew heated. Barack Obama sought a second term on a platform of, essentially, having saved General Motors and killed Osama bin Laden. Republican Mitt Romney, the former Massachusetts governor, began taking aim at Obama's support of billions in loans for green companies, including the lifesaving loans that Tesla was granted in 2010, which were helping pay for the Model S. During the first debate between the two men, Romney went on the attack for Obama's support of Tesla, Fisker, and Solyndra. "I had a friend who said, 'You don't just pick winners and losers; you pick the losers,'" Romney said.

Solyndra was an especially juicy target. The solar panel darling, also located in Fremont, had received $535 million in loan guarantees under the Obama administration. But as a glut of Chinese-made solar panels flooded the market and a reduction in subsidies in Europe cut demand, Solyndra had found itself in trouble, filing for bankruptcy almost a year earlier.

Fisker appeared to be struggling as well. The Department of Energy had issued the electric car startup $529 million in loans under the same program as Tesla but put a hold on doling out the money in late 2011 amid concerns about the viability of the business. Fisker had delayed the release of its luxury sedan, the Karma, then sold fewer than promised under the terms of its loan agreement with the government. When the Karma eventually hit the market, Fisker was hamstrung to address its quality problems and parts shortages; it had leaned on suppliers for almost everything. It was on the road to eventual bankruptcy. The takeaway was clear: the company had ceded too much control of its business to others.

Tesla's relationship with the DOE was tense as well. But its problems were different: Musk had bet

on controlling as much of the Model S as possible.
The delays at Fremont meant that the revenue Tesla
had been betting on wasn't rolling in yet. During
the late summer of 2012, as Passin's team struggled
to ramp up production, the company was left with
only enough cars to sell half as many as it had forecast for the season. It had increased production to
a rate of 100 cars a week by the end of September.
Passin need to do much better than that in order
for the company to deliver 5,000 Model S sedans
to customers in the final three months of the year,
as promised to investors.

The effect of the slow start could be seen in the
company's finances. Operating costs were soaring
while revenue was $400 million worse than what
CFO Deepak Ahuja had forecast at the beginning
of the year as a **worst-case** scenario for the company. The problem for Ahuja was that the business
plan had bet on cars going out of the factory and
into the hands of buyers at a rapid pace, bringing
in the money needed to pay for parts as the bills
for them came due. Instead, the factory was struggling to use the parts they had even to make the car.
Cash on hand had dwindled to $86 million. To save
money, the company was letting bills pile up—its
accounts payable had doubled from a year earlier.

It wasn't a sustainable situation. Musk had to
raise money again. He needed to sell more shares
to keep things going until the factory was cranking out the Model S. Unlike in 2008, when he was

able to search for financing in private, Tesla's actions would be under the microscope. The week before the Obama-Romney debate in Denver, **The New York Times** ran a blistering story about Tesla's troubles, which came on the heels of the company announcing plans to sell more shares to raise cash and lowering its revenue targets. The company now expected to deliver between 2,500 and 3,000 Model S sedans to customers in the final three months of 2012—a staggering increase for the little company and still well shy of its previous estimates. "Tesla's story is starting to show some serious cracks," an analyst told the **Times.** "This shows that capital raising is a necessity, not a luxury, as the company had maintained."

Before the debate, Musk tried to do damage control, posting a blog on the company's website that said the media had misunderstood his motives for raising money, likening it to being prepared in case of a natural disaster. He noted that one of Tesla's suppliers had recently suffered a flooded factory, which affected parts getting to the carmaker.

Despite Musk's putting on his best public face, it was obvious that Tesla was in trouble. The company found itself spoken of in the same breath as foundering companies, and Musk was having to deflect, telling reporters that Tesla had more money than it needed to complete the Model S.

"The most you could say is that Solyndra executives were too optimistic," he told reporters at the

Press Club in Washington, D.C., subtly and unin-
tentionally showing his own cards. "They presented
a better face to the situation than should have been
presented in the final few months, but then, if they
didn't do that, it would have become a self-fulfilling
prophecy of—as soon as a CEO says I'm not sure if
we'll survive, you're dead."

CHAPTER 15

ONE DOLLAR

Deepak Ahuja, the soft-spoken chief financial officer, finished his calculations. It was only a few days into 2013, after a year in which Tesla had delivered just 2,650 of the Model S—having fallen far short of fourth-quarter prognostications. Things were looking up in other areas, though. They'd successfully raised some money to buy themselves time. By the end of 2012, Gilbert Passin's team in Fremont had hit a weekly build rate of 400 cars. To keep track of all of the problems, Jerome Guillen, the Model S program director, kept a giant spreadsheet of every car and its problems. Engineers were assigned to the issues and twice a day he would check on the status until they worked their way through the logjam. They were months later than they'd hoped to be in reaching the 400-car milestone but proud that they had pulled it off.

Now that the factory was churning out cars at a reliable rate, Tesla needed to sell them. And yet, despite a list of thousands of pre-orders from people who had put down a refundable $5,000, the sales team, again, wasn't closing deals. Musk publicly attributed it to the challenges of delivering cars to customers during the Christmas holiday. Whether or not that was true, the fact was that for the first time in its life, the company had hundreds of unsold vehicles on its hands.

Early in the year, Musk told Wall Street that he expected Tesla to turn a slight profit in the first quarter. The words "slight profit" were totemic in their significance. After losing more than $1 billion, Musk was suggesting, the company now sat on the razor's edge of profitability. If they could muster those sales, they'd reach a milestone of incredible importance. It would signal that Musk's dream was possible, that Tesla wasn't just a cash-eating machine. Another quarter like what they'd seen in late 2012, however, and the company would again be out of money. This time it wasn't clear that there would be any recourse. His dream of a mass-market electric car would almost certainly be over.

Ahuja considered the numbers. If Tesla delivered 4,750 Model S sedans in the first three months of 2013, nearly as many cars as the factory could crank out, they would turn a profit of $1. They set out to do precisely that.

Musk turned to George Blankenship, now the head of global sales, with clear instructions: "We make a dollar, we have a company. We lose a dollar, it's another losing quarter—we do not have a company."

The toll of years of working to bring out the Model S could be seen on many of those involved. Rik Avalos, the recruiter tasked with building out the company's staff, had lured many families to Silicon Valley to work at Tesla. He watched as the pace ground away at them. At the holiday party, he'd meet unhappy spouses and be greeted with icy responses: "Oh, you're the reason we're here?"

"Whether they got divorced, whether they quit and left. There was a lot of family disruptions," he recalled. "That was really hard on the heart."

Many had taken pay cuts to be there on the bet that Tesla could be bigger than it was. Avalos had offered them a vision of a $50 share price. As they struggled with the Model S, however, the company's stock provided little comfort. Avalos recruited one lawyer who took a nearly 70 percent salary cut when he joined. The shares would really have to soar to offset that loss.

Despite the challenges, Musk had a way of rallying his company. Gathered for cake one day, he told them they needed to keep pushing, they needed to get the Model S out, and then came the next generation. **That** was the one that could make it all go mainstream.

"I know I've asked a lot out of all of you and I know you all have worked really hard," Musk told them. "I wish I could tell you that we didn't have to work harder, but we're gonna **have** to work harder. If we do not do this—we will fail, and we will go down in a ball of flames." But, he added, "If we do this, this company could be worth $200, $250 a share."

Avalos looked over at his manager, who shrugged back as if to say "He's fucking crazy." The recruiter would be perfectly happy with $50 a share, a price he'd all but promised to so many families. As 2013 began and shares hovered around $35, such a rise would mean 50 percent growth. A share worth $250 sounded like a hallucination. That would suggest a market value of $28 billion, fully half the valuation of Ford Motor Co. It was unimaginable.

If riches or a rewarding corporate culture couldn't sustain them, one thing could: the car. Tesla had gotten a staggering jolt of good news the prior fall. The venerable **Motor Trend** magazine, after reviewing twenty-five makes of vehicles for its annual Car of the Year award, shocked the automotive industry in November by picking the Model S. While the likes of GM and BMW had fought hard to woo judges, Tesla had been given the kind of accolade that told gearheads Musk's car wasn't one to take lightly.

The writeup was glowing: a cover that included Musk, the Model S, and a headline of the January issue proclaiming "Electric Shocker!" In the review, they touted the performance, the handling, the interior comfort and exterior looks, concluding: "The mere fact the Tesla Model S exists at all is a testament to innovation and entrepreneurship, the very qualities that once made the American automobile industry the largest, richest, and most powerful in the world. That the 11 judges unanimously voted the first vehicle designed from the wheels up by a fledgling automaker the 2013 **Motor Trend** Car of the Year should be cause for celebration. America can still make things. Great things."

The critical acclaim came just as the business press was detailing Tesla's troubled finances. The juxtaposition of praise and doomsaying would become a hallmark of publicity for Tesla for years to come.* Still, it was the kind of adulation that Musk had aimed for from the beginning when he pushed the team to not only make the best electric car but the best overall car on the market.

---

* It was also a period of improvement in Musk's personal life. A few months after his divorce from Talulah Riley, she was back in his life that fall, telling **Esquire** that her role was to keep him from going "king-crazy." Elaborating on the English phrase, she added, "it means that people become king, then they go crazy." They'd remarry in 2013.

At an event for customers in New York City, Blankenship and Musk celebrated the win. An emotional Blankenship talked about how Tesla's mission had once been seen as impossible, but that the company was on the cusp of something big. "We're not doing this for this year or next year or the next two years," he said, as his voice cracked with emotion. "What's happening here is bigger than that, it's bigger. What we're doing this for is your children, your grandchildren.

"I believe tonight is the catalyst. I see it as something that's going to help us as a company go from maybe crawling to walking, getting ready to run."

Blankenship had done a marvelous job expanding Tesla's footprint of stores across the globe. By the end of 2012, after less than two years' worth of work, the company had thirty-two stores with plans for as many as twenty more in the next six months. During the final three months of 2012 alone, he had opened eight stores, including dazzling new spots in Miami, Toronto, and San Diego. His team counted more than 1.6 million visitors to Tesla stores in the period—almost as many people as had visited the company's stores during the first nine months of that year. It reflected a stunning degree of interest for a company that hadn't even spent money on TV ads.

But while the stores were getting people to kick the tires, for some reason they weren't converting. Sales managers looked at the rate of cancellations and it was grim. The company was facing negative sales growth—more cancellations than orders for the Model S. Even for some Tesla employees, who believed in the mission and thought the vehicles were superior, buying the car could be oddly terrifying. For many people, a car is one of the largest purchases they'll ever make. All the more so for a performance series Model S, a battery-powered car from a company with a limited history that retailed for $106,900 (a number that was staggeringly more than the $50,000 price tag Musk had promised when he first revealed the vehicle in 2009). If their car broke down, would Tesla even still be around to fix it?

What Tesla needed, as with the Roadster, was an army of salespeople swooping in to alleviate the inherent fear, to help turn those lookie-loos into buyers. Again the company turned to its all-purpose fixer, Tim Watkins.

He organized an all-hands-on-deck sales drive: Employees in recruiting were deployed to call potential customers who had inquired about buying a Model S, so-called hand-raisers. Human resources processed orders. Meanwhile, Blankenship managed delivery, tracking every car on its way to the customer. Corporate accounting dictated that a car couldn't be counted as sold until the owner took possession of it. He kept a whiteboard tracking cars in transit. When a truck flipped over in the Midwest carrying half a dozen cars, Blankenship got a call with the bad news ten minutes later. He told an assistant to erase the cars from the board—they wouldn't be delivered this quarter.

Luckily, many of the customers lived in California, making for speedier delivery. Every night just before midnight, Blankenship sent Musk an email with the tally.

"OK, more," Musk would respond one day.

The next day: "Too little, too late."

On the Tuesday before the quarter ended, Blankenship realized they were on pace to hit their number. He emailed Musk the latest.

"This looks promising," he replied.

Blankenship's emails evolved into hourly updates, distributed to more and more people throughout the company. In the office on the final Saturday of the quarter, the 4,750th car arrived at 3 p.m. He pushed send on an email. Exhausted, he turned on the **Rocky** theme song from his computer and cranked up the sound as far as it went.

An aide turned to him: "George, we're just getting started."

Between that afternoon and the end of Easter Sunday, they delivered another 253 cars. Sales soared to $329 million that month alone, more than what Tesla brought in during the entire year of 2011 and 80 percent of the total revenue in 2012.

Turning on the sales faucet had a multiplier effect. Other automakers that failed to meet the strict requirements of California and other states for selling zero-emissions vehicles could lessen their financial penalty by buying credits from companies with a surplus of qualifying sales. Every sale Tesla made was qualifying. In the first quarter, the company was selling credits for about half of its sales, generating $68 million from those other car companies—pure profit. Put another way: Tesla earned 68 percent more from the sale of **credits** in the first three months of 2013 than it had during all of the previous year.

It was the kind of boost that pushed Tesla over the line to turn its first quarterly profit ever. The company hoping to net a single dollar instead reported an $11 million gain. Musk couldn't contain his excitement, tweeting out a press release late that night. He followed up with a tweet to note that he was in California, where it was still late on the final day of March. This wasn't an April Fools joke.

Tesla's stock began to soar in response, rising just about every day: $43.93. $44.34. $45.59. $46.97. $47.83. Then on April 22, the stock closed the day at $50.19. Avalos, the recruiter, could hardly believe it, rushing outside for fresh air, tears swelling; a sense of relief washed over him. They'd done it. He hadn't led those people astray.

The public market was clearly rewarding Tesla. The ability to build a car and turn a profit, even with the help of regulatory credits, signaled that maybe everything else was promising—an affordable electric car for families—was doable.

But the experience of the previous months seemed

to have rattled Musk. Again, Tesla had come close to running out of money. To make matters worse, the natural challenges of birthing a car seemed to unnerve investors in a way that Tesla hadn't experienced in 2008 when the Roadster launch upended things. Musk complained to his staff about having to cater to the whims of the market; that every move was taken out of context and overreacted to; that their focus was on the next quarter when he was thinking about the next decades.

There was heightened scrutiny from regulators, too. Musk's friend and board member Antonio Gracias, along with his firm Valor, found themselves under investigation by the Securities and Exchange Commission for selling Tesla shares about a year earlier, just before another large shareholder unloaded stock, unnerving the market and sending shares down at a time when there was already concern stemming from Peter Rawlinson's unexpected departure. While the firm defended its actions, Valor's timing looked too coincidental to some. The SEC would eventually opt not to pursue any claims against Valor, but only after Tesla endured a round of bad press.*

All of it left a bad taste. Musk seemed to have second thoughts about public markets in general. At

---

* Later that year, the SEC informed Valor in a letter that it wasn't going to recommend any enforcement action against the firm, according to three people familiar with the matter.

almost 1 a.m. on June 7, 2013, he sent out a memo to his employees at SpaceX, which was still privately held and where some workers undoubtedly were looking forward to the payday associated with an IPO. It wasn't to be, Musk announced. At least, not anytime soon. "Given my experiences with Tesla and SolarCity, I am hesitant to foist being public on SpaceX, especially given the long term nature of our mission," he wrote.

For those who hadn't been paying attention to Tesla's dramas, Musk spelled out his thinking. "Public company stocks, particularly if big step changes in technology are involved, go through extreme volatility, both for reasons of internal executions and for reasons that have nothing to do with anything except the economy. This causes people to be distracted by the manic-depressive nature of the stock instead of creating great products."

What Tesla had done was a massive achievement. But now the rubber was, quite literally, meeting the road. All of the years of sweat and sacrifice would be tested. Cars were in the hands of customers. Everything was out of Tesla's hands as they held their breath, waiting for owners to speak and, hopefully, for more positive reviews to follow.

If Musk was right, buzz would begin, launching a word-of-mouth, viral marketing campaign. No better endorsement could come than a positive review by **Consumer Reports,** the not-for-profit publication from the Consumer Union that helps guide readers on everything from cars to washing machines. Unlike **Motor Trend, Consumer Reports** prided itself on its secrecy and independence, eschewing free test vehicles and secretly buying its own cars at random locations, then putting them through the gauntlet. It performs 50 different tests, creating mounds of data on each. Automotive executives' careers have been made and broken on the magazine's results. Detroit car companies have long complained of bias toward Japanese vehicles, even as they took great pains to improve their rankings.

So the industry took note when **Consumer Reports** announced that the Model S had scored 99 out of 100 points—a truly stunning result. Only one car had ever scored so high in the magazine's history, the Lexus LS large sedan. The review was uncharacteristically rapturous, saying the Model S was "brimming with innovation, delivers world-class performance, and is interwoven throughout with impressive attention to detail. It's what Marty McFly might have brought back in place of his DeLorean in 'Back to the Future.'"

The **Motor Trend** endorsement the previous fall had helped Tesla with the gearheads, but this was significantly more important. This signaled to mainstream buyers that the Model S wasn't some science experiment (such as Fisker's Karma, which scored 57), but rather a vehicle that rivaled that of global automakers. Importantly, the review also told buyers that the car had "ample" capacity for running errands and taking "the long, winding way home."

That last point could only have reassured range-anxious customers. The company was already rushing to build its own charging stations along major highways in California and across the U.S. with the goal of being able to say that road-trippers could go from Los Angeles to New York City without having to worry about finding a spot to charge. Better still: Tesla was giving away the electricity for free.

Tesla was riding high on the day that Musk summoned Blankenship to see him, the company having seemingly just pulled off a sales miracle. Despite the undeniable success, both men agreed it was time for Blankenship to give up the reins of the sales organization. Musk's relationship with Blankenship had become strained after the near collapse in the first quarter, leaving the CEO harboring a resentment that he'd been kept in the dark by his sales chief about the full extent of the trouble. "For Tesla to succeed you need the best person you can get at every single position," Blankenship said. "As far as sales, I'm not that person." Blankenship was tired, ready to return to retirement with a sense of victory.*

---

* Blankenship would hang around for a few more months working on opening stores overseas before retiring—again.

—

Musk may have been enjoying the moment, but he didn't want to get caught off guard again. Tesla in 2013 wasn't the company it had been in 2009, when it had just barely weathered financial collapse. Musk had set out to rebuild Tesla in his image, in the hope of ushering in the Model S first, on the path to the Model 3. He'd built an all-hands-on-deck culture of risk-takers, their ranks swelling to almost 4,500 people.

But with growth came distance: He no longer had a finger completely on the pulse the way he once did. His direct control was slipping. In a series of emails to employees, he laid out his expectations that managers shouldn't be blocking the flow of information. "When I say that managers will be asked to leave if they take unreasonable actions to block the free flow of information within the company, I am not kidding," he wrote. In another message, he assured workers they could talk to him directly. "You can talk to your manager's manager without permission, you can talk directly to a VP in another dept, you can talk to me, you can talk to anyone without anyone else's permission," he wrote in another note. "Moreover, you should consider yourself obligated to do so until the right things happen. The point here is not random chitchat, but rather ensuring that we execute ultra-fast and well." That may have been so, but there was a message lurking beneath it all: Just because Tesla was bigger, that didn't mean Musk didn't want to be involved at every level.

To better ensure control of the company's finances, Musk pushed forward with a plan to cut ties with the Department of Energy. If the loans were going to be a political lightning rod, Musk wanted to get away from them as soon as possible. Plus, he'd grown tired of the restrictions placed on him by government aid, the constant need to seek approval for business decisions, the limitations on how he could spend Tesla's money. Shares in Tesla had more than tripled in value so far that year. Investor enthusiasm gave him a chance to raise a record amount of money, and he did just that, pulling together about $1.7 billion through new debt and selling new shares. The proceeds were used to pay off the government loan years early. Musk announced that the rest, about $680 million, along with cash that he expected to generate from sales of the Model S, would be enough for Tesla to bring out the Model X, the SUV built on the Model S platform, and, finally, its third-generation car—his car aimed at the masses, the one that he had promised would cost around $30,000 and change the automotive landscape. He expected that BlueStar, as the consumer-level car was dubbed internally but which would soon be known as simply Model 3, would cost $1 billion to develop.

It all added up to a rare moment of unqualified

triumph for Tesla. Little known, and perhaps more surprising, was how easily it could have gone the other way. It was just months earlier that Musk had, in a darker moment, considered handing away ownership of the company he'd bled for. He had quietly reached out to his friend, Google co-founder Larry Page, and offered to sell Tesla to Google—potentially for about $6 billion, plus another $5 billion in expenses. As part of the deal, Musk sought the $5 billion to expand the Fremont factory, and he would remain running the automaker for eight years, to ensure Tesla successfully turned out its third-generation, mass-market car. His role at Tesla after that would be anyone's guess.

But once first-quarter deliveries came through, talks with Google were quickly scuttled. Tesla didn't need a buyout. Musk had done it again.

CHAPTER 16

A GIANT RETURNS

Dan Akerson did not want to be spotted driving around Detroit in a Tesla Model S. But by the middle of 2013, the chief executive of General Motors needed to know what all of the fuss was about over the electric sedan named **Motor Trend** Car of the Year—the title the Chevrolet Volt, GM's own answer to Tesla, had won just two years earlier. Despite critical success, the Volt was a sales laggard whereas the Model S had fueled Tesla's first quarterly profit and was giving new credibility to the idea that Elon Musk could bring out a next-generation electric vehicle for the everyday driver. Tesla ended the first half of the year having sold 13,000 Model S sedans and valued at $12.7 billion by the stock market, or more than three times more than when the year began.

Elon Musk was talking about sales of the Model S reaching 35,000 a year in 2014.

Akerson, a former Navy officer and telecom executive, had joined the automaker's board as part of the company's bankruptcy reorganization in 2009, in the wake of the Great Recession. He became CEO ahead of GM going public again in the fall of 2010, weeks after Tesla's IPO suggested an improving market. From his earliest days on the board, it was clear Akerson wasn't impressed with what he found at GM. While the bankruptcy had wiped away billions of dollars in debt and put GM on firmer financial footing, Akerson was convinced the company needed an injection of new thinking. The managers who were left over seemed too inward-looking, too slow to adapt to a world that no longer moved at the carmaker's plodding pace.

Akerson saw the Model S being greeted in Detroit with the kind of derision that the American car giants had once reserved for the likes of Toyota, back when Japanese automakers were the upstarts. He saw how that had turned out for GM; he knew it was time to investigate.

The list of reasons, among car experts, for why Tesla was a one-trick pony and was destined to fail was lengthy. Yes, Tesla brought out an amazing Model S, costing on average $100,000. But it had taken Musk and his team years of sole focus to get it to market. The next product would be more constrained, with the company under greater pressure to get it out quickly, all while facing the challenges of continuing to build the Model S. Could Musk keep all those balls in the air and not go broke while trying? Unlikely.

Still, Akerson was convinced that GM's R&D team had become stale, spending time on projects that had no future for the automaker. It couldn't sweep away its modern history of failing to industrialize on its ideas. There was the EV1, for example. GM had conceived of placing the batteries on a flat, skateboard-like frame beneath the car, but the idea was discarded by GM's engineers for the Volt in favor of a T-shaped battery pack that sat inside the cabin, eating up back-seat space. Tesla had seized upon the idea of a skateboard battery pack, using it to create the Model S's spacious cabin. Why should Tesla reap the benefit of innovations GM had arrived at first?

Shortly after becoming CEO, Akerson made a point to visit the R&D operation in the automaker's spacious campus north of Detroit, in a suburb named Warren. He found that the majority of the team had master's degrees or PhDs; they celebrated each newly received patent as a career high, accompanied with a bonus from GM. The automaker ranked among the largest holder of new patents in America each year—$7.2 billion was spent in 2013 alone on R&D. Nominally, Akerson was there to congratulate the new patent holders. He posed for a photo with one engineer who'd already racked up several patents while at GM. But their innovations weren't making their way into GM vehicles. This burned at Akerson. How could that be?

Another issue sure to eat at Akerson was his perception that GM had failed to take proper advantage of the cellular phone technology built into its vehicles. The company's response to the rise of personal technology in the 1990s had been the development of OnStar, a phone signal transmitted by the car that allowed drivers to ask an operator for help or directions. Musk had proven that such a connection could be used for so much more. The Tesla Model S could have its software updated remotely that way, or through the user's home wireless internet. This allowed engineers and programmers to make improvements to the sedan after it was sold, without necessitating a burdensome trip to the shop. A part, for example, might be wearing out, but it could be saved by programmers if they changed the car's code to reduce the torque applied to it.

This ability became critical for Tesla in the fall of 2013, when a string of Model S fires began raising concerns about the safety of a vehicle with thousands of lithium-ion cells onboard. The first fire occurred in October, near Seattle. A Model S ran over debris on the road that damaged the underside of the car, puncturing the vehicle's battery pack (legitimizing Peter Rawlinson's earlier concern, back when he and Musk had fought over millimeters of car height). No one was injured but the fire department struggled to extinguish the flames—an effort caught on cell phone video and circulated on the internet. A second Model S caught fire in Mexico, then a third in November, in Tennessee, piquing the interest of the National Highway Traffic Safety Administration. The reports fed into a long-held concern about the safety of using of lithium-ion batteries to power cars—a concern Tesla's own engineers had worked to address from the start.

It was the kind of controversy that GM engineers had feared for their own cars; their worries seemed well founded. Tesla's share prices plummeted. To top it off, that fall, actor George Clooney, whose early interest in Tesla had been used in publicity by the company, complained to **Esquire** about his Roadster's reliability. "I had a Tesla. I was one of the first cats with a Tesla. I think I was, like, number five on the list. But I'm telling you, I've been on the side of the road a while in that thing. And I said to them, 'Look, guys, why am I always stuck on the side of the f—— road? Make it work, one way or another.'"

Tesla's engineering team got moving. As they studied the fires, they realized that the car's low height off the ground increased the statistical likelihood of running over something that would puncture its batteries. Thousands of other cars might pass over the same road debris, but because their underbodies sat fractions of an inch higher, the odds of damage dropped dramatically. Tesla engineers calculated that if they used the car's suspension to lift the body up just a smidge, the chances of striking debris would be reduced. They changed the software and sent it out to the fleet that winter. It worked, buying them a few months to come up with a thicker plate to protect the battery pack. In the meantime, reports of car fires quickly disappeared.

GM hadn't proven as nimble with its electric vehicles. Even though the Chevrolet Volt had been a win for GM, at least in securing funding from the government and in the world of public relations, Akerson looked at the car with frustration that it hadn't caught on better. He owned one and thought it was impressive, bragging to his golf buddies about how he rarely bought gasoline, yet could still take a road trip. GM had made something unique, yes; but they had also brought to market a $41,000 (before the $7,500 federal tax credit) family sedan that was full of compromises, including looks, performance, and roominess. The back seat only fit two people—it might be able to take a road trip, but not a very enjoyable one.

Not so the Model S. Its exterior was on par with a Porsche, its interior on the level of a Mercedes E-Class. And the real kicker: Tesla's battery range was almost as good as the Volt's range **with gasoline.**

Musk's vision was generating justified questions from Wall Street about whether traditional automakers could compete with Silicon Valley.

As Akerson drove the Model S, he couldn't deny he was impressed. "It's a damn good-looking car. We ought to build that one and put a combustion engine in it," he said the first time he saw one. "We'd do well." Convinced this could be the next threat to GM, Akerson quietly created a team within the company to study how Tesla might bring down the behemoth automaker.

Akerson's task force, made up of about a dozen high-potential managers in their thirties and forties representing different parts of the company, could tell Akerson was different from other GM executives they'd dealt with. He came from the world of telecom. He knew the world could change in an instant when the right technology emerged. "That newly crowned leadership group saw a world of change," a task force member said. "They knew it was coming quickly."

The engineers remained dismissive of Tesla's battery technology, raising concerns about the types of cells being used and the risk of fire they carried. They also worried about the giant touch screen in the middle of the dashboard, saying it posed a danger of distracting the driver. The managers also questioned the legality of Tesla's sales strategy of selling directly to buyers and avoiding franchise dealers.

There was also this: It cost $100,000. GM didn't have a single vehicle that started at a price comparable to that of the Model S. In fact, few carmakers in the U.S. did. The 2012 Mercedes-Benz S-Class, the German automaker's top-of-the-line large sedan, started at $91,850 and ranked as the best-selling vehicle at that price point, with 11,794 sales in 2012 in the U.S. Tesla had long promised a $50,000 Model S, but actually delivering a car at that price was going to be nearly impossible with the inherent cost of batteries. As Tesla announced its first profitable quarter in 2013, it quietly killed off plans for a base Model S, one with a smaller, 40 kWh battery pack. The company justified the decision by saying only 4 percent of orders were for the cheaper version of the car anyway. Those customers' orders would be honored with a Model S with a larger 60 kWh battery pack, though its range would be limited by software.

As 2013 neared an end, Tesla was on pace to sell almost 23,000 Model S sedans—more than the 20,000 it had promised. In the U.S., the Model S was outselling the Mercedes-Benz S-Class, an arguably more luxurious and better-made car. The Model S was redefining what luxury **was** to a certain buyer. Tesla was creating a new kind of market segment—one for buyers excited about technology and the perceived virtue that came from driving a "green" vehicle. Musk was predicting a more than 55 percent sales increase in 2014, to more than 35,000 units, as Tesla pushed Model S sales into Europe and Asia.

GM engineers watched all this and grumbled. They, too, could have delivered a $100,000 sedan; their product planners had never even imagined such a market existed. Within the task force, some thought that Tesla was destined to remain a niche, something wonderful for wealthy Californians but impractical for wide swaths of the world. They questioned whether the startup was ready for mass production of vehicles, which would amplify any quality problems it faced in its factory.

Rather, Tesla might be a harbinger of another threat. Might well-funded Chinese automakers use the Tesla playbook to enter the U.S. market? For years, industry insiders had fretted about the Chinese gaining the manufacturing might to compete in the U.S.—it was seen as only a matter of time. Several had already made pledges to be in the U.S. by certain dates, but those plans were seemingly missing a key component: distribution. Established automakers had a huge moat around them, with thousands of franchise dealerships selling and servicing their cars. If Tesla could prove that you could sell cars directly to customers through a handful of company-run shopping mall stores and a zippy website, why couldn't a Chinese car company do the same and kill Chevrolet on pricing?

Tesla's stores became a key interest for a segment of the task force. Spies from GM were sent to observe the shopping experience. They took note that stores might have one or two vehicles on hand for a test drive, but that Tesla was sending buyers to a computer to build out what their order might look like. As they watched, it seemed many shoppers were looking at the Model S as maybe a third car—not a daily driver. Compared to competitors, Tesla was among the best with visual aids in its stores, but it ranked among the worst in such basic sales functions as asking a prospect's name, offering a test ride, or discussing financing options.

They also questioned how Tesla planned to service a growing fleet of customer cars without dealers. Despite glowing marks from the media, several owners were reporting problems. Edmunds.com, which provides car shoppers with reviews and sales data, bought a Model S in early 2013 and kept a running log of issues that it encountered, including seven unscheduled visits to the Tesla service center and one breakdown that stranded the driver. Two visits involved the car's drive unit, which included the motor and battery pack—pricey fixes. They weren't alone; one owner claimed five drive unit replacements in the first 12,000 miles. A telltale sign of the problem was a milling noise that came during acceleration, interrupting what normally would've been a relatively quiet ride.

"When I first sat down to write this post, I

was all fired up, as I tried to picture myself in an owner's shoes. If I had to replace the engine on my car twice—hell, even once—I would swear off the brand forever," the Edmunds reviewer wrote. "But after talking it over with some colleagues, I was reminded that the people who buy Teslas aren't just buying basic transportation. They are early adopters and willing beta testers of a shiny new piece of tech."

To GM's task force, it seemed like Tesla was going to need to adjust its sales and service strategy as it grew into a broader mass market. Musk had already told the world that Tesla's third-generation vehicle would have a range of two hundred miles or more, at a fraction of the price. The task force had begun to hear industry gossip suggesting Tesla might be trying to bring the car to market in 2016. They harangued GM's battery engineering team. How's Tesla going to get the propulsion cost down so it can sell a car for $35,000 and make money?

"They can't," the engineers replied. "They can't make a $35,000 car any more than we can."

GM was, if anything, better positioned to try. GM should be able to get better pricing on parts, since it bought so many across the company, and the team could reuse components from other vehicles. For the previous two years, teams in Michigan and Seoul had been toiling away on their own next-generation vehicle, and they were excited to have achieved a range of about 150 miles.

"If you can't get this to 200 miles don't bring it out because you're going to embarrass yourselves," Steve Girsky, GM vice chairman, told them.

That fall, GM announced that the company was working on an electric car that could go 200 miles on a charge and cost $30,000 (at a loss to the manufacturer). The message was clear: Detroit had called Musk's bet.

CHAPTER 17

INTO THE HEART OF TEXAS

Seventy years old and dressed in a suit and tie, Bill Wolters arrived at Tesla's headquarters in Palo Alto after a trip from his home in the Texas state capital of Austin. The longtime lobbyist for franchise car dealers wanted to meet directly with the man who seemed hell-bent on overturning generations of norms in the business of selling cars. Wolters came with the goal of convincing Elon Musk that it was time to use franchise dealerships to sell his hot Model S sedan.

Wolters had watched Tesla open its first gallery location in the Houston mall in 2011 and its second in Austin. When he'd talked about Tesla's plans with Diarmuid O'Connell, Musk's deputy who had been lobbying in Texas, Wolters had been dismissive. "Good fucking luck, son," he said. By mid-2013, however, Tesla was turning a profit and Musk was showing up more often in Texas, in part because SpaceX was making plans to expand there. That spring, he'd flown in for a hearing with the state senate and spoken at the annual South by Southwest festival.

His profile had grown far beyond Texas, of course; he had gained broader cultural currency. Robert Rodriguez included him in a brief scene in the 2013 movie **Machete Kills.** The movie also included Amber Heard. Though Musk didn't share a scene with the actress, he began trying to meet her through Rodriguez. "If there is a party or event with Amber, I'd be interested in meeting her just out of curiosity," Musk wrote in an email to the director that was later leaked to the trade press. "Allegedly, she is a fan of George Orwell and Ayn Rand . . . most unusual."

That summer, however, Musk remarried Talulah Riley, the actress who would eventually be cast as a sexy robot in HBO's **Westworld.** Their relationship was fraught, the turmoil often carrying over into Tesla, workers said. Some said they tried to anticipate Musk's mood by following news of his personal life, even tracking Riley's hair color, believing Musk was happiest when her locks approached platinum.

All of this might have seemed a million miles away from Texas car dealerships, but the increased attention fed on itself, helping Tesla make noise. General Motors spent $5.5 billion in advertising and promotions in 2013—an amount about

$2 billion shy of its R&D budget. It and other car-makers were among the biggest TV advertisers in the U.S.; their franchise dealers, too, spent large amounts of money with local newspapers and radio and TV stations. Musk had long eschewed advertising, chalking it up as contrived and inauthentic. The quality of Tesla's cars would be enough to sell them, he argued. He could say that, in large part, because he was able to generate so-called free media. Just like a politician benefits from a drumbeat of news coverage, Musk and Tesla were benefiting from their own attention. His Twitter account could stir the pot and generate excitement. As Tesla rushed to open more stores, local media dutifully wrote stories about the new showrooms.

Franchise car dealers could no longer afford to write the carmaker off. Dealers in Massachusetts and New York had filed lawsuits trying to stop Tesla from selling directly, while state lawmakers in Minnesota and North Carolina were exploring changes to their laws to do the same. To Wolters, it didn't make sense that a company like Tesla would want to shoulder the cost of opening its own store network. Why not push that cost onto dealers? That's what had led the likes of GM and Ford to franchise their stores generations ago.

Wolters was greeted at Tesla by O'Connell, who gave him a tour of the headquarters and took him through the battery lab. They eventually ended up in a small conference room, where he was introduced to Musk. "I really admire what you've done to create this new product and we want to help you be successful," Wolters recalled saying. "And we want to work with you in any way possible to help you do the things that you want to do in our state through the franchise system."

Wolters came at his position with bias, of course. He'd been president of the Texas Automobile Dealers Association for a generation. He'd joined the association after starting his career working for Ford in Texas and elsewhere, a role that had him dealing with the automaker's franchise dealers. It had been a long career and one that shaped his thinking about the fabric of the small towns across his native Texas. As big-box stores and shopping malls rose to replace familiar downtowns, in many places the car dealer was one of the last remaining locally owned businesses. Yes, the customers might be buying a Chevy, but they were buying and getting it serviced regularly from a recognizable family name. "When I was a kid in Lewisville, Texas, there were 2,000 people that lived in that small farming community and there were 40 Main Street merchants—all locally owned and operated," Wolters would later say of his motivations. "Today, one of those businesses exists and it's Huffines Chevrolet. Everybody else has been replaced by a big-box store because they didn't have laws that prevented them from being terminated."

The typical franchise car dealer generated money

on a mixture of new and used car sales, as well as servicing those vehicles. Overall, the average dealer that year made about $1.2 million in profit before taxes, selling 750 new vehicles and 588 used ones, according to the National Automobile Dealers Association. The service end of the business remained where the profit was. (On average that year, a new vehicle netted the dealer just $51.) Like so much of the auto industry, scale was the key to success. Franchise dealers had once mostly been family-owned enterprises. But as with many things, that was becoming less common. Large corporations now owned dozens, if not hundreds, of franchises. AutoNation Inc., a publicly traded car seller based in Florida, was the largest with 265 franchises in the U.S., selling everything from Chevrolet to BMW. In 2012, AutoNation employed 21,000 people (compared to the 2,964 full-time workers at Tesla). Its largest shareholder was Microsoft co-founder Bill Gates, who invested $177 million in the business that year. The company generated $8.9 billion in revenue off the sale of more than a quarter of a million new vehicles.

While the evolution of commerce might be a little more nuanced than Wolters described, he came to his views with an unwavering resolve, along with the backing of more than 1,300 franchise dealerships in 289 cities and towns across Texas. Those dealers netted more than $1 billion a year in profit and represented one of the largest payrolls, tax bases, and sources of civic involvement in the state.

"I didn't take it lightly," Wolters later reflected of his trip to see Musk. "I was there to come to an agreement that we could work together."

It wasn't to be. Musk wasn't interested, not remotely. He talked about surveys that showed that the majority of people wanted to buy their cars directly from carmakers. Wolters disagreed. "We registered 2.8 million new and used cars last year at our franchise dealerships and not a single person said, 'Oh, I wish I could have bought from the factory.'"

Musk wasn't buying that. He showed very little interest in dragging out the meeting. "I'm going to spend a fucking billion dollars to overturn the dealer franchise laws in America," he declared.

Wolters was taken aback. "The quality and safety of every Texan in my state, 28 million people, depends on the franchise dealer network."

Musk only stared at him in reply.

"So, this is just about you then?" Wolters asked.

Few people talked so directly to Musk anymore. To him, Wolters represented everything he wanted to change—a stodgy older man who felt entitled to keep the legacy system humming along for the benefit of the people who had been lucky enough to inherit a cash-printing business, one that took its customers for granted. Musk couldn't contain his rage. He jumped up and left the room, slamming the door and yelling, "Get that guy the fuck out of here!" as he walked away.

Musk would later tell people that Wolters accused him of being un-American. Wolters denied that. The two men were at opposite ends of the spectrum when it came to how they saw the future of America. Wolters was fighting to keep a system in place that he believed was helping families across his state, a way of doing business that was the very fabric, as he saw it, of today's America. Musk held a more typically Silicon Valley disruptor's viewpoint. He saw a better way of doing things and didn't want to be constrained by old rules. For the previous couple of years, Tesla had been trying to work around state laws in establishing its stores. It was time to change that approach, time to go on the offensive. If he couldn't work around laws, he was going to change them. Texas would be his first fight.

Texas became a regular haunt for Elon Musk that year. He approved turning on money to fund a Tesla lobbying campaign, spending up to $345,000 to hire eight lobbyists in the state. It was more aggressive than a parallel campaign by Musk's SpaceX, which was also trying to get state law changed to allow for a commercial space port to be built in south Texas.

That number paled in comparison to the money

flowing from Wolters's shop. The Texas dealers association hired almost three times as many lobbyists, for up to $780,000, and that was on top of the car dealers' political donations that had swelled to more than $2.5 million during the previous year's elections. Musk could feel the influence dealers had. During a visit to the statehouse, a senator approached Musk. "I love what you're doing with SpaceX," the man told him. "I hate what you're doing with Tesla." Musk maintained his composure but inside he was seething.

Despite the challenges, Tesla was able to find sponsors in the house and senate for bills that would narrowly change state law to allow Tesla to own its own stores. Musk wanted to make a big showing to help the legislation get traction. He testified before a house committee in April and sent a company-wide email urging employees to reach out to anyone they knew in Texas to rally at the statehouse:

It is crazy that Texas, which prides itself on individual freedom, has the most restrictive laws in the country protecting the big auto dealer groups from competition. If the people of Texas knew how bad this was, they would be up in arms, because they are getting ripped off by the auto dealers as a result (not saying they are all bad—there a few good ones, but many are extremely heinous). We just need to get the word out before these guys are able to pull a fast one on us. For everyone in Texas that ever got screwed by an auto dealer, this is your opportunity for payback.

The rallying cry worked. Local owners parked their Model S sedans in a neat row outside the capitol and packed the room of the house committee on business and industry to show support for Musk and watch as he made his pitch. Dressed in a dark suit, Musk wasn't the cavalier showman from so many product reveals. He talked in a more measured way about how Tesla needed to reach a new kind of buyer, outside the usual car-buying experience. He believed that traditional franchise dealers had a conflict of interest in selling electric cars because it undercut their business selling gas cars.

He faced skepticism from lawmakers. One questioned whether Tesla would ultimately need a franchise network to handle financing and trade-ins, once it got past the early adopters and aimed for more mainstream buyers. Musk suggested Tesla might one day need to add franchise stores, but what it wanted now was options. "Really what we're trying to do at Tesla is make sure we have the maximum chance of succeeding," Musk said.

Separately, he held a press conference with a little more flair. "Everyone told us when we were getting into this that we'd get our ass kicked. Well, I guess there's a good chance that we will get our ass kicked. But we'll try."

The committee that day moved the bill forward, but it failed to go much farther as the state's 2013 legislative session wound to an end. It was dead. Tesla promised to return in 2015 for the next session, but executives knew it faced an uphill fight. In Texas the galleries would have to suffice for now.

With plans ramping up for their third-generation car—its name by now cemented as Model 3—it wouldn't be enough for long. Tesla had surprised the automotive industry with the success of the Model S, but what got it this far wouldn't get it where Musk was aiming. Every part of the company's game would need to improve, scale up. The next evolution was to the big time. What Musk wanted was nothing less than to take his technology startup and turn it into a real car company. The road was littered with wrecks of companies who'd had the hubris to think they could get there too.

PART III

A CAR FOR EVERYBODY

CHAPTER 18

GIGA

With Tesla's plan for a luxury electric sedan nearing its completion, JB Straubel had gone on to oversee the creation of a charging network from San Francisco to Tahoe and Los Angeles to Las Vegas, to alleviate concerns from Californians about running out of charge while on road trips. A similar network of so-called Supercharger stations was being deployed across major interstate highways in the U.S. Now, aboard Elon Musk's private jet in 2013 on the way to Los Angeles, Straubel thought through the ramifications of his boss's next ambitions for Tesla.

Ever since the electric-car maker acquired the former GM-Toyota factory in Fremont, Musk had been convinced it could once again make 500,000 vehicles in a year, a milestone the facility had almost achieved at its height many years earlier.* Musk had told Wall Street that he thought there was global demand for 50,000 Model S sedans each year and was targeting as many as 50,000 Model X SUVs as well. That left room to maybe build 400,000 of their forthcoming Model 3s at the old factory—a vertiginous number for a company that had struggled to make Model S sedans at all the previous year, and that was still having a hard time increasing its production.

The bottleneck was largely batteries. Tesla was dependent solely on Panasonic for the thousands of lithium-ion cells that were packaged together to create a battery pack in each of its vehicles. Straubel's back-of-the-envelope math suggested that to produce cars at the factory's peak, Tesla would need a yearly supply of batteries equal to roughly what was being built in the entire world at that point. The bigger problem: price. At the current battery rates, Tesla couldn't afford to sell a $30,000 electric car. Despite all of the work by Straubel and Kurt Kelty to get the battery costs down, it was still keeping Tesla from becoming mainstream. The cells were costing an estimated $250 per kilowatt hour, down from $350 per kWh in 2009. That meant that the cells in a 1,300-pound battery pack with a capacity

* The Fremont factory's production peaked under GM-Toyota in 2006 with about 429,000 vehicles, according to a California study on the history of the facility.

of 85 kWh, a reasonable goal for a sedan that could compete with its combustion-engine counterpart, would cost about $21,000. That already represented a huge chunk of the Model 3's projected sticker price. Analysts believed automakers needed the cost to drop to around $100 per kWh before the cost of making an electric car would be comparable to that of a traditional one.

As Straubel discussed his math with Musk on the airplane, the two quickly became convinced: they were going to need a factory to churn out batteries just for Tesla. It was the only way they could scale up the way they wanted to. That could cost billions, and while Tesla was seeing success with the Model S, its cash balance after raising money earlier that year was just shy of $800 million. That was supposed to fund the Model X and the Model 3, and it was already becoming clear inside Tesla that the SUV was going to be more costly than Musk had promised investors. Even with its own factory, Tesla needed Panasonic's help if they were going to crank out billions of battery cells a year. That wouldn't be an easy ask.

Moreover, Musk's standard playbook wouldn't get the job done. The Model S was a hit with critics and tastemakers but it failed abysmally in meeting Musk's stated goal of costing $50,000. For now, that was OK; going forward would be different. He needed to build a machine, figuratively and literally, that could make his new car affordable. Scale would be his friend in fighting cost, spreading the price of building vehicles over as many hoods as possible. The bulk of battery cost remained in its manufacturing; scale could cut away at that line item. But huge scale was in turn going to require rapid sales growth. Tune those two elements—batteries and deliveries—and it would be 2008 and 2013 all over again, Tesla taking huge, quantum strides into the market.

The challenge wouldn't be in crafting some ingenious plan; the way forward was clear enough. The struggle—and it would be greater than anyone in 2013 could have foreseen—would be in the execution.

Straubel's role in developing the battery pack technology that made Tesla cars work had been critical, so much so that Musk (and the company's literature) viewed him as a co-founder. The engineering was impressive, of course, but it had also relied on Straubel's ability to convince Panasonic to work closely with Tesla. It had been a rocky relationship from the beginning, and pulling it together had depended on persistence and good fortune, not to mention Straubel's critical discovery in 2006 of Kurt Kelty from Panasonic.

But Tesla had pretty much been left using off-the-shelf batteries for the Roadster, and Straubel had wanted to tweak the chemistry and structure of the cells to make them more robust for the broader automotive world as he eyed the Model S. The demand for cells was going to be large enough that Panasonic would need to commit additional resources to it. Musk's timeline was also quicker than the Japanese company was accustomed to. As the two sides worked out the details of their battery agreement, following the announcement of Panasonic's investment in 2010, Musk had become unhappy with the battery supplier.

The issue came down to price—as it always did. After a particularly disastrous meeting in Palo Alto in 2011, the partnership looked to be in jeopardy. While Musk had humored CFO Deepak Ahuja by wearing a tie for the introductory meeting with the CEO of Toyota Motor Corp. years earlier, he began showing less interest in entertaining the formality of Japanese companies such as Panasonic. When Panasonic proposed a price for its cells for the Model S, Musk grew irate. "This is insane," he told them. He got angry and stormed out of the meeting. As he walked away, his aides, including Straubel, tried to direct him to an all-hands meeting, where several hundred Tesla employees were gathered for an update from their CEO. He could be heard muttering to himself, "This is a disaster. We're not doing this." He turned to Straubel, told him to do the all-hands meeting without him, and walked away.

It added up to a new assignment for Straubel

and his team: Tesla was considering a move into the battery business. If Panasonic was going to charge too much for and move too slowly with its cells, Tesla would try to make its own. Musk began reaching into Straubel's team himself, reassigning members to the task of building a battery factory. It was daunting prospect; some new hires weren't even sure if Musk was serious.

"Yes, I'm fucking serious!" he bellowed from his desk one day.

But beyond the question of cost, Panasonic had spent years and years developing their manufacturing procedures for making cells, which were highly volatile, requiring clean rooms and special suits to protect the material from contamination. Tesla was still trying to figure out how to use carts to move Model S bodies around an old factory. Some on the Tesla team felt it was a hopeless errand. After months of work and as projected costs mounted, Musk eventually dropped the idea. They weren't ready to go head-to-head with the Panasonics and Sanyos of the world. But he didn't forget the notion entirely.

Out of Musk's sight, Kelty quietly worked toward an agreement with Panasonic that both sides could live with. Musk's anger eventually fizzled (it was unclear to Straubel's team if he'd been persuaded by the agreement's logic or simply lost interest, or if his ire had been a negotiating tactic). In October 2011, Panasonic announced a deal to produce enough cells for Tesla to build more than 80,000 vehicles over the next four years. It guaranteed that it would deliver enough cells in 2012 that Tesla could make more than 6,000 Model S sedans that year. To meet the demand, Panasonic would expand from one assembly line to two.

That timeline would prove more ambitious than Tesla initially required. By 2013, however, it was clear the Model S was a hit not only for Tesla but for Panasonic. It was the kind of win the Japanese company desperately needed after struggling through 2012. A decade or so earlier, the company had made large bets on mobile-phone handsets and flat-screen televisions. But competition from cheaper Chinese competitors led Panasonic to fail in those ventures, leading to billions of dollars in losses. A painful, multi-year restructuring of the company went into place, culminating with Kazuhiro Tsuga becoming president in 2012. He abandoned the TV screen business and cut tens of thousands of jobs. But cutting wouldn't be enough; he knew he needed to point the company toward new growth areas.

Tesla's demand for more batteries for the Model S in 2013 came right on time for Tsuga. He wanted to transform the company's automotive business into one of the company's main units going forward—

a high-profile partnership with Tesla could serve as a calling card to other automakers being nudged into the electric car market.

Tsuga was eager to expand the relationship. Panasonic even pushed Tesla to etch the supplier's name on the back window in exchange for $50 million in cash. Musk wouldn't hear of it. Tsuga installed new leadership over his battery unit who traveled to Palo Alto for a meeting with Tesla.

Musk believed the Panasonic visitors were coming to discuss a price cut. It made sense.

Demand for the Model S was so strong that Tesla's nascent assembly line was now facing a slowdown because it lacked the batteries to keep pace, and Musk was pushing for more and more cells. More business should mean a discount. Instead, the Japanese visitors wanted to **raise** prices. Perhaps it was a negotiating tactic to impress the new CEO back home but it backfired horribly— to put it mildly. Musk controlled himself only inasmuch as he didn't curse out the visitors, something he had done to other suppliers. Instead, he plotted revenge.

The following day, a Saturday, he summoned his team to Tesla headquarters with a familiar order: Tesla was going to make its own batteries. This time would be different from any previous attempt, however. If the Model 3 was going to be a success, Tesla could no longer be wholly dependent on a third party, as it had been with Panasonic on the

Model S. What he and Straubel had envisioned on the private jet was a go.

But just because Tesla wasn't going to be buying Panasonic's cells, it didn't mean they didn't want their **money.** Just as Tesla had benefited from Kelty's hiring in 2006, it was poised to benefit from a new leader within Panasonic's ranks. Yoshi Yamada, a senior Panasonic executive, had a more Western view on business than some of his contemporaries, and he was eager to break out of the Japanese company's old ways. He had helped turn around Panasonic's U.S. operations with a more modern management approach and spent time developing relationships in Silicon Valley. As he neared retirement, he had taken up marathon running at age sixty in 2011; he spent vacations traveling around the U.S. visiting Revolutionary War–era battlefields.

A move brought him back to Japan, where he was given control of the unit that included batteries. With it, he had inherited the Tesla relationship. As a leader of the U.S. organization in 2009, he had visited Tesla when Panasonic was first exploring a partnership. Before that he had been hosted by Kelty several times on visits to tech companies, back when the American executive still ran the Japanese company's Silicon Valley office. So when Musk's demands in 2013 threatened to derail talks, Yamada personally intervened to put them back on track. Until that point, the Tesla relationship had been managed like a regular part of the battery business; no major executive from Japan was involved. Yamada thought more oversight was needed.

That fall, they announced a contract extension. To address Tesla's needs for the Model 3 and Panasonic's lack of capacity to handle the new demand, Yamada proposed a joint-venture between Panasonic and Tesla. Musk despised such arrangements—essentially a 50-50 company, with both sides jockeying for control. Anyone who had worked for Musk knew that he wasn't into power sharing. But by then, Kelty and Straubel were cooking up something to allay their boss's concerns.

It was a season of change for Straubel, who had met Boryana, a young woman in Tesla's HR department, a self-described nerd who shared his affinity for data. They married that summer. In late 2013, he was crafting slides to make the case for a massive battery factory. It was an audacious plan: a factory to be built in phases, adding capacity as needed. It would cost as much as $5 billion, comprising 10 million square feet under a single roof (making it larger than the Pentagon). The facility would require as much as 1,000 acres of land and employ 6,500 people.

The proposal had more in common with how Henry Ford had envisioned his business a hundred years earlier than with how most automakers operated today. Some on the team were against the idea of taking on responsibility for batteries, as Musk was pushing for. Kelty feared the complexity was too great, and that it would be pushing his counterparts at Panasonic into a novel solution: In Straubel's grand plan, Panasonic would essentially be setting up shop inside a Tesla factory to build cells at one end, while Tesla assembled battery packs at the other. It would be a vertically integrated facility, with raw material for the cells coming in and battery packs coming out, for use in the Model 3s to be assembled in Fremont.

Build it in the U.S. and the savings in shipping alone could amount to 30 percent. But Tesla was going to need even more savings if it was going to lower the cost of its cars and make a mainstream electric vehicle. Gasoline cars were still about $10,000 cheaper than comparatively sized electrics.

Straubel's Gigafactory, as Musk called it, came about as the chief technology officer increasingly needed a new focus. Musk was looking to expand his senior leadership team, and he became convinced that manufacturing needed to be put under one person instead of the two warring units of vehicle and batteries. Straubel advocated for elevating his own manufacturing leader, Greg Reichow, who had overseen the rollout of the battery pack line at Fremont with relatively little drama. Musk agreed, but in a surprise to Straubel, Reichow would join Straubel in reporting directly to Musk, given orders to create a new assembly line for both the Model S and Model X. Straubel's influence waned.

Likewise, Musk had been courting Doug Field, a high-ranking Apple engineer, to join the team. As Musk led Field around the factory on a tour, senior leaders recognized that Field was being considered for more than just advanced engineering. Field represented a new kind of executive for a new era of Tesla. Unlike those engineers recruited by Peter Rawlinson, he wasn't someone who had spent a career at a traditional automaker and was looking to escape for a fresh start with a small startup. Nor was he a recent Stanford grad looking to jump-start a career in Silicon Valley, as Straubel and much of his team had been. Field was a seasoned corporate warrior, one who oversaw thousands of people at Apple and was responsible for carrying out the iconic Mac computer's engineering. Field's hiring would be a statement to Silicon Valley that Tesla could play with—and poach from—the big boys.

Field was the right guy to usher Tesla into a new, more professional era. He had cut his engineering teeth at Ford Motor after graduating from Purdue University in 1987, then departed in frustration with the automaker's culture. He'd been assigned to study competing Lexus and BMW cars and in doing so realized it would be a long time before Ford could compete. He'd gone on to Segway, where he oversaw an electric scooter that was before its time, before he was recruited to Steve Jobs's Apple.

As Straubel watched Musk courting Field, he knew Field would be brought in with the promise that he could oversee Model 3 development. Straubel would need something new. The Gigafactory, then, would allow him to build his own empire. Just as he had solved the critical problem that made a lithium-ion-powered electric car work, he would tackle the biggest issue keeping electric cars from going mainstream: cost.

So, working together, Musk and Straubel concocted a high-stakes gambit. They conceived of a way to pressure both their suppliers and local governments into helping Tesla fund the new facility. The plan called for suppliers to contribute to the development, estimating that half of the price tag could be paid for by companies such as Panasonic. It also depended heavily on winning government incentives; whichever state housed the factory would reap the benefit of thousands of high-paid, high-skilled jobs. Diarmuid O'Connell, who had navigated the company through its lifesaving DOE loans and in its fight with franchise dealers, began reaching out to states with the goal of pitting a few against one another in a bidding war. Texas was appealing because it might help give Tesla the kind of leverage it needed to win approval in the statehouse to overturn franchise dealer protection laws. Musk preferred California because it was closer to home.

A location outside Reno, in Sparks, Nevada, however, was increasingly looking like the perfect fit. The state seemed eager to make them welcome, and the location was a mere four hours' drive from the Fremont factory. It was only a little farther on Musk's private jet than flying from LA to the Silicon Valley.

Government officials were invited to Tesla's headquarters to hear about the plan. They arrived to find their competitors in the same room. They quickly realized the cost of winning Tesla's presence in their state would be steep.

By late February 2014, to turn up the pressure, Tesla went public with its plans, seeking to raise $1.6 billion in new debt. The money, the company told investors, would be for a massive battery factory, the third-generation vehicle, and other corporate matters. Speculation immediately turned to whether Panasonic was on board.

Back in Japan, Yamada was facing resistance on the project. It wasn't clear to many in the auto industry what the true demand for electric cars might be. Consumers weren't rushing to embrace the Chevrolet Volt or another new entrant into the market, the Nissan Leaf. Panasonic officials also didn't like the idea of setting up a costly shared factory that Tesla would own. It had never done such a thing.

To win over Panasonic, Straubel needed to convince Yamada that Tesla meant business. He had a plan. It harked back to how Tesla had cajoled early investors into taking a chance on a little startup. To stoke Daimler and Toyota for the Model S, years before it was ready, they had worked up mule cars—dummies that were close enough to the real thing to give their audience a taste of what was to come. Tesla needed something they could showcase, a mule **factory.**

Blueprints for a factory, however, failed to capture the kind of excitement their prototype cars had. The Tesla team became convinced that they needed to demonstrate to Panasonic and other suppliers how serious they were about the project. Quietly, they forged a deal with landowners in Sparks and began preparing the site for construction. They called bulldozers and earth movers from around the state, erected massive lights, began moving tons of dirt. The bill was enormous, climbing to $2 million a day. Straubel wanted to have a site prepped for a demonstration to Tesla's would-be partners. He had to make it convincing enough to suggest that Tesla was charging forward—with or without them.

It was a risky gamble. If word got out, it might look like Tesla had already selected a location without any need for state money. Musk was looking for a state's help to the tune of 10 percent of the cost of the project, or $500 million. Why would Nevada lawmakers, or any other state, approve incentives for a factory that was already under way? When the local newspaper caught wind of the massive work under way and broke the news, suggesting it might be for Tesla, the company deflected, saying it was one of two locations being prepared so it could be ready to go—an excuse that was hard to swallow given the extravagant costs and the company's limited finances.

But those details didn't matter. Musk and Straubel were creating an illusion, a Potemkin factory. At the Sparks site, Straubel built a raised viewing area that gave an expansive view of the construction. He invited Yamada for a visit. As the two stood looking over the site, Straubel arranged for giant earth movers and dump trucks to rush by on cue for dramatic effect. He wanted Yamada to know the future was under way, with or without him.

Straubel was excited; he expected Yamada to be as well. Instead, the executive was quiet. He looked pale. Sick to his stomach, maybe, over what he was seeing. The trick had worked better than Straubel had even intended. Tesla hadn't just enticed Panasonic, it had backed them into a corner.

Yamada, who by now had bought into Tesla's vision for the future, would return to Japan, and within weeks Straubel and Musk would fly to Japan to meet with Tsuga for one final dinner. After brief small talk, Musk cut to the chase: Are we doing this?

Tsuga agreed.

After dinner, Straubel boarded Musk's jet to return to California. In Tesla's short life, the company had inked many lifesaving partnerships—Daimler's supplier agreement helped keep the lights on, Toyota's disused factory let them make the Model S. The deal with Panasonic was something else, maybe something even more; it suddenly gave them the potential to scale up by an order of magnitude. If Straubel and Musk were right, they had just opened up the bottleneck, ushering in the era of the affordable electric car.

CHAPTER 19

GOING GLOBAL

Far from California, Tesla was gaining a cult-like following in a seemingly unlikely spot: Norway. In mid-2012, as Satheesh Varadharajan, an IT entrepreneur in Oslo, began looking to replace his used BMW X5 SUV, he stumbled upon an online video from earlier that year of Musk revealing the Model X. He was intrigued by the SUV; he quickly read as much as he could and took a trip to a newly opened local Tesla store, where he saw the Model S for the first time. The SUV was still years away from production, but he could get into a Model S within a few months.

"This thing was packed with things you'd never seen before," he recalled. The large screen, the acceleration. He was sold. The price helped; the effective cost with government incentives in Norway was around $60,000, roughly half what he had paid for his used BMW. Others took notice, too, with sales in Norway soaring to become Tesla's second-largest market behind the U.S.

China held even greater promise. Facing polluted cities and clogged roads, the government was also pushing for the adoption of electric cars. In places such as Shanghai, where most cars were limited to which days they could be driven, electric cars were exempt from such restrictions. Nationwide buyer tax breaks would lower the effective cost. The expansion of the electric car market in China was projected to fuel much of the growth of the electric car market overall. Global luxury carmakers were already raking in money from the explosion of Chinese buyers for BMWs and Mercedes. The idea of a luxury electric car seemed like a sure shot—which could fuel Tesla's growth ahead of the Model 3 and, in turn, create an even larger market for a mainstream car. Just as Tesla needed to build out its ability to produce billions of batteries, it needed to enter a market where millions of buyers might help boost its sales.

With George Blankenship's departure, Elon Musk turned sales and service over to Jerome Guillen, a Ph.D. in mechanical engineering from the University of Michigan, who had already proven himself to Musk when he was tasked with shepherding the Model S into production in the wake of Peter Rawlinson's departure.

The French national's arrival at Tesla in the fall

of 2010 had come as a surprise to some in Europe; the then thirty-eight-year-old's career had been on the rapid rise at Daimler AG. German media had speculated that he was a strong candidate for promotion to Daimler's powerful executive board someday. Daimler chief executive Dieter Zetsche had plucked him from the company's commercial truck division, where he had overseen the development of a new generation of semitrailer truck, to be the founding leader of Daimler's newly formed business innovation department. The high-profile role had him building a team to evaluate new business opportunities for the company, which was considering the way technological advances might change its future. One of Guillen's successes was the launch of a car-sharing business called Car2Go, which allowed users to rent Smart cars on an hourly basis.

Guillen had two sides to his personality, according to those who worked alongside him. With Musk, Guillen was deferential. In the trenches, he was hard-driving, quick to yell and threaten underlings. Human resources received complaints about his management style and tried to coach him on how to deal with people less caustically. Musk publicly touted a "no-asshole policy" for hiring, but it seemed to many that added consideration was given to some difficult personalities as long as they delivered results (and, importantly, weren't an asshole to Musk himself). It was all a matter of perspective, of course. One man's asshole is another man's straight shooter. To Musk, Guillen was a problem-solver, the guy called in when things went sideways. Guillen's advice to his underlings was to seek his input on Saturdays, when he had time to think, and to never interrupt him during lunch—his only moment to himself. His approach ran in stark contrast to that of his predecessor, George Blankenship, who had reveled in rewarding stores for their sales results with pizza parties, and who was always quick with kind words.

Guillen's native Europe was Tesla's next beachhead, but the most prized market—by far—was going to be China. Musk began 2014 telling **Bloomberg News** that Tesla's China sales could equal those in the U.S within a year. ("It's not my firm prediction—it's more like a low-fidelity guess.") Guillen hired an executive from Apple, Veronica Wu, to oversee the growth. She'd helped build out the tech company's presence in China from almost nothing into a major growth engine for Apple, on its way to rivaling the company's sales in the U.S. Her part of the business had been less glamorous but no less significant than selling directly to retail buyers. Wu had been responsible for education and enterprise sales.

During her job interview, Guillen asked her to write up her thoughts on the Chinese market to share with Musk before meeting him. She stressed Tesla's lack of brand awareness in China and its need to position itself locally. She warned that foreign companies tended to run into problems when they assumed that what works in the U.S. will automatically work in China. Lastly, she told Tesla that it needed to think carefully about government relations. It wasn't just a question of regulation, such as in the U.S.; the government in China could make or break Tesla if it chose to. She cautioned that survival in China was almost Darwinian, that it often came down to a company's ability to adapt, rather than just its strength or smarts.

Her number-one job after getting hired was to seek government approval to get Tesla vehicles into the country and qualified for EV subsidies. In Shanghai, she found the city government welcoming, even offering to qualify Tesla's cars for EV license plates before receiving qualification from the national government for EV subsidies. City officials wanted to know if Tesla had plans to establish a China-based factory; if so, they wanted it in their community. The promise of a new car factory and the jobs it would create proved an enticing carrot for local Chinese leaders, whose careers could be boosted if they were seen to have grown their municipalities' economic fortunes. Shanghai had always been more welcoming to Western automakers than the rest of the country. But a foreign carmaker couldn't just set up a factory to build cars locally. It needed a local partner to own about half of the operation. General Motors and Volkswagen both headquartered their China operations in Shanghai and were partnered with the same local joint venture.

Musk, however, was resistant to the idea of going into China with a partner—he was concerned about losing control of the brand and the technology. Without a local partner, foreign car companies couldn't set up local manufacturing, and vehicles imported from overseas were greeted with tariffs of 25 percent.

Though it could derail any dreams of selling the Model 3 with a lower price tag, the lack of local production wouldn't be Tesla's biggest initial hurdle. The more immediate concern was a lack of charging infrastructure. Many buyers in the country's major cities lived in high-rise buildings, without parking or a place to charge their vehicles. This put on pressure to rush the opening of Supercharger stations around Beijing and Shanghai.

Despite the challenges, Wu was making headway in her first few months. In late April, Musk came to Beijing to celebrate the handing over of the first Model S sedans to local customers at a charging station that had been set up in an industrial park. Tesla was finding that Chinese customers were motivated differently from those in California, where buyers largely bought into the idea of Tesla as much as into the car itself. The Chinese consumer buying a $120,000 car expected to experience luxury through their purchase—not a bare-bones showroom without any of the perks found at the country's franchise dealerships, such as lounges and snacks. Customers staged publicity stunts over their perceived grievances with the brand; one man gathered reporters outside the Beijing store to watch him bash his newly purchased Model S's windshield with a sledgehammer in protest of its late delivery.

Wu argued to Guillen that Tesla needed to expand its sales in China through the use of partner retailers, much in the way Apple had (unlike its U.S. stores, which were Apple-operated). It was a non-starter for Guillen, who knew Musk's position on franchising. Tesla was built on the idea of controlling its own sales experience, and it wasn't going to give that up as it ventured into a new, major market. Despite the challenges, sales in the third quarter in China were picking up steam.

Just as the business seemed to be finding its footing, though, Wu was caught with a huge, unexpected problem. Musk performed his marketing magic back in California in the fall of 2014, and Chinese customers, watching on the internet, began to revolt.

It started that October when Musk, after using his Twitter account to tease that an announcement was coming, revealed at an event that Tesla would soon be coming out with a **new** version of the Model S—one with a dual motor. Musk promised quicker acceleration—a blazing 3.2 seconds from 0 to 60, rivaling a McLaren F1 super car. It would begin rolling out software called Autopilot that promised to use artificial intelligence to handle some of the car's driving. The vehicle's range would also improve.

"This car is nuts," Musk said. "It's like taking off from a carrier—it's just bananas. It's like having your own personal rollercoaster that you can just use at any time." Drivers would be able to select from their touch screens drive modes called "normal," "sport," and "insane." Tesla said customers could begin ordering in the U.S. right away, with an expected delivery in December in North America, followed by Europe and Asia.

The China team was caught on its heels, especially when customers with pending orders for the older Model S began demanding the new version instead, rather than the one they'd already configured. It took about two months for a car to leave the Fremont factory and make it by boat to China, then through customs, before arriving in the customer's hands. That meant Wu's team had hundreds of cars about to be delivered that were immediately outdated. Even worse, it wasn't clear when the new vehicles were going to arrive in China or what their new pricing was going to be. When she and her colleagues reached out to Palo Alto, they found a U.S. team that wasn't ready with answers.

Car companies around the world have long had to deal with model changeovers, as this year's cutting edge becomes last year's out-of-date. Most companies try to manage their inventory so that dealerships aren't left sitting on too many older cars. Model year close-out sales are well-known to car buyers, and August is often a good time to get a deal on outgoing models. Such discounts went against Musk's ethos for selling. Palo Alto wanted to hold firm on its Chinese sales.

But Wu's team was seeing orders getting canceled at an alarming rate. Tesla's arrival in China was on the verge of being stillborn. Sales dropped 33 percent in the fourth quarter compared to the third, and according to an equity research firm, almost 50 percent of the vehicles Tesla had imported into the country went unregistered by year's end. Guillen blamed Wu for the decrease, not Musk and his theatrics. She was summarily ousted. As sales in China fell even more in the beginning of 2015, Musk sent out an email with a clear threat: he would fire or demote managers if they were "not on a clear path to positive long-term cash flow."

Fully a year after ordering a Model S in Oslo, Varadharajan received his car in mid-2014. The Norwegian IT executive had no regrets, marveling at the ease of charging it at home and enjoying the perks of never having to visit a gas station again. The customer experience was also unlike anything he'd seen before. On a road trip in Europe, his car broke down. He called the service center back in

Oslo and they told him to fly home, that they'd ar-
range to take care of the car for him. The service
center even offered to pay for his flight, he said.
He'd heard similar stories from other customers in
Norway. In June 2014, he became president of the
Tesla Owners Club Norway, an effort to build out
a network of charging stations to augment those
operated by Tesla.

In his first year, he'd admittedly had a couple of
other issues with the Model S. The door handles
stopped working, a common problem, and he experienced some rattles. He found it easy enough to
get an appointment for fixes, with little wait time
for repairs. By 2015, however, that was changing.
It might take several weeks to get an appointment,
followed by several days, if not weeks, for repairs.
Other club members were reporting much longer
times, he recalled.

In Palo Alto, Guillen's team knew there was a
problem. Their data showed that the average time
to get a vehicle repaired in Norway was sixty days,
and some customers on internet forums were complaining about waits that were much longer than
that. (In California, the turnaround time was more
like a month.) In many ways, Norway was the ca-
nary in the coal mine for Tesla. Because of government subsidies aimed at fostering electric car sales,
Tesla's business had boomed there. Unlike in the
U.S., where, according to a 2014 survey of new
vehicle buyers, the average Tesla purchaser owned

two vehicles, in Norway the car was a daily driver for many.

That was the case for Varadharajan. If Tesla didn't do something, matters in Norway would only get worse as the Model X arrived, and worse still when the third-generation vehicle landed after that. It was exactly the kind of troubles forecast by franchise dealers in Massachusetts, Texas, and other states, who were still trying to fight Tesla's direct sales approach. If delays spread to the U.S., Tesla would be handing easy wins to its competitors, not to mention scaring away new customers just as the brand was taking off.

Bonnie Norman, the Roadster owner in California, had begun to see a change in the kinds of messages, and their tone, on the Tesla club website that she helped moderate. New buyers were complaining about issues with their vehicles, venting that they hadn't expected to encounter such problems with such an expensive car. She was a believer in Musk's road map to bringing electric cars to the mainstream, but she began to worry that more was needed to help welcome those converts into the world of EV ownership.

She wrote a memo to Musk deputy Diarmuid O'Connell, suggesting the company develop a more robust education strategy. The loyal owners that had spawned clubs in several cities, including the one she would create for Sacramento and Lake Tahoe, could be a powerful force. While Tesla had eschewed traditional advertising, it had begun trying to incentivize buyers to become de facto brand ambassadors. The company created a referral code system for owners to pass along to potential buyers, who would get a benefit for using the number. In turn, the owner would earn points, almost like a commission for a successful sale, toward other Tesla products, including future car purchases. Norman was on her way to getting a special Model X with her referral codes.

"You have an owners group ready to do just about anything for you—they make their own signs for events, they form parades, some have barbecues for others who drive cross country—how can you use that passion and energy to start educating the Model 3 market so that when they start to buy their first Tesla they do not overwhelm Tesla with questions and non-problems?" Norman wrote. "How do you aim this amazing gun of Tesla owner passion at a problem that Tesla is sure to have in the future?"

The challenge of the global sales and service operations was getting to Guillen. Musk had begun 2015 with the ambitious goal of increasing Tesla's annual sales to 55,000 vehicles, a 74 percent jump from 2014. By the end of the first half, however, Tesla was again falling short of the pace it would need to meet that goal. The sales team was struggling—including its leader. Burned out, Guillen stepped away from his role, eventually leaving the company in what was described as a sabbatical. It was another period of turmoil for the sales department.

That spring, Musk took direct control of sales himself, and board member Antonio Gracias and his partner Tim Watkins returned to Tesla to dig into the problem. They had helped build up the sales department in its earliest days, when the Roadster was sputtering, then again when the Model S needed a more aggressive push. But at this point, even Gracias and Watkins were struggling with new ideas to juice sales. Increasingly, some on the senior executive team wondered if they had reached the ceiling of demand for the Model S. Musk turned to his cousins at SolarCity for help, asking that the solar panel company's top sales guy, Hayes Barnard, help diagnose the problem.

Part of the issue, Barnard found, was that the sales force took weeks to close a sale, a holdover from when the organization thought of itself as primarily about educating customers, avoiding hard sales techniques. Musk wanted to revise that approach, to have the team focus on its pitch instead. Barnard brought in Tesla's top-performing sales members from around the U.S. and videotaped their approach, as part of a training program that would go out to the global sales force.

Musk decided he needed an executive to handle non-engineering tasks, the things that bored him and that he had handed over to Blankenship years earlier. He approached Sheryl Sandberg, the chief operating officer at Facebook Inc., about becoming his COO at Tesla. She demurred, instead recommending Jon McNeill, a friend of her late husband.

McNeill was different from many of the other big hires in recent years. He was an entrepreneur who understood the kinds of risk-taking required in a startup. **Fast Company** magazine, almost a decade earlier, had included McNeill among its annual list of the most innovative entrepreneurs, for his work growing an auto repair business called Sterling Collisions Centers Inc. to forty locations and annual sales of about $120 million. He did so by improving the typically inconvenient process of getting body work done after a car crash. McNeill would bring to Tesla an acute understanding of how to use data to improve the customer experience. Even before officially taking over as president of sales and other non-engineering efforts, as McNeill traveled, he began stopping in to Tesla stores to better understand the sales process. He'd take a test drive and leave a different email address at each visit. But after a few weeks, none of the stores had reached out to him in an attempt to close a deal. From its early days, Tesla understood that the test drive allowed a buyer to see what was different about an electric car over a traditional one. Gracias and Watkins had built the sales process around that test drive; it was supposed to be the moment when the salesperson latched on and closed the deal. But if no one was even following up to **try** to make a sale, it was clear that the sales team had lost its discipline.

The newly hired McNeill went to Norway and met with Varadharajan and members of his club's executive council at a Tesla location. He told them Tesla was rushing to make things right. What wasn't said was that McNeill was going to have to find ways to improve the customer experience beyond just hiring a bigger staff and opening more stores and service centers. Tesla didn't have the money for what would be truly needed once the Model 3 arrived. Instead, his team began looking at the data coming off the Model S and realized they could identify 90 percent of the service issues remotely and fix 80 percent at an owner's home or office, including seat replacements and brake repairs—basically everything but replacing batteries and drivetrains. Instead of spending millions to build more service bays, McNeill's team would put hundreds of technicians into service vans to make house calls.

As Varadharajan looked around the Oslo service center, he believed McNeill's commitment: While they chatted, he could see a couple dozen people interviewing for jobs.

CHAPTER 20

BARBARIANS AT
THE GARAGE

Sitting on the fourteenth floor of Trump Tower in midtown Manhattan, Lawrence Fossi had the unique job of managing the private office of one of New York City's most colorful billionaires. Though then in his late sixties, his boss, Stewart Rahr, had been a larger-than-life character around the Manhattan party scene ever since he'd sold his pharmaceutical company and divorced his wife. His exploits with models and celebrities were detailed in the **New York Post.** When not in newsprint, he had developed a habit of email-blasting tidbits of his adventures (along with photos of him and celebrities, such as Leonardo DiCaprio, or with a bevy of beautiful women in various states of undress) to a list of hundreds of celebrities, journalists, and fellow billionaires. A 2013 article in **Forbes** carried the subhead: "The

Unhinged, Hedonistic Saga of Billionaire Stewart Rahr, 'Number One King of All Fun.'"

Fossi's life was more sedate. He was the guy behind the guy. It had been a far climb for Fossi, who had grown up with six siblings and was the first in his family to attend college. He picked Rice University in Houston because it seemed about as far from his home in Connecticut as possible. Born in 1957, he graduated during the Watergate era and went to work at a tiny weekly newspaper in Wilton, Connecticut, the kind of job that let him do a little bit of everything, from reporting on local governments to laying out the front page. Most important, it allowed him to write. A year later, he was off to law school at Yale University followed by a job at Vinson & Elkins LLP, a big law firm in Houston.

There he focused first on business but eventually developed an expertise in commercial litigation. That's how he became connected with Rahr. In 1999, Rahr hired Fossi as part of a lawsuit against a waste disposal company that Rahr had invested in and felt defrauded by. The company's executives had taken the almost $12 million of his investment and siphoned it into their own pockets, according to the litigation. Rahr won his lawsuit and kept in touch with Fossi, turning to him for his sundry legal needs. When it was time to sell off his pharmaceutical company, Rahr leaned on Fossi again.

After the sale, Rahr asked Fossi to manage his family office in New York City. Fossi was hesitant at first. He didn't think of himself as an investment guy. He didn't have a business degree and had very limited accounting in his background. His experience in business was more like that of a pathologist. He'd spent a career wading through corporate litigation, diagnosing the causes of problems, whether fraud or malfeasance. But he had something critical for the job: Rahr's trust.

So in 2011 Fossi moved to New York. It was in service to Rahr that, in 2014, his interest in Tesla was piqued. Fossi had a passing familiarity with the company. He knew Rahr had been an early fan, buying several Roadsters years earlier. But as Fossi began paying more attention, things about the company just didn't add up to him. He watched an online video of Musk from the previous year, in which Tesla announced its plans for battery swapping.

To combat the slow charging times of its electric vehicles—and in hopes of making the technology appealing to the mainstream—Tesla was proposing a plan for its drivers to simply swap out battery packs when they reached the end of their charge while on a road trip. Tesla's loan from the U.S. government provided for this kind of plan, and it would make the company eligible for the increased tax credits given in California to encourage the development of a fast recharging system. Musk's idea might have seemed simple enough, but to anyone familiar with the state of the electric vehicle, it would have been anything but.

Standing onstage that summer, Musk put on a signature performance before a large crowd. Dressed in a black T-shirt, jeans, and a velvet jacket, he promised a battery swap that was quicker than filling up a gas tank in a typical car. He described future Tesla Stations, where owners could choose between either a free charge or a faster battery swap, for a cost of maybe $60 to $80. "The only decision you need to make when you come to one of our Tesla Stations is do you prefer faster or free?" Musk said to laughs.

To prove it, a red Model S pulled onto the stage below a giant video screen that showed an employee pulling up to a gas station to fill an Audi. The company's signature club music began thumping. A giant timer was projected. Beneath the Model S sat a contraption that supposedly would pull out the battery pack and replace it with a new one, though the audience couldn't see exactly what was going on beneath the car. A camera person followed the Audi driver, meanwhile, as he picked up the gas handle and began filling his tank. Musk stood on the side of the stage, arms crossed as he watched.

After a little more than a minute, Musk chimed in: "So we have automated nut runners, these are the same nut runners that we use in the factory, and they find the spot where the bolts are and automatically torque the bolts to the exact specification that each bolt needs so it's torqued to the battery specification every time there's a battery pack swap."

A little more than thirty seconds later, the Model S was done and drove offstage to cheers while the Audi continued being filled. Musk looked up at the screen as the timer counted onward. "Yeah, looks like we've got some extra time," he said. "Let's do another one." A white Model S pulled into position amid wild cheers and laughter. More seconds ticked by; the second Model S pulled away after about ninety seconds. "Almost done with the gas station," Musk told the crowd. More seconds racked up. "Sorry, I don't mean to keep you waiting—my apologies," Musk said as the Audi driver began to finish up. The crowd watched him climb back into the car and pull off almost a minute after the second Tesla was done.

Musk returned to center stage with a mischievous grin and shrugged, to cheers again from the crowd. He thanked them for their support, noting, as he often did, that Tesla wouldn't be where it was without them. "What this is about is convincing the people that are skeptics—there are just some people who take a lot of convincing," he said. "So what we really wanted to show is that you can be more convenient than a gasoline car. . . . Hopefully this is what convinces people, finally, that electric cars are the future."

That may have been Musk's intention, but as Fossi watched the video from his Trump Tower office, his own skepticism grew. Maybe it was because, with the benefit of hindsight, he could watch that video knowing that Tesla's boisterous plans for a battery swap had never panned out. Tesla found little interest from owners; some worried about their battery pack being replaced with a faulty one. The whole show seemed driven, as much as anything, by a desire to help the company qualify for regulatory credits.

What really caught Fossi's attention, though, was the crowd's enthusiasm. "It struck me that this was like a religion," Fossi recalled. He marveled at how Musk had forged a narrative of himself as a great tech visionary—landing rockets, disrupting industries, making the world cleaner. "This guy— he's a prairie preacher," Fossi said. "He's got the revival tent up and these people are in it."

Since Tesla had gone public, there had been those who questioned Musk's plans. The stock began attracting the kinds of investors who used market maneuvers such as short selling to bet against Tesla—people who judged that its stock was overvalued and would ultimately fall in price to match its real value.

A typical investor buys a company's stock for a price of, say, $100 a share, with hope that the value will increase over time, to, say, $105, and can be sold for a $5 profit. Short sellers do the opposite. They borrow the stock at $100 a share and sell it right away, betting that it will drop in value, perhaps to $95, at which point they'd buy the stock back, return it to its original owner, and pocket the $5 difference. It's a complicated game, and it comes with huge risk. The most a long seller can lose in such a scenario is their $100 investment. If a shorted stock rises not to $105 but to, say, $1,000, the short seller would have to buy it back at that higher price, eating $900 on the sale. In theory, there was no limit to how much they could lose.

About 20 percent of Tesla's shares were being traded by short sellers in 2015, which only added volatility to an already tumultuous stock. A graph plotting Tesla's trajectory from its first profitable quarter, in 2013, through to 2015 was like a roller coaster that generally crept up and up, but that was punctuated by dramatic dips. For example, if an investor had held a share of Tesla stock from the start of 2014 through that year's end, it would have risen almost 50 percent—a remarkable growth. But it would have been a hair-raising ride. The year began with share prices falling 7.4 percent to a closing low in mid-January of $139.35. By September, the price had bounced back, more than doubling to a closing high in September of $286.05. But by the end of the year, it had plummeted again, this time by 22 percent, finishing at $222.40.

Depending on when a short seller makes its bets, it stands to profit from those periodic drops. Over the course of time, though, as Tesla shares trended upward, short selling was proving a losing proposition. An estimated loss, on paper, of almost $6 billion had accumulated from short positions in the years between Tesla's IPO through 2015. Despite that, many short sellers remained confident that Tesla's number was bound to come up. In the meantime, 2015 was proving to be another gut-churning ride.

Short sellers like to work around a company's events, such as its release of quarterly financial results or the release of a new product. They're looking for a piece of news that's perceived as bad by the market, and that triggers a sell-off. But many are living on borrowed time, especially if they see a stock continuing to soar. Some short sellers, feeling the pressure, will begin to gang up together to attack a company—through the media, online investor forums, and, increasingly, on Twitter—all as part of an effort to change the narrative of a company, to highlight the negatives around its business or reveal a failing that normal investors may not recognize. Essentially, they're trying to scare investors and drive a targeted stock price down.

One of the most legendary short sellers on Wall Street in recent generations cemented his fame by predicting the demise of Enron Corp. Jim Chanos was first attracted to Enron in the fall of 2000 after reading an article in the Texas section of **The Wall Street Journal** that said the company was boosting its profits by reporting unrealized, non-cash gains on long-term energy deals—essentially adding to its balance sheets revenue it wouldn't actually see for some twenty years. His analysts dug further, and Chanos came to the conclusion that Enron was a "hedge fund in disguise," using energy trading, rather than energy distribution, to generate the bulk of its earnings. And by his calculations, Enron wasn't very good at being a hedge fund. It required more and more capital to eke out modest results. His math suggested Enron was generating a 7 percent annual return, but at an expenditure of more than 10 percent.

He talked about his views on Enron at a conference for other short sellers in early 2001, hoping to stoke interest. Bethany McLean, then a reporter at **Fortune,** eventually called and he helped her piece together the story. As pressure grew on Enron in 2001, the beginning of the end was marked by a conference call between CEO Jeffrey Skilling and industry analysts. One asked him why Enron didn't produce a balance sheet along with its earnings reports the way other companies did. Skilling called the analyst an "asshole." Eight months later, Enron filed for bankruptcy. **Barron's** described Chanos's short as "surely the market call of the decade, if not the past 50 years."

Chanos's rise was no less improbable than Fossi's. The son of a second-generation Greek-American who owned dry cleaners in Milwaukee, Chanos had gone to Yale, where he studied economics, then landed a job in Chicago at the brokerage firm

Gilford Securities. He gained attention for making a sell recommendation on a company named Baldwin-United, a call that defied conventional wisdom from competing analysts, who were bullish about the annuity company. Chanos called the company a "house of cards" that was doomed to collapse because of too much debt, wonky accounting, and negative cash flow. A little more than a year later, he was vindicated when the company filed for Chapter 11 bankruptcy, wiping away $6 billion in market value. He was hailed in **The Wall Street Journal** and other media for his gutsy call. He went on to set up his own fund and found continued success into the early 1990s, shorting regional banks and other financial institutions that had exposure to the collapse of real estate in Texas, California, and New England. He also found success in shorting Michael Milken's junk-bond kingdom. His run saw his fund rise at twice the rate of what the S&P Index did in the same years, more than quadrupling in value until his luck ended in 1991, when the overall market turned sour. He had made some bad bets over the years. He took a short position on McDonnell-Douglas in the '90s. He also bet wrong in interpreting America Online's balance sheets as those of a company in trouble. America Online rode a wave of internet success in the '90s.

But on balance he had proven a truly extraordinary, prescient trader. The 2008 global financial collapse was good for his business, Kynikos Associates, named after the ancient Greek word for cynic. The firm peaked in 2008 with almost $7 billion in assets. **New York** magazine profiled him in a lengthy story, full of anecdotes about his tussles with Goldman Sachs and others, calling him the "Catastrophe Capitalist." Others likened him to the LeBron James of the short-selling world. And as his reputation grew, the mere announcement that he was taking an interest in a company with a short-sell position could roil a stock.

In his zeal, Chanos could strike a righteous tone about his place in the ecosystem of Wall Street investing. He told a reporter he was "convinced to the deepest part of my bones that short-selling plays the role of real-time financial watchdog. It's one of the few checks and balances in the market."

By 2015, however, his firm was facing new challenges. A large stakeholder withdrew its money from the fund; Chanos opened it up to outside investors. Also that year, Chanos began making a lot of noise about Musk's businesses. During a CNBC interview that January, he questioned Tesla, noting that the shares were valued on projected 2025 earnings while the company struggled to forecast its next quarter. "The guts of this product [the battery] is made by Panasonic," he said. "[Tesla is] a manufacturing company. It's an auto company. It's not a changing-the-world company."

That's not how Musk was painting Tesla's future

that winter. In a February call with analysts, Musk laid out a path for how Tesla could be worth $700 billion, or about the same as Apple at the time. "We're going to spend staggering amounts of money," he said. "For good reason and with a great ROI." Tesla was going to see a growth rate of 50 percent annually for a decade, he said, with an operating margin of 10 percent and a price-to-earnings ratio of 20. His math suggested unheard-of growth in the auto industry, as well as an unbelievable valuation for most companies. It just didn't make sense to skeptics, especially for a company that had turned a slight profit once, for one quarter in 2013. At best, Musk's claims sounded like bravado. At worst, he was seriously overplaying a weak hand.

In August, Chanos returned to CNBC, where he announced he was shorting SolarCity shares. In the web of his businesses, Musk saw SolarCity as a complementary product and service, whose solar panels would generate the energy to power Tesla cars. Whether or not that vision was plausible, the far more immediate issue for the company was its fundamentals. The company installed solar panels on homes and businesses, and the core of its business was going door-to-door making sales, but also financing the purchase over the course of twenty years. That timescale could prove challenging because newer technology was bound to emerge. Chanos equated the business to a subprime financing company for the way the panels were financed, treated essentially as a liability attached to the property. "You basically lease the panels from SolarCity. They put them on your house and they collect the lease payments. So in effect, if you put on the panels you have a second mortgage on your home because you hope it's an asset but in many cases it turns into a liability." SolarCity was burning through cash and racking up debt. "That is a very scary proposition," Chanos said.

As he'd surely hoped, SolarCity shares plummeted that day.

If SolarCity was in trouble, so was Musk's business empire. The solar company, along with Tesla and SpaceX, represented a pyramid that in Musk's mind he sat atop of. If part of that pyramid began to crumble, the whole thing could collapse. Musk's personal finances and his companies' were tied in a complicated knot. Since taking Tesla public, his personal finances had only grown more muddy as his borrowing increased. He had taken out personal loans, pledging 25 percent of his Tesla shares and 29 percent of his SolarCity holdings as collateral. He had a credit line from Goldman Sachs and Morgan Stanley totaling $475 million, some of which he had used in past years to purchase stock in Tesla or SolarCity to bolster those companies. If SolarCity's stock cratered, Musk would likely have to pony up money or more stock to make the banks happy. He'd long loathed the idea of parting with any Tesla shares, doing so only on rare occasions, such as when he'd repaid the loan from SpaceX that he had used to stave off Tesla's bankruptcy in 2008. His position as the largest Tesla shareholder helped him keep strong control of the automaker. The more his ownership stake dwindled, the more vulnerable he might be to a takeover or ouster as CEO. After more than a decade, he'd been able to avoid losing control of Tesla the way he had at PayPal. Tesla's success in raising the money it needed to pay for expansion seemed increasingly intertwined with his public persona.

The stakes of all the volatility in SolarCity's stock could be felt by Musk and his family that fall; it threatened their whole house of cards. Tesla board member Kimbal Musk was operating his finances not unlike his older brother. In late October, as shares of SolarCity fell to half their value from the start of the year, Kimbal faced a margin call from his bank, a request that he deposit more money to cover the losses he'd accumulated. "I was nervous watching today," his financial adviser, Karen Winkelman, wrote him about SolarCity's stock. Kimbal was in a pinch, as he wanted to expand a restaurant business that he had invested in, but he was low on funds. He told his adviser that he would seek a loan from Elon.

The elder Musk, however, didn't take kindly to the request. "You know that I don't actually have cash, right?" he wrote. "I have to borrow."

The Musk family had been professionally

intertwined for years. Musk had helped his cousins Lyndon and Peter Rive found SolarCity in
July 2006, the same month Tesla introduced the
Roadster at the airport in Santa Monica. He mentioned at that event, though this was largely overlooked by the public, that his vision for Tesla was to
be tied with generating power from the sun through
solar panels. And in his master plan for Tesla released the following month, he talked about future
partnerships between the automaker and SolarCity.

In theory, the solar business was simple. Homeowners and business owners had two choices when
it came to acquiring a solar system. They could plop
down the $30,000 or so it typically cost to put one
on a house, and they'd be eligible for any federal tax
credits that came with it—at the time 30 percent
of the system's cost. Or a homeowner could lease
the system. If they went this route, they'd get a low
monthly payment but not the tax credit. Instead,
the tax credit would go to the organization that financed the purchase of the system.

In practice, as Chanos alluded to on CNBC,
SolarCity had evolved into a complicated financial
operation that was essentially two businesses working together: a business that sold and installed solar
systems; and a business that created investment vehicles, selling the rights to tax breaks and other benefits associated with those solar systems.

The setup required a lot of cash. SolarCity, which
went public in 2012, almost two years after Tesla,

had never turned a consolidated profit in that time. It lost $1.5 billion from 2009 through 2015 and raised money through stock sales and issuing debt. SolarCity subscribed to Musk's philosophy of running tight on cash, which he believed forced executives to run their operations more efficiently and to find creative solutions—and avoided further diluting his ownership stake. But things were on the cusp of running **too** tight. As short sellers, such as Chanos, began targeting SolarCity in 2015, Musk used SpaceX to help improve SolarCity's finances, purchasing $165 million in bonds from the solar company. It marked the only time SpaceX had ever invested its cash in a publicly traded company.

In all his talk of SolarCity, Chanos hadn't forgotten about Tesla. He had a theory that high executive turnover was a sign of trouble inside a company. Musk had burned through several general counsels in the past few years, eventually settling on his former divorce lawyer, Todd Maron, to handle corporate affairs; Deepak Ahuja retired as CFO in late 2015 as production was supposed to ramp up on the Model X, and several leaders of Autopilot, the company's self-driving initiative, had quietly departed. A few months after his SolarCity comments, Chanos appeared on Bloomberg TV. He noted the dramatic difference in market value for BMW compared to Tesla. BMW sold 2 million vehicles a year while Tesla was saying it planned to do 55,000 in 2015. Investors, however, had pushed the value of Tesla's stock so high that the electric-car company was worth about half of BMW.

"It's an overpriced car company," he said of Tesla. He also cautioned that other automakers were making plans for electric cars and would catch up with Tesla soon. "They have to become a car manufacturer and becoming a car manufacturer is a lot more difficult then becoming a high-tech darling."

Back in Trump Tower, Larry Fossi had his own suspicions. At home at night and on the weekends, he began putting his thoughts on Tesla into words. He had been toying with the idea of publishing his thesis about Tesla on **Seeking Alpha,** a website that published content from investors. But he wanted to do it anonymously. He also wanted to feel like he had skin in the game, leading him to short Tesla himself.

He needed a pseudonym. Fond of Montana, where he planned to retire, he went with Montana Skeptic. He picked a drawing of Galileo Galilei—the astronomer who had been condemned as a heretic by the Catholic Church for his (it turned out, accurate) view of the sun as the center of the universe—as his avatar. In late 2015 he published his first piece entitled "Why Tesla Will Fall Far Short of Elon Musk's Model X Delivery Forecast." What followed was a nine-page analysis critical of Musk's ambitious production goals and his history of missing targets. A few weeks later, Fossi published another analysis as Montana Skeptic warning that the Model 3 was on the road to trouble.

CHAPTER 21

LABOR PAINS

A scraggly mountain range that gains two thousand feet of elevation sits above Tesla's sprawling assembly factory in Fremont, California. During the wet winter months, the treeless hills turn an emerald green, a stark contrast to the factory's bright-white paint job and stylish gray Tesla lettering. The factory had never looked as inviting during the almost two decades that Richard Ortiz worked there, back in its previous life. Even so, he had dreamed of working in that car factory since he was a child. It was a dream that had eluded his father, and that had looked unlikely for him when he reached high school.

General Motors opened the facility in 1962, four years before Ortiz was born, as part of the Detroit automaker's strategy to build cars and trucks near where customers bought them, to save on shipping costs. Generations of families' lives were tied to the middle-class life that the factory had provided. But by the 1980s, that system was in jeopardy. U.S. automakers were facing increased competition from Japanese competitors such as Toyota, while also enduring years of mismanagement that left their vehicles suffering by comparison.

GM shut down several facilities, including the Fremont factory, in 1982. It was seen by GM as one of its worst performing, with a reputation for a tough, organized labor force under the banner of the United Auto Workers (UAW). At the time, workers called the drab greenish-tan factory the "battleship," where endless conflicts raged for generations before its eventual closure. The workers used every trick in the labor books to flex their might for the GM managers—sick-outs, work slowdowns, wildcat strikes. Daily absenteeism reached 20 percent. More than six thousand work grievances remained backlogged in GM's system when the factory closed.

Even as a teenager, Ortiz understood the power of the UAW. As part of a school project, he read a book about the life of Walter Reuther, the union's founder. He dreamed of one day becoming the president of a UAW local.

The factory found salvation in 1984. Toyota sought to set up shop in the U.S. as it faced increased protectionist fears. A confluence of needs led GM and Toyota to consider partnering on a manufacturing facility. GM was eager to learn about

Toyota's fabled manufacturing system; Toyota didn't feel confident it could work with U.S. workers. So a deal was reached to reopen the Fremont factory together, drawing from many of GM's former workers. Those who returned found a very different environment. The factory was renamed NUMMI, for New United Motor Manufacturing Inc., jointly owned by GM and Toyota. The Japanese automaker brought in hundreds of trainers from its home country to retrain the California workforce in their methods, emphasizing continuous improvement, respect for people, and standardized work. The system expected managers to make decisions that were best for the long term, not just to fix immediate problems. Work was designed to be done with the fewest motions possible. The assembly line moved at a constant speed, with the expectation that each job would occur within a sixty-second increment. Workers were empowered to pull a cord hanging from the ceiling to stop the line if they saw problems. The mantra was to do the job correctly the first time to avoid the chance of developing defects down the road.

Academically, it all made sense. The challenge was making it work, especially when managers faced pressure to meet their daily production quotas. Senior managers had to live by the principles. In 1991, when managers felt consistency was slipping, they hung banners and handed out buttons to encourage workers to focus on quality. That year, the factory won J.D. Power's prestigious award for the quality of its vehicles.

Ortiz was hired in 1989. His lack of seniority in the union meant he was first sent to the paint shop instead of becoming a welder, which he had previously trained for. The paint shop was a tough job, one that opened his eyes to a world beyond Fremont. Japanese management's strong belief in training gave him an expansive look into the business, including an instructive trip to Toyota City in Japan. He also learned how to operate within the UAW, rising to the level of a committeeman of the local, a powerful position tasked with ensuring that the union's contract was being followed. He got his family members jobs at NUMMI, bought a house, and grew a family. He was pleased to learn that he had made his father proud. "All he does is brag about you," family and friends told him.

After almost twenty years, in 2006, he left. He'd grown frustrated with the politics of the place, and he was going through some tough times with his wife, whose likeness he had had tattooed on his right arm. After they split, he sought training to become certified in collision repair. In many ways, it was fortuitous timing: The auto industry was heading into its 2008 and 2009 declines, and GM's bankruptcy reorganization included shedding its participation in the NUMMI factory. Toyota said the factory's finances didn't work without GM, and it began unwinding it in 2010. The end of NUMMI was painful for its workers, many of whom held union leadership to blame for its woes. The long-held promise of Toyota's involvement was that they'd have jobs for life. The average tenure of the almost 5,000 workers at the plant turned out to be 13.5 years, with an average age of 45. Ortiz watched his family members struggle to find new jobs.

When Tesla acquired the factory in 2010, it didn't need thousands of workers with its smaller production runs and its emphasis on automation. Plus, Ortiz wasn't much interested in returning to the factory, where the personal memories were still too fresh. In the years that followed, he fell and hit his head, resulting in the detachment of his retinas. He awoke from the fall blind and wondered if he'd ever see again. He underwent a surgery that required a grueling recovery, including holding his head in a specific position for hours at a time. As he healed, his vision returned. "When I woke up after the surgery, I could see," he said. "All of the problems I had, they were problems no more."

As he adjusted to life after the accident, he rode past the factory with his son one day in December 2015. "Why don't you apply?" his son asked. That night, Ortiz went online to do so. Again, it was fortuitous timing. Tesla was in desperate need of workers amid a struggle to increase production of the Model X. Within a few days, Oritz was heading into the factory again for the first time in years.

When he entered, he could see that things had

changed. While the bones of the building were the same, it was unlike anything he had seen there before—in so many ways. Gone were the dark recesses and dingy walls. The floors were painted white and everything looked so fresh and bright. New windows had been installed. During orientation they talked about how Tesla was revolutionizing the auto industry, focused on building cars better than ever. "It was an autoworker dream," he recalled.

As he settled into work on the general assembly line, Ortiz began to feel at home. He picked up the duties quickly, reassigned to harder and harder tasks. The pay wasn't as good as he remembered. He was making $21 an hour compared to $27 when he left the GM-Toyota factory. He also sensed that despite the shiny paint and happy talk, things weren't as cheery as they seemed. He detected an anxiety radiating off his colleagues.

He didn't know it yet, but Ortiz had arrived at Fremont at a perilous time. All of the adulation surrounding the Model S had puffed up Tesla's belief in its abilities. Elon Musk had sold investors on a dream of what was ahead with its third-generation car—a dream that largely assumed that everything was in order with the Model S and Model X, when things were still far from in order.

Long before Ortiz arrived at the Tesla factory, the seeds of trouble were being sowed. Just as Peter

Rawlinson's team had struggled with knowing when to stop work on the Model S—to hand over the designs and for the factory to get building—the Model X proved equally chaotic. With the Model S, an external pressure ultimately forced their hands: Tesla was running out of money. They needed the Model S in production to create revenue. In 2014, Tesla found itself in a new position. Amid rave reviews, the Model S was generating cash and, for a time in 2013, profit. Musk grew confident—maybe overly confident. He wanted the Model X to be an even more amazing car. The engineering and manufacturing teams would visit Franz von Holzhausen's design studio in Hawthorne, where seemingly every new idea was allowed to blossom. Concerns about a supplier's inability to produce or the factory's limitations quickly ran into the buzzsaw.

Musk particularly disliked hearing excuses about time constraints. He'd dig into the notion, for instance, that a part couldn't be made in the proposed time. On an individual case basis, he might have been right. Why **couldn't** this supplier speed things up? But as the requests to do the impossible grew, and as more and more of the vehicle depended upon unheard-of timelines, the risk of failures stacked one upon the others.

There was the front windshield, which was much larger than that of other cars, sending the team searching the world for a supplier. One was found in South America that met the requirements.

The second-row seats were a challenge. To have a chance of withstanding crashes, they had to meet certain load requirements. Most cars have them bolted to the floor at a seat's four corners. The seat belt is often latched to a structural pillar as well. But Musk's vision for the Model X included an effortless entry into the third row. He wanted the second-row seats to seemingly hover in their spots. He didn't want cumbersome seat belts stretching from top to bottom across the entrance to the back. All of this required a specially designed seat, one that was attached to the floor through its own pillar. Getting that correct was proving harder than imagined.*

All of these challenges, however, paled in comparison to the rear falcon doors. Peter Rawlinson had warned Musk in late 2011 about them, but those words were long forgotten. By spring of 2015, the team was grappling with how to get them to open upward like a bird's wings in flight. The hydraulics weren't standing up to testing; they were leaking on passengers. Sterling Anderson, an MIT researcher who had attracted attention for his work in autonomous cars, was hired in late 2014 as the program manager of the Model X and was quietly engineering a new design for the doors, one that would use

---

* Musk would eventually grow so unhappy with the seats that he ordered the company to begin manufacturing its own—a costly and time-intensive process that would eventually lead to a factory down the road.

a less complicated electromechanical system. Musk liked it and ordered the last-minute change.

Changing the door design so late in the game was risky. It would require adjustments to the body of the vehicle, requiring new dies to be created. Normally, this might take nine months, followed by three months of tuning the tools to get the proper parts made. But they didn't have a year; production was supposed to begin in just a few months. All of this was occurring as the factory team was trying to transition from Gilbert Passin's innovative cart system to an actual assembly line, in order to handle the increased demands for production. Again, Tesla was trying to build the airplane as it was taking off.

The factory went ahead, using the old tools, and made a few dozen early versions of the SUV for final testing, before starting official production. The cars were almost unbuildable. They looked horrible, with wide gaps between pieces. But the team didn't have a choice. They needed to assemble the first ten SUVs to be handed over to customers at an event to celebrate the start of Model X production, scheduled for late September 2015.

The vehicles were taken to a backroom and disassembled so that teams of designers and engineers could rebuild the SUVs by hand. Parts had to be remade. Workers took knives to hand-carve door seals. They worked around the clock for two weeks just to get the first SUVs ready. The morning before the event, the cars were still plagued with snags.

Their dress rehearsal was a disaster—most vehicles' doors were buggy. Software programmers were using laptops to figure out why the doors wouldn't open. Musk was cool under the pressure, encouraging the team to do whatever needed to be done for the show, assuring them they could address the real problems afterward.

The night of the show, Musk emerged in his signature black velvet jacket, jeans, and shiny shoes. He began his presentation, as was becoming the norm, by recapping why Tesla was doing what it was doing. "It's important to show that any type of car can go electric," he said. "We showed that you could make a compelling sports car with the Roadster and that could go electric, and we showed that you could do it with a sedan and now we're going to show that you can do it with an SUV." Then to the sound of electronic dance music, the SUVs drove out.

The team backstage held their breaths when it was time to reveal the doors. The key feature was being tested like it had never been before. Failure could turn all of their hard work into lost jobs.

The doors worked.

They had pulled off another close call, but just like in 2012, when the company had celebrated the start of Model S production, Tesla wasn't remotely ready for the kinds of production figures Musk was promising. He told Wall Street that Tesla would make 15,000 to 17,000 vehicles in the final three months of 2015, or roughly 1,250 to 1,400 a week.

It fell to Greg Reichow, head of manufacturing, and his deputy Josh Ensign, a former Army officer who had joined Tesla in 2014, to figure out how to make that a reality. As they chased Musk's goal, the factory began pumping out vehicle after problematic vehicle. The doors wouldn't shut, the windows didn't work. Soon the parking lot was full of hundreds of cars that had problems, and they weren't sure what was causing them.

These issues weren't surprising to workers such as Ortiz. It was as though when Tesla repainted the factory, they'd also done away with Toyota's manufacturing spirit. Instead of thinking about long-term effects, Ortiz found his managers focused on short-term fixes. As with the Model S, they weren't standardizing work at the different build stations; processes were left up to chance. Instead of choreographed steps, Ortiz was running around the body of the vehicle to place parts; there wasn't the focus on efficiency he'd learned. Sometimes the parts themselves seemed off. He'd flag for a supervisor a bin of cracked door parts, for instance. He might be told to go through the pile and find good ones. Other times he was told to use them anyway, he said, good or bad. It didn't feel like his voice mattered.

Whereas Toyota would've stopped the line to fix problems like these, Ortiz had to send vehicles through, to be reworked at the end of the assembly line. There, the crews might remake a part by hand. Other times, they manhandled the pieces together, forcing a fix. It was time-consuming, backbreaking work. The effects could soon be seen on the workers; executives began to notice the rate of injuries climbing.

In 2015, Tesla recorded 8.8 injuries per 100 workers, exceeding the industry average of 6.7. Many were related to repetitive movement, according to managers; Ortiz was noticing the kinds of injuries to backs and arms that could be avoided through proper ergonomics. Unlike Toyota, Tesla hadn't spent time designing the SUV to be built by workers. Those fancy second-row seats may have been conceived to spare customers the awkwardness of strapping in a car seat, but they required workers to bend awkwardly into the cabin to bolt them to the floor. Bolts would strip and workers would be forced to do the work manually.

Even with these challenges, though, there was a sense of camaraderie. Tesla was fighting to prove the world wrong. As the year came to an end, Ensign sent out a message that they needed to plan to work the next twenty-one days straight to meet their annual target. Musk insisted that they show the world they could do it—or at least look as if they could.

In the end, Tesla's overall production fell about 1,000 cars short of their goal, but total sales—almost all because of the Model S—fell within the range of what Musk had promised. Tesla made just 507 SUVs that final quarter of 2015. Most came in the last days of the year. That's when the crews pushed to reach a weekly rate of about 238 vehicles in seven days. Musk began telling investors they were on track to make 1,000 a week by June. His executive team, however, was pushing him to back off—to let the workers catch their breath. But Musk was worried that a slow quarter would spook the market. He told executives that Tesla's success was based on the perception that demand was greater than supply. The moment it appeared otherwise, they were doomed. His patience with the factory was waning. He wasn't interested in how the mess had been created—design decisions made months earlier that had cascading effects on the assembly line. He just wanted it cleaned up—and now.

As 2016 progressed, his fuse seemed to grow shorter. His investment in SolarCity was struggling, with quarterly reports suggesting the company was running low on cash amid an industry-wide downturn in the solar business. His marriage to actress Talulah Riley was ending—for a second time. He was quietly spending more time with actress Amber Heard, who was married to actor Johnny Depp. That spring while Depp was out of the country, Musk visited the couple's downtown Los Angeles condo building, according to workers there. He would arrive late at night and leave early in the morning.* Musk's sleeping habits—or lack thereof—were well

* Musk has said their relationship didn't begin until Heard filed for divorce in May 2016. They were friends before that.

known among senior staff. Middle of the night emails were the norm. He seemed to be one of those people who naturally didn't need much sleep. But as his relationship with Heard gained steam, Musk seemed to be stretching his already vexing schedule to accommodate her, executives said. He was taking seemingly spur-of-the-moment trips to Australia aboard his jet, where she was filming the movie **Aquaman,** and returning after spending very little time on the ground. He was spotted by the tabloids with Heard at nightclubs in London and Miami. He commented during staff meetings that the travel was getting to him. It was clear to those around him that he was trying to see her as much as possible.

"Lack of sleep is not a thing for him, he operates on a fine level with a couple hours of sleep a night," one longtime aide said. "The thing that was more taxing on him was all of the travel and the various time zones."

Despite Tesla's struggles, it had opened many eyes to the possibility of an electric vehicle, erasing the notion that traditional automakers held the lone key to carmaking success. It should come as no surprise then that Silicon Valley's biggest player, Apple, had taken note. In 2014, the company had quietly begun work on its own electric car program, hiring a slew of experienced hands to launch the effort, dubbed Project Titan. But it quickly became clear that developing a car was harder than they first thought. So, with Tesla's stock fallen from its previous highs and with its well-publicized struggles over the Model X, Apple CEO Tim Cook apparently saw an opening.

Tesla and Apple had developed a complicated relationship. Musk admired Apple's achievements and was eager to hire people with Apple on their résumés. His stores were explicitly modeled after Apple's; his designs took inspiration from the iPhone. Both companies had developed cultlike fans, some of whom saw synergy potential in the two. At the Apple shareholders meeting in early 2015, Cook faced some investors eager for a marriage. "Quite frankly, I'd like to see you guys buy Tesla," one of them told Cook, followed by laughs and cheers from the audience.

Among the first signs that Apple was working on its own car came when Cook's team began trying to poach Tesla executives, with promises of 60 percent salary increases and $250,000 signing bonuses. All of which was too much for Tesla to grin and bear. Musk began taking shots back at Apple, claiming that Apple hired away those who weren't tough enough for Tesla. "We always jokingly call Apple the 'Tesla Graveyard,'" he told German newspaper **Handelsblatt** in 2015.

For all the sparring, Musk was interested to hear what Cook was thinking, and a call was arranged. According to people who heard Musk's version of events, Cook tested the waters about an acquisition, and Musk showed some interest, but with a condition: He wanted to be CEO.

Cook quickly agreed that Musk would remain CEO of Tesla under Apple. No, Musk allegedly responded. He wanted to be CEO of **Apple.** Cook, who since Steve Jobs's death had shepherded Apple into the most valuable publicly traded company in the world, was gobsmacked by the request.

"Fuck you," Cook said, in Musk's telling, before hanging up. (Apple declined to comment on the record.)*

Whether or not this was an accurate recounting, it's hard to imagine Musk was serious about wanting to be CEO of Apple. Rather, the story played into Musk's vision of Tesla becoming on par with Apple. It also served a more immediate purpose: It told those senior managers hoping for salvation from Apple to think again. They had to fix the mess at the Fremont factory or else.

None of these distractions were enough to take Musk away from the demands of the factory. Just as in 2008, when the Roadster's rescue had been seen as the result of Musk's personal intervention,

---

* Cook told a **New York Times** podcast in 2021 that he had never spoken to Musk. However, Cook had been photographed sitting next to Musk at a meeting held by Donald Trump in 2016, and both executives served together on a business school advisory board in China.

he jumped again into the weeds. Musk set up an air mattress on the floor at the end of the Fremont assembly line. He began talking publicly about how he was sleeping at the factory.

An especially perplexing issue was the Model X's passenger windows, which made a horrible screeching sound when going up or down. Many knew that Musk was sensitive to smells and sounds—so much so that crews working around him on the factory floor were instructed to disable safety beepers on carts when he was nearby. (He also disliked the color yellow and insisted on replacing the typical safety color with red whenever possible.)* Late one evening, he gathered the team doing rework on the Model X for a pep talk. As he meandered his way through his speech, he told them he understood the sacrifice they were making. He paused; his eyes swelled with tears. He knew firsthand the toll it can take on a family, he said.

It was those moments of emotional transparency that helped motivate many. Musk wasn't asking anyone to work harder than he was working himself. Even Ortiz admitted that Musk's presence seemed to kick the team into high gear. A telltale sign that Musk was visiting the factory was the arrival of popcorn machines for snacks near his desk. Managers also seemed more on edge. It reminded

* Musk has denied this, including in a May 21, 2018, tweet: "Tesla factory literally has miles of painted yellow lines & tapes."

Ortiz of his days at NUMMI, when OSHA would show up for an inspection.

Still, the work was getting to the men and women on the line. They liked the overtime pay but resented the unpredictable hours, especially getting called into work on the weekends only to stand around because of delays.

Facing a revolt, Ensign promised the workers that each month they'd get advance notice of a weekend off. The problem, however, was that when it came time to have that first weekend off, Musk wanted the factory humming. A problem had arisen with a change to the headlamps; the supplier in Mexico was struggling to make them, sending small batches by airplane every day. The domino effect was a disaster. Hundreds of SUVs couldn't be completed. Musk was unhappy that not a single Model X had yet to be built flawlessly.

At the end of the assembly line one evening, Ensign and Musk got into a heated debate about giving the workers time off. The factory boss made the case that the team needed to rest. Ensign's entire argument seemed to set Musk off. "I can be on my own private island with naked super models, drinking mai tais, but I'm not," Musk bellowed. "I'm in the factory working my ass off, so I don't want to hear about how hard everyone else in the factory works."

And with that Musk stomped off. Later that night, as Musk continued to walk the line, he came across a worker struggling with a window. It made that irritating screeching sound. Musk's anger seemed to boil over—until an hourly worker on the line spoke up. "I know how to fix this," he said. He explained that if he made an incision along the door seal, it would eliminate the sound. Musk told him to demonstrate and the worker dutifully did so. Sure enough, the sound went away.

The result was unlike any reaction the worker likely suspected. Musk erupted at Ensign: **How the fuck do you have somebody in your organization that knows the solution?**

Ensign hadn't wanted to point out in front of the entry-level worker that, in fact, the engineers had long ago tried the suggested approach, only to find that after a few weeks the issue frustratingly returned. No need to embarrass the worker, who was feeling like a hero to Musk. But Musk was in a rage. The "idiot bit," as some employees dubbed it, had been flipped, as if a switch had been turned on that signaled to Musk that an employee was unworthy. Musk continued: **This is totally unacceptable that you had a person working in your factory that knows the solution and you don't even know that!**

The team watched as Musk and Ensign's boss, Greg Reichow, went off to a conference room for a heated conversation. The results were swift: Ensign was fired. Reichow quit. Musk found himself on the verge of launching the Model 3 without the heads of his manufacturing operations.

A few weeks later, Musk would celebrate the first Model X to roll off the assembly line without any defects. It occurred at around 3 a.m. Even with the push, Tesla failed to reach its goal of 1,000 cars a week by quarter's end. Still, Musk told investors the company was on track to make 50,000 vehicles in the second half of the year—or as many as Tesla had made the entire previous year.

The push to produce Model Xs may have helped goose Tesla's stats for Wall Street, but for customers it only kicked problems down the road. The car would be plagued with complaints. National sales of Tesla's vehicles hadn't yet reached the annual volumes to merit a J.D. Power review of initial quality, a benchmark study of car performance. (Tesla had been able to block its own inclusion by rejecting J.D. Power's request to call its customers in certain states, such as California and New York—states that require an automaker's permission.) But enough vehicles outside those states gave J.D. Power data to draw directional insight. It showed that Tesla's initial quality was the worst among luxury brands, and among the worst in the industry across the board. Only Fiat and Smart had more overall problems, according to the Tesla-specific study, which wasn't released broadly in detail at the time because the numbers were so raw.*

---

* J.D. Power wouldn't begin including Tesla in its study publicly until 2020.

The list of complaints was long, from struggles with excessive wind noise, to misaligned body panels, to difficulty operating seat belts. Despite that, the study found that Tesla cars had their advantages, with their large touch screens and the performance of their electric powertrain. Essentially, Tesla outshined competitors in the areas that automakers traditionally didn't understand, while Tesla was doing horribly at the decades-learned part of carmaking.

And while customers were reporting a high number of problems to J.D. Power, the appeal of the brand wasn't hurting for it. It scored higher for brand excitement than any competitor. J.D. Power concluded its private report with a warning that initial quality concerns "may grow as sales increase and a different type of buyer base emerges."

Ortiz understood the disconnect. His son seemed to idolize Musk as a tech hero, so much so that he tried to get his kid an autograph from his boss one day. Yet he could see the way Tesla was poorly managed—perhaps dangerously so. He often thought that the UAW would have a field day with the factory.

By summer, he got to put that idea to the test. He received a mysterious invitation for a weekend get-together with some old union buddies. He walked in to find organizers from the UAW sent from Detroit. They wanted to know if he wanted to help organize Tesla, an opportunity he had dreamed of.

It would be easy, he thought to himself; the workers were ready. He wanted to hark back to UAW's glory years, when its efforts shut down GM factories.

"I wanted to walk them out—that was the end goal," he said. "I was going to walk everybody out of that building—the old-fashioned way. No arguing, no talking, no vote counting. You recognize us and we'll go back to work. You don't, we'll stand right here." He thought maybe he'd end up the president of a UAW local after all.

CHAPTER 22

CLOSE TO S-E-X

Elon Musk settled on the compact car's name after first wanting to go with Model E—for electric, but also because it would allow his company's vehicle lineup to spell S-E-X. His senior leadership team laughed out loud at the suggestion, but it had to be tweaked when they realized Ford held the trademark to a Model E. So "E" was reversed to "3," which was apt for a third-generation car, but which also kind of allowed him to maintain the joke.

That's not how Apple would have named its next iPhone. But a lot of life at Tesla was different from what Doug Field had experienced at his previous employer, where he'd overseen thousands of engineers developing the newest Mac computers. It quickly became clear to him: There were things that needed to change at Tesla. If the car company was going to become mainstream, it couldn't keep making the same mistakes it had with the Model S and Model X. The size of the company was growing too large, the stakes too great for the kinds of errors his team had been guilty of. Those delays on a mass-market car might kill the company for good. His hiring signaled that it was time for Tesla to grow up, to shed its life as a startup and mature into a corporation.

As Field settled into the company's Palo Alto headquarters, the company's naivete was evident. A divide that went back to the Peter Rawlinson days existed between the car guys and the tech guys. (They were all largely men.) The vehicle and manufacturing engineers—the car guys—had grown out of Rawlinson's organization in LA, eventually relocated to Silicon Valley. They typically arrived from automakers, often European ones. Several had British accents. They ditched their suits, untucked their button-downs, went without ties. They were in their forties and fifties, living in tony Bay Area suburbs such as Pleasanton or Walnut Creek.

The tech guys came to Tesla through JB Straubel. They resembled a Silicon Valley startup, many a generation younger than their counterparts in cars, often (inevitably) Stanford alums. They favored T-shirts and fancy sneakers, lived in San Francisco or Palo Alto.

Tensions ran deep between the two factions. The car guys felt the tech guys lacked respect for the auto industry's hard-earned lessons, for the proper way to make a car. The tech guys felt the car guys lacked their engineering skills, relying on the past as a crutch. As one executive observed about the groups: "There was not a single cultural axis on which you'd say there was alignment."

Things needed to change. Field needed them to work together if they were to going to find a way to take everything that everyone loved about the Model S and make it at a fraction of the cost. It was going to take a change of perspective. Years earlier, when engineers working on the Model S mentioned budgets to Musk, he'd lose his temper. He had wanted them to focus on making the best car possible, costs be dammed. Not now: Field had a different mandate. Tesla had proven it could make the best car. Now, Tesla's challenge was to make an everyday car that made money—a car for everybody. That was the only way that Tesla could evolve from a wild idea into a reality.

During an early meeting with his team about the Model 3, Field put up a slide to get them focused. The Model 3 needed to achieve a selling price of $35,000, a range of more than 200 miles on a charge, and to generate something close to the customer experience and passion of the Model S. If they did that, he told them, "We're going to change the world."

The starting price of $35,000 was key. It was the sweet spot for sedan sales. The BMW 3 Series and

Mercedes C-Class began at that price, while the Toyota Camry reached that price on the high end. In the U.S. in 2015, the average new vehicle sold for $33,532, according to Edmunds. A fully loaded Model 3, of course, would sell for more, adding to the profit they could make.

But the challenge of it continued to be the same as it had been a decade prior, when Martin Eberhard first began thinking about ways to convert AC Propulsions' tzero into a production car: battery costs. Straubel's idea for a Gigafactory and his partnership with Panasonic were a start. Tesla estimated the location of the factory, near Reno, Nevada, would help reduce costs by close to a third. But so much more needed to occur. Field needed a way to motivate a team that hadn't traditionally been accustomed to thinking about return on investment.

He crafted a chart from the sales data. It showed how the cost of a car affected sales. For every dollar they could cut from the cost of the car, they'd gain a hundred sales a year. "This is what a dollar means," he told them. "It means a hundred more families, a hundred fewer internal combustion engines on the road, a hundred safer people, happier people that get the joy of owning a Model S." In short, their mission was to make a cheaper version of the Model S, one that cut costs while keeping all of the exciting parts.

Beyond manufacturing efficiencies with the

battery, another way to reduce cost would be to pack more energy into the batteries, and thus to use fewer cells. The Model S used lithium-ion cells dubbed 18650. Chief Technology Officer JB Straubel's team envisioned slightly larger cells (21 mm by 70 mm), which would allow the engineers to get greater internal volume and pack more energy. They wanted to find a way to get the same distance as the Model S but using 25 percent less energy.

To do that, Tesla needed to focus on improving the vehicle's overall efficiency. Field's engineers were given two budgets for their parts. He wanted them to think about the cost of their decisions not just in money but in terms of the load it put on the car's electrical system. So for instance when the team presented the cost of proposed brakes, they listed the dollar cost but also the cost of the energy associated with accelerating and decelerating. A plastic strip under the vehicle's body might be $1.75 to make and 25 cents to install, but the improvement to aerodynamics meant it improved the range as much as an additional $4 worth of batteries. That was a win. As the team whittled away at weight and finessed the aerodynamics, they improved the car's range to 335 miles, well above its original goal.

Other ways to get the cost down came from moving to a steel vehicle frame, instead of the pricey aluminum found in the Model S (though the doors remained aluminum). Inside the car, they found huge savings by replacing the typical dash of gauges

with a single expanded flat screen in the center of
the car, covering all of the car's information.

They developed a vent system that was much
cheaper than what was typically used and relied on
fewer parts. Car designers for years had dreamed of
doing away with the circular or rectangular vents
in a car's dash. They envisioned replacing them
with something more elegant. None had achieved
it. Taking a hack at it for Field's team would be
an engineer named Joe Mardall, who had previously worked at McLaren Racing—improving the
aerodynamics of race cars—before joining Tesla in
2011. He had worked on developing an air filter
system for the Model X in 2015 that Tesla claimed
removed at least 99.97 percent of fine particulates.
His new task required him to figure out a way to
use streams of air to act like the gates of a vent—
to aim airflow without the usual, cumbersome
openings. He crafted a sleek and unobtrusive slit
in the dash to direct the air. Gone were the vent
circles. It was emblematic of what Tesla had been
trying to do since the Model S: throw together a
small group of the smartest engineers and let them
come up with out-of-the-box ideas to fix seemingly
intractable problems.

Cutting cost was one thing. The nightmare occurring at the Fremont factory in the spring of
2016 underscored the need to improve not just the

economics but the buildability of Tesla's cars. Field and Greg Reichow, before he quit, were in agreement on that. The key problem to starting production of the Model X had been last-minute changes to the design and engineering of the SUV. The factory needed time to work with parts suppliers to set up the tools and test and tune them to make the vehicles, before production was under way.

To get the car developers and factory engineers on the same page, Field gathered them all together for a meeting away from the Palo Alto headquarters and the Fremont factory. On a Friday when Musk would be in LA, they quietly took the managers and directors, about fifty in total, for a daylong event at the Presidio, the former military fort in San Francisco with views of the bay. The day began with Franz von Holzhausen's design team and the technical program managers presenting the design for the Model 3 and its proposed user experience. They compared it to the Model S but also to competitors in Germany and Japan, going through specs ranging from dimensions to performance. They were targeting the BMW 3 Series, the Audi A4, and the Mercedes-Benz C-Class. With an understanding of what the car was going to be, they turned, that afternoon, to the plan for building it. Key product design engineers took them through the process. They had ideas for using a newly acquired die factory to reduce the amount of time needed to turn around stamping tools and presses.

Field framed it through a simple metaphor: he wanted his engineers to think of themselves as farmers. The team could avoid a lot of issues, he argued, if they took the time now to seed the fields with engineering work, to make sure the Model 3 would be ready when it was time to harvest. Since its beginning, Tesla had struggled to focus on more than one thing at a time. It was evident in its execution of the Roadster, then the Model S, and even now as they struggled to turn the Model X out of the Fremont factory. Field warned his engineers that they needed to avoid finding themselves in a similar situation when the Model 3 came out. In any product development process, he told them, the engineers had an opportunity to make the most influence on the project at its beginning. That's also the time, he told them, when a CEO is often paying the least attention. By the time a product nears completion, when a CEO's attention burns brightest, engineers have very little ability to make improvements or course corrections. This is what was happening with the Model X. Musk was walking the line at the factory, unhappy about engineering and design decisions made years earlier and demanding painful changes.

Musk may not have been at the off-site meeting with the team to take in the message, but he was demonstrating a newfound appreciation for the complexities of manufacturing. As von Holzhausen held design sessions in Hawthorne for the Model 3, the managers noticed a change in tone from the

Musk they had seen years earlier during Model X development. He could still be hands-on, but instead of offering pie-in-the-sky visions, he would drill purchasing and manufacturing managers about the effects of their decisions. To some managers who'd been around long enough to have seen the old Musk, he didn't seem as deeply engaged with the Model 3 as he had been with the Model X. He was spending more time at the factory, of course; but he was also carrying the pain of troubles at home and elsewhere.

He didn't **need** to be as involved: He had a senior management team that was clicking. Field seemed to be on top of the Model 3's development. Meanwhile, the manufacturing team in Fremont was working on an ambitious plan to build the car. They envisioned creating a new assembly line just for the Model 3. They were toying with adding more automation, too, in an effort to improve quality. They began making plans for production to begin at the end of 2017, with the goal of making 5,000 Model 3s a week by the summer of 2018. At that pace, on a yearly basis with typical downtime for maintenance, they'd be cranking out about 260,000 compact cars a year, a major milestone. A second phase would allow them to add more robots and automate more parts of the assembly line, to reach the goal of making 500,000 vehicles a year in 2020.

It was aggressive, they all agreed. But the

two-stage approach would allow them to tackle the complicated process of adding more automation, which takes time, while production was under way. It would also let Tesla generate revenue from the Model 3s rolling off the assembly line, helping fund that automation.

The key to it all, though, was avoiding the mistake of the Model S and Model X: They couldn't build the manufacturing line at the same time that the factory was trying to begin production. Tesla was already reworking thousands of Model Xs that suffered from an assembly line rushed into production before it was ready. A factory pumping out more than three times as many cars a week in need of similarly heavy rework could have disastrous consequences for the company.

As Field was trying to unify the company, some members were looking to chart their own path. JB Straubel's team at the Gigafactory had been debating how to transport battery packs from Sparks to the Fremont factory, and one idea thrown around was developing an electric train powered by battery packs. The conversation morphed into developing an electric semitrailer truck, as the costs made more sense. The team grew excited; Straubel quietly authorized them to begin working on a prototype to see if it was possible. They envisioned building a small fleet of electric trucks shuttling between factories. In his fashion, he had the team acquire a gas-powered Freightliner that they could hack into an EV. He assigned Dan Priestley, a young engineer, to tear the entire vehicle apart and build a proto-type, using half a dozen Model S battery packs. The team began taking it on test drives, charging it at Superchargers, and marveling at the acceleration.

After a while, though, they faced a problem. They'd developed an entirely new product with-out telling anyone. Field's team developed vehicles; Straubel's was supposed to be building a battery factory. Plus, Musk didn't like surprises—unless he did. Straubel bet that the new semi would be so impressive that all would be forgiven. He invited Musk to see something behind the Fremont fac-tory one day, offering a demonstration. The truck his team had made had blazing acceleration—more like a sports car than a lumbering semi.

Musk was suitably impressed. But he decided the project wasn't going to sit with Straubel (who did have a giant factory to focus on). Musk had other ideas for it: He wanted to use the promise of a new vehicle program as a carrot to lure back Jerome Guillen, who had earned Musk's trust by bringing out the Model S, before burning out while running the sales department.

Since departing in August 2015, Guillen had been decompressing. He traveled across the U.S. on an epic camping trip, sending occasional photos of himself in far-off places to friends back at Tesla.

They couldn't remember seeing him so happy. When Musk raised the possibility of returning, he seemed unsure. It wasn't even clear there was a place for him. Field was already overseeing Model 3 development and Jon McNeill was running sales and service. What did they need him for?

Musk's latest pitch had some appeal. Guillen had worked on semis early in his career, before Tesla. With this new vehicle, could there be a totally new market for the company? Persuaded, he returned in January 2016 to oversee the project, a nice quiet backwater at Tesla. It also meant that even as they wrestled the Model X into production that winter, they were developing not one but two new vehicles. It was a lot of capacity to manage, but it was also the kind of flexibility Tesla would need in the future if it wanted to move beyond its niche.

On a breezy March night in 2016, Elon Musk, Doug Field, Jon McNeill, and the other senior leadership stood backstage at the Tesla design studio, where hundreds of customers and supporters waited for a look at the car Musk had long promised: the Model 3. On Twitter, days earlier, Musk had announced the reveal, going so far as to take pre-orders for a car that hadn't even been seen in the wild.

In the wings, the executives watched videos on social media showing lines of people waiting at

Tesla stores around the country. They also kept an eye on a screen that kept a running total of $1,000 deposits coming in. They couldn't believe it. They'd had an informal pool going for how many deposits they'd take in. When Tesla first revealed the Roadster, they hoped to get 100 buyers willing to put down $100,000. It took a few weeks to do that. The Model S collected 3,000 pre-orders in the months after it was revealed. But this was unlike anything they had expected. Tens of thousands of people were signing up for the car, sight unseen. Then hundreds of thousands. The team was floored, searching for words. It was all so jarring: If you computed Tesla's anticipated build rate for the Model 3, the reservations alone that night ate up the first few **years** of planned production. They knew from experience that many of those reservations wouldn't convert into actual sales, but still, it signaled huge interest. Field thanked Musk for the opportunity to be there.

Musk walked onstage to cheers. "You did it!" someone in the crowd screamed. At long last he revealed the Model 3. In the crowd was Bonnie Norman, the Roadster (and now Model X) owner who acted as a volunteer brand ambassador, and Yoshi Yamada, the Panasonic executive who had championed his company's involvement in the Nevada factory. Both were in a celebratory mood.

As the Model 3 rolled out, the most striking thing about it was not the freshness of its aesthetics.

In many ways it resembled the Model S, though smaller and with proportions that were slightly less elegant. (The effect was most noticeable in the nose, the short front end giving it an almost turtle-like profile.) To many, the differences between the Model 3 and the Model S were hard to spot. And that, in itself, was the stunning thing: A car with a promised starting price of $35,000 might easily be mistaken for its $100,000 predecessor. The night was a spectacular coup.

Many executives at Tesla were bonded by the experience of **managing up** to their CEO—or, in other words, managing their manager. Musk continued to meet weekly with the senior team, at a fraught time where a misstep could put anyone under his hot glare. One way to handle Musk was to suss out his mood with his chief of staff, Sam Teller, who traveled with him and acted as gatekeeper to all things Elon. Teller was one of the few people beside Musk who could see beyond the silos of Tesla, SpaceX, or SolarCity, who could caution when another part of the empire was giving Musk grief. When new executives were hired, his advice to them was simple: Problems are not actually problems. They're normal. Musk spends all of his time solving things that were fucked up. **Surprises** are the problem. Musk doesn't like to be surprised.

JB Straubel saw potential to help defuse

internecine rivalries at these meetings by using Teller as a conduit, pulling together key deputies for pre-meetings, so they could hash out their differences in private before putting up a united front to the boss. It seemed to be working—until Musk found out. He was predictably unhappy. He ordered that stopped. Deputies weren't to plot behind his back; he wanted to see them clawing at each other in front of him. The implied message: Musk wanted to retain decision-making authority, not to have decisions foregone by the time they came to him.

Regardless of such hiccups, Musk arrived jubilant at Tesla headquarters the week after the Model 3 reveal. The executive team sat in the glass conference room near his desk marveling at the number of reservations still rolling in, well beyond their wildest expectations. But the excitement soon turned tense. With all the anticipation, Musk made the call that onlookers must surely have dreaded: He decided to accelerate production for the Model 3.

The plan they had crafted over the past few months had called for production to begin at the end of 2017, after months of setting up a new assembly line, testing it, validating that each workstation could bang out job after job every few seconds.

Under that plan, Tesla's now CFO, Jason Wheeler, who had taken Deepak Ahuja's role in late fall 2015, projected that revenue would more than double in 2018 over 2016. He forecast that Tesla would achieve its first year of profit in 2017, making $258.9 million, followed by almost $900 million in 2018. By 2020, when Musk promised to deliver 500,000 vehicles in a single year, Wheeler predicted revenue of $35.7 billion, with a profit of $2.19 billion. Tesla was on a path to becoming much bigger, all thanks to the Model 3.

All of a sudden that wasn't good enough for Musk. The existing plan meant that the first few years of production would be feeding pre-orders alone. By the time the company worked through its backlog, the car it was selling would be outdated. The orders also signaled to the world that there was huge potential demand for electric vehicles, which would surely motivate competitors to get serious with their fledgling plans.

What Musk didn't say directly, but many believed to be a major part of his motivation, was the company's cash burn. The delay in the Model X was eating money. The company finished March with $1.5 billion of cash on hand, a level that would've been much lower if it hadn't tapped a revolving credit line (essentially a corporate credit card). It dwarfed the numbers that had been in Tesla's coffers just a few years prior, but all the same, at its current pace the company would be out of cash by year's end. Tesla either needed to boost Model X production or the company had to raise more money from investors.

The cost of ramping up the Model 3 would

inevitably add to the financial pressure. So the faster Tesla could reach 5,000 cars a week, the faster it could generate the kind of cash it needed to keep the business going. But the team was divided on the wisdom of hastening their plan. The Model X assembly line debacle was a case in point. Field, in particular, was vocal in his opposition to the idea. Straubel, too, was concerned, saying he wasn't sure he could get the Gigafactory up to speed in time to meet demand.

Musk wouldn't budge. It wasn't a discussion. He had made up his mind.

Field kept his opinions about the changes to himself. He gathered his team to rally for what was ahead. "You are now working at a different company," he said. "Everything has changed."

With Reichow and Josh Ensign gone, manufacturing reported directly to Musk while sales reported to Jon McNeill. A cohort of managers were directly experiencing their mercurial CEO, without any buffer. Shen Jackson, a manufacturing engineer, proposed a three-phase schedule, ramping up from 5,000 a week to 10,000 to 20,000. Musk told him it wasn't fast enough. He wanted to begin Model 3 assembly the following summer, about six months ahead of the original plan. Not only that, Musk wanted to reach a rate of build much more aggressive than had ever been planned.

The manufacturing team raised the idea of managing bottlenecks at different workstations. They'd feather in buffers, a few cars being made at once. If a critical workstation went down, they could still keep downstream workers cranking without shutting down the line entirely. A sensible idea: Musk rejected it. If the engineering was perfect, there was no need for such diversions.

Some managers began pushing back, including a paint shop manager who told Musk that what he was proposing wasn't possible. Musk told him to find another job—he was fired. He was one of many who would learn to keep their doubts to themselves if they wanted to keep their job.

Suppliers, too, began expressing concerns. Panasonic executives were shocked at the accelerated schedule. They were still struggling to set up in Nevada, where the facility had many new bugs—the power kept blinking out and a new workforce needed to be developed in a part of the country unaccustomed to manufacturing. Musk was undeterred.

On May 4, he took away the safety net. He announced to investors, through a letter to shareholders and a conference call with analysts, that Tesla was accelerating its plans to make 500,000 vehicles a year to 2018, from 2020. Tesla would make **1 million** vehicles in 2020.

It was a stunning projection. When the media was given an opportunity to ask questions, Phil LeBeau, the longtime automotive correspondent from CNBC, asked for clarification. "Is that a production target, a production goal, or hypothetical?"

Tesla would do more than 500,000 vehicles in 2018, Musk said, followed by roughly 50 percent growth the next year, until reaching 1 million in 2020. ("I mean," he said, qualifying it a bit, "that's my best guess.")

Musk told investors that suppliers were being instructed to be ready to begin volume production on July 1, 2017. He conceded that it might take a few months after that deadline to hit full capacity, as the company dealt with whatever issues might arise. "In order for us to be confident of achieving volume production of Model 3 by late 2017, we actually have to set a date of mid-2017 and really hold people's feet to the fire internally and externally to achieve an actual volume production date of late 2017," Musk said. "So as a rough guess, I would say we would aim to produce 100,000 to 200,000 Model 3s in the second half" of that year.

To customers, he had a piece of advice: Now was the time to order. "You don't have to worry about [. . .] placing your order and receiving it 5 years from now. If you place your order now, there's a high probability you will actually receive your car in 2018."

He had proclaimed it now to the public. The Model 3 would be churning through factories and humming through streets in as little as a year. The die had been cast.

CHAPTER 23

CHANGING COURSE

Forty-year-old Joshua Brown of Ohio was driving along a divided highway south of Gainesville, Florida, a little past 4:30 p.m. on a Saturday in early May 2016 when his Model S went straight into a semitrailer crossing the roadway. His vehicle, traveling 74 miles per hour, didn't slow down; it went beneath the trailer, shearing off the roof of the car. It continued into a drainage culvert and through two wire fences before hitting a utility pole, rotating counterclockwise and coming to a rest in the front yard of a house. Brown died on impact. The truck driver was uninjured. While officials would say the truck driver failed to yield the right of way, the immediate question was: Why didn't Brown, traveling a straight road without any obstructions, make any apparent attempt to stop or even slow down for a giant truck crossing the roadway? The fact that the vehicle kept going suggested that there could be a fatal flaw in a new, high-profile feature that had begun rolling out.

It would take officials months to dig into the crash, but Tesla was able to quickly retrieve the data and soon knew that Autopilot, its driver-assist software, was, in fact, in use at the time. For most of his forty-one-minute trip from the state's west coast, Brown had had the system on, applying enough torque to the steering wheel seven times, for a total of twenty-five seconds, to convince the system that he was in control. The longest he'd gone between alerts that he needed to touch the wheel was nearly six minutes. He'd received almost two minutes' worth of warnings to re-engage with the wheel during his trip. The system had last sensed his movement of the wheel two minutes before impact. It made no attempt to stop.

The crash came at a horrible time for Musk. Following the excitement around the Model 3 weeks earlier, he was racing to raise money to pay for the car's production. He was also quietly working with his cousin Lyndon Rive, CEO of SolarCity, to figure out a way for Tesla to buy the struggling solar company, which had long come under the attack of short sellers who questioned its business. The confluence of these three events would threaten Tesla's ability to fulfill Musk's years-long effort to bring out the Model 3.

For years, Musk had somehow managed to

balance the demands of running Tesla and SpaceX while also serving as the chairman of SolarCity. It hadn't been easy, especially in 2008 when Tesla and SpaceX both struggled. As winter became spring in 2016, his elaborate financial house looked in jeopardy again, a by then familiar state. Tesla's reveal of the Model 3 gave incredible hope for what was ahead for the automaker, but the months-long effort to churn out the Model X had left the company financially wounded. All the while, in the background, SolarCity's business had grown weaker.

Early in 2016, Musk surprised his new CFO at Tesla, Jason Wheeler, with an early Saturday morning call. It sounded noisy in the background, like maybe Musk was on his airplane. The instructions were clear, though. Musk planned to call an emergency meeting of Tesla's seven-person board of directors. Surprisingly, he wanted Wheeler to be prepared to give a business pitch for what an acquisition of SolarCity by Tesla might look like. He had forty-eight hours.

Unbeknownst to Wheeler, Musk was spending the weekend in Lake Tahoe with Rive. SolarCity was in trouble, and if SolarCity was in trouble, so was Musk's business empire. Musk's call to his Tesla CFO that February morning came after a month of plotting by Musk and his cousin about ways for SolarCity to conserve cash. The company ended 2015 with $383 million on hand. Buried deep in the fine print of one of the revolving loans that was helping the company do business was a requirement that SolarCity keep an average monthly cash balance of $116 million. If it fell short, the company would immediately be in default, which could trigger defaults across the company's other debts and would have to be publicly announced. That could threaten SolarCity's ability to remain solvent. And since its finances were intertwined with Tesla's, such a default could make it harder for Tesla to take out new loans, which it would need with the delay of the Model X and the mounting costs for bringing out the Model 3.

Concern about taking unneeded risk as they prepared for the Model 3 weighed heavily on Wheeler, who had joined Tesla months earlier from Google, where he had been vice president of finance. He had replaced Deepak Ahuja, who opted for retirement after a grueling seven years next to Musk. Wheeler and his team worked all weekend to put together a report on the ramifications of Tesla buying SolarCity.

The result wasn't pretty: SolarCity was hard to see as anything other than a money loser. Merging the companies would almost double Tesla's debt load and could prove a risky distraction. Wheeler put his numbers for the board into a presentation he entitled Project Icarus. The short version: A deal would hurt the value of Tesla's stock. The board agreed that it wasn't the right time, especially as the Fremont factory continued to struggle with ramping Model X production. The idea was tabled.

By May, however, wheels were already in motion for Tesla to go to Wall Street to raise its own money, amid the excitement following the Model 3 reveal. The growing number of customers putting down deposits would help Musk make an easy case for further investment: Tesla needed to fund the car's ever-expanding production. His senior executives were united in pushing him to raise twice as much as he wanted—an amount they believed would give them enough cushion to bring out the car on the accelerated timeline that he was now demanding. But Musk didn't want to raise that much (and in so doing dilute his own stake). Instead, Tesla filed on May 18 to sell shares to raise $1.4 billion.

Less than two weeks later, at the private quarterly meeting for Tesla's board, Musk again raised the idea of buying SolarCity. This time the board, which had several members with interests in SolarCity's success, was open to it, authorizing Tesla to assess a potential purchase and hire advisers to help out.

For years, Musk and some of the board members, such as early investor Antonio Gracias, had talked about putting the two companies under one roof, especially as Tesla and SolarCity were increasingly working together. In 2015, JB Straubel, Tesla's CTO, developed yet another new line of business for the automaker, selling large battery packs for home and commercial spaces, which appealed to solar customers eager to capture the energy they were creating by day to use at night. The idea of adding a solar component to Tesla's business had been proposed to the automaker's board way back in 2006, before SolarCity was even founded, but the board rejected it at the time. Kimbal Musk, Elon's brother and fellow Tesla board member, had chalked it up to the other board members being too shortsighted.

The idea took on a new, greater significance in early 2015, as the Tesla board visited the construction site for the Gigafactory in Sparks, where Musk and Straubel planned to work with Panasonic to build millions of battery cells and battery packs. As they looked at the massive site, the gravity of it hit many of them. If Tesla was positioning itself to build battery packs and sell them as part of solar panel systems, they should control the entire customer experience. To board members such as Gracias, it crystallized the understanding that they were moving into a new era for the company: the electricity storage business.

On June 20, less than a month after getting permission to investigate a deal, Musk called a special meeting of the board—this time they met in the early afternoon at the Fremont factory.

As they discussed a potential acquisition, CFO Wheeler again expressed concern. He worried that the merger would result in Tesla having to pay more to borrow money in the future. According to notes taken by consultants, among the issues considered was the effect that short sellers, and Jim Chanos specifically, might have on the stock.

The board decided that one of their ranks, Robyn Denholm—who had just joined Australia telecom giant Telstra as chief operating officer, in line to become CFO—needed to get out in front of investors to make the case for why the two companies made sense together. She was seen as one of the few board members without any ties to SolarCity (which counted Gracias as an investor and board member, too). As talk turned to what Tesla should pay for SolarCity, Musk wanted to put a price on the table that was publicly defensible. As for negotiating tactics, Musk told the group, "I don't negotiate."

He left the room along with Gracias, a concession to their conflicts of interest. The group settled on a range of $26.50 to $28.50 per share—a valuation of as much as $2.8 billion for SolarCity. The company quickly sent a formal letter to Rive at SolarCity and ran a blog post on Tesla's website the next day, after the markets closed in New York, announcing the proposal.

"Tesla's mission has always been tied to sustainability," the post began. "It's now time to complete the picture. Tesla customers can drive clean cars and they can use our battery packs to help consume energy more efficiently, but they still need access to the most sustainable energy source that's available: the sun."

Investor reaction was immediate and unmistakable. SolarCity surged 15 percent, a hopeful sign for a company whose shares had been down more than 60 percent over the previous twelve months. Tesla shares fell off a cliff. The car company's valuation dropped by $3.38 billion.

With news of the SolarCity deal and Tesla's tumble, CNBC knew just who to turn to. The business channel interrupted its afternoon coverage with a blaring "breaking news" banner. Jim Chanos had sent a statement lambasting the proposal: "The brazen Tesla bail-out of SolarCity is a shameful example of corporate governance at its worst." Of course, Chanos had a dog in the fight. While Tesla was tanking, SolarCity shares were soaring—bad news if, like Chanos, you were betting on SolarCity's collapse.

But the bad reviews were coming from more than just short sellers. Tesla's consultants, a firm called Evercore, began collecting a string of negative industry analyst notes, used by investors to gauge the market. "Absent a detailed explanation (at this time) we are struggling to see brand, customer, channel, product, or technology synergies," a J.P. Morgan analyst wrote. "We believe this integration is possible, but likely challenging and fraught with financing risks," an Oppenheimer analyst concluded. And the most stinging came from Morgan Stanley's Adam Jonas, the analyst whose enthusiasm for Tesla helped inspire his return to the U.S. from London: "We believe Tesla's most valuable asset may be the trust it has built with its providers of capital. Even a rejected deal could 'leave a mark' in terms of investor questions of corporate governance, altering the price that investors may demand on Tesla's capital."

Amid the market onslaught, the National Highway Traffic Safety Administration, the federal regulator that oversees cars, equipped with the power to force painful and costly recalls, announced it was opening an investigation into the fatal Florida crash, making public for the first time its concerns about Tesla's Autopilot feature and setting off critical media coverage of the system. **Fortune** magazine's Carol Loomis questioned why Tesla hadn't disclosed the crash when it raised money eleven days later.* That very question raised concerns with the Securities and Exchange Commission as well, putting Tesla under the microscope of the powerful regulatory body that oversees public equities.

As he pushed back against Autopilot criticism, Musk and the board seemed caught off guard at just how much the market was rejecting their SolarCity idea. Denholm was getting an earful directly from Tesla's largest shareholders, the institutional investors that manage billions in holdings, and which would carry important sway in approving the

---

* Musk defended Tesla's action in the story by saying the issue wasn't "material to the value of Tesla." He told **Fortune**: "Indeed, if anyone bothered to do the math (obviously, you did not) they would realize that of the over 1M auto deaths per year worldwide, approximately half a million people would have been saved if the Tesla autopilot was universally available. Please, take 5 mins and do the bloody math before you write an article that misleads the public."

proposed deal. T. Rowe Price—one of the largest institutional shareholders—was unhappy that Tesla had just done a public offering, mere weeks earlier, and hadn't disclosed the potential acquisition. Jeff Evanson, Tesla's vice president of investor relations, was getting a lot of blowback, too.

"Honestly, we hate being public," Kimbal Musk complained in a text to a friend about the dramatic drop in Tesla's share price following the SolarCity announcement. It was a subject he and his brother had discussed before, bemoaning how much easier it was to run the privately held SpaceX. It wouldn't be the last time the lament was aired.

The proposed SolarCity deal didn't just roil the markets; it didn't go over well inside the company either. Tesla had taken risks before but always in the pursuit of something bigger. It may have seemed mawkish to outsiders, but many longtime managers subscribed to the company's environmental mission—that they were working toward bringing out an electric car that could help reduce pollution and improve the planet. By 2016, some found it hard to believe their shares could rise any higher in value. For them, there was a common bond helping justify the long hours, personal sacrifice, and the erratic behavior of their CEO.

This felt different. It seemed now as though Musk was acting in his own blatant self-interest. "Elon was risky but he usually made pretty good business decisions," one longtime manager said. "But this just made no sense: Why would you acquire **that** company, **that** business, **that** brand? Except for the fact you're trying to bail it out."

Evanson in investor relations confided to one of the Tesla board members that the deal was disliked within the ranks of Tesla leadership. "Lousy sentiment outside and inside the company," he said in an email. "Senior execs need to shut up and get onboard."

Though Musk was chairman and the largest single investor of both Tesla and SolarCity, the two companies' cultures had diverged greatly. The basic means of selling their products were miles apart, for instance. Musk despised hard sales practices, which had contributed to Tesla's stores acting more like education centers. SolarCity was all about the hard sell, with salesmen traveling door-to-door and using call centers to pressure potential customers. The sales force was highly incentivized for such practices, too. Tesla disliked the idea of having its salespeople working against each other, shying away from the commissions that SolarCity embraced. The differences showed in the books. Wheeler's team calculated that SolarCity had spent $175 million in commission payments in the previous twelve months, compared to just $40 million at Tesla.

Even seemingly little things were misaligned. At Tesla, Musk had resisted doling out fancy job titles, dismissing job candidates that he thought were seeking titles rather than interesting work. At SolarCity they had an abundance of title creep, if perhaps to compensate for salary, which was generally lower. Both companies employed about 12,000 people in the U.S. Sixty-eight people held the title of vice president at SolarCity and were making average salaries of $214,547, compared to twenty-nine people at Tesla who were making on average $274,517.

Tesla's engineering leadership wasn't happy as it looked over its new potential workforce. Michael Snyder, an engineering manager in Tesla's energy unit, believed SolarCity's engineering talent was subpar, telling Musk that they'd rate 2 or 3 on a scale of 1 to 10. He went so far as to say there was only one person he'd consider hiring. Musk reassured him: "We are going to part ways with a lot of SolarCity personnel."

To be sure, Tesla was changing from the company many had joined during the days of the Model S and Model X, just as the company had evolved after its near bankruptcy in 2008.

To address the shift, much in the way that he'd set out his vision for the company's path from Roadster to the Model 3 in 2006, Musk posted another blog entry that summer detailing what was ahead for the company—a clear bid to change the narrative on Wall Street about the floundering deal.

This one was entitled "Master Plan, Part Deux." In it, he worked to better articulate the ecosystem that he, his brother, and Gracias had been talking about together all of those years. He wanted to, he wrote, "create a smoothly integrated and beautiful solar-roof-with-battery product that just works, empowering the individual as their own utility, and then scale that throughout the world. One ordering experience, one installation, one service contact, one phone app."

But Musk went beyond explaining the benefits of buying SolarCity. He laid out a vision for the car business. After the Model 3, he saw a compact sport utility vehicle (smaller than the Model X and built on same chassis as the Model 3, known as the Model Y), and a new kind of pickup. He imagined a semitrailer truck for commercial customers. Tesla likely wouldn't need to develop a cheaper car than the Model 3 because, Musk wrote, he expected robot taxis to displace car ownership. He envisioned a time when a Tesla car owner could deploy a vehicle into a robot taxi fleet "just by tapping a button on the Tesla phone app and have it generate income for you while you're at work or on vacation, significantly offsetting and at times potentially exceeding the monthly loan or lease cost."

It was classic Musk. He was painting the kind of vision for the future of cars that Silicon Valley had fantasized about for generations. But somehow, when he said it, it seemed plausible. (Trying to catch up, General Motors in March announced it was acquiring a little-known San Francisco startup called Cruise to jump-start its own autonomous car program. What attracted everyone's attention was the headline number for the price of the deal: more than $1 billion.) Musk didn't lay out how he would pay for his future or belabor the timelines. He didn't need to. His pronouncement had one convincing bit of history in its favor: Ten years earlier, his talk of bringing out an all-electric luxury sedan and compact car had seemed far-fetched. And look at Tesla now.

As if to punctuate the point, Musk pulled a lever that Tesla had rarely used before as it focused its efforts on dramatic growth instead of the bottom line. It started drumming up profit. In November, the company announced the July–September quarter had been profitable by $22 million, only its second profitable quarter ever. The increased deliveries of the higher-priced Model X helped, but like in 2013, Tesla also benefited mightily from sale of regulatory credits to competitors who didn't meet emissions targets in places such as California, and who'd otherwise face fines for their failure to do so.

Then Musk went to work the way he knew best, giving investors the shiny objects they needed in order to be persuaded of a combined Tesla and SolarCity. SolarCity, he said, was going to create a new kind of solar panel. Instead of sitting atop the roof, the panels would **become** the roof. After meeting with skeptical investors, Musk dashed off a note to his cousin, Peter Rive, at SolarCity and JB Straubel: "Latest feedback from major investors is very negative on SolarCity. We need to show them what the integrated product looks like. They just don't get it. Needs to happen before the vote." At Universal Studios, Musk in late October revealed his vision, using the roofs displayed on the houses from TV's **Desperate Housewives.** None of the solar panels actually worked, but that was beside the point. Musk promised to make roofs sexy. Weeks later, shareholders approved the deal.

Musk was working just as steadily to change the perception of Autopilot. It had attracted a lot of attention when it was announced in 2014 and rolled out in earnest in late 2015, further burnishing Tesla's credentials as a car company of the future. It had been used by Musk as an example of how the automaker was on the road to deploying fully self-driving cars, the kind of technology that was gaining increased attention in Silicon Valley with advances made by Google and others.

Tesla's underlying technology came from a parts supplier named Mobileye, which had developed a camera system to identify objects on the roadway. Tesla's team had worked to push the boundaries of the system through clever software programming. The effect was that a user could turn on a series of systems such as lane-keeping and adaptive cruise control, which kept the vehicle centered in its lane and following behind traffic at a certain distance. It wasn't foolproof, as the Florida crash indicated. It sometimes didn't recognize road hazards, which is why drivers were instructed to remain vigilant at all times. But it worked well enough that users became complacent.

The Autopilot software team, led by Sterling Anderson—who had also helped shepherd the final development of the Model X—had proposed different ideas to ensure driver engagement. They had been monitoring the torque applied to the steering wheel (a less than reliable measure) and the team was exploring building in a sensor that could detect if a driver's hand was even on the wheel. Musk had initially balked at the idea of making the system anything less than seamless, or else contributing to a feeling that the car was nagging you. After the Florida crash, he approved changes that included shutting off Autopilot if users kept disregarding warnings about keeping their hands on the wheel. He also wanted to push forward a new generation of the system replete with extra features.

As his team got closer to announcing its developments that fall, Anderson grew increasingly concerned over Musk's habit of overpromising, according to people familiar with the team's work. The last thing they needed was for Musk to go out and tell the world that the next version of Autopilot was fully self-driving.

Anderson expressed those worries to Jon McNeill,

head of sales and marketing, among others. It just wouldn't be accurate to say that the system could possibly be capable of taking control of the car without a person sitting ready behind the wheel, watching the road as a backup. Tesla's legal and PR departments had already been losing an uphill battle with Musk on messaging, according to those privy to the pushback. For the past year, they had stressed the importance of drivers **keeping their hands on the wheel,** working to ensure that any official Tesla communication demonstrated such use. But when Musk took TV reporters out for rides, he quickly demonstrated Autopilot—with his hands off the wheel.

In October, it came time for Musk to announce the new hardware. He said that it was shipping in all cars, and that buyers could upgrade to the enhanced features. He also proved Anderson's concerns well founded when he declared that the system's hardware was capable of fully self-driving, claiming that by the end of 2017 he would demonstrate a car able to travel from Los Angeles to New York City driving itself. Customers could purchase the feature with their new cars, to be installed as soon as it was available, with the minor caveat that regulatory approval might still be required.

His comments rubbed some of the engineers the wrong way. They didn't think what he was proposing was possible. Others shrugged it off—Musk said it, maybe it **was** possible. After all, they hadn't expected to squeeze so much functionality out of the previous system. That was what was so exhilarating about working with Musk. He'd pushed them this far, cleared hurdles that other automakers would've thrown up.

Those closer to Musk, however, didn't take his pronouncement so lightly. They knew not only that his timeline wasn't tenable, but that he'd be looking for somebody to blame soon enough. His promises struck some as crossing a new line. In the past, he had shocked them with bold pronouncements. Now he was truly promising the impossible.

As 2017 began, Anderson and Wheeler would both exit Tesla. On his final conference call with analysts, Wheeler welcomed the return of Deepak Ahuja as CFO in comments that seemed to foreshadow the troubles he saw on the horizon. "With Deepak's history here, on the verge of bankruptcy and everything he's gone through, he's well positioned," Wheeler said before Musk cut him off.

As concerned as Wheeler may have been, Musk's gambles seemed to be paying off as the new year began. First, the NHTSA in January 2017 closed its six-month probe into Tesla's Autopilot, announcing they'd found no "defects in the design or performance." In dense language, the government said it had reviewed Tesla data from the system and found its vehicles' crash rate **dropped** by almost 40 percent after "autosteer" was installed. Musk was ecstatic—he quickly highlighted the 40 percent findings on Twitter. Some members of the team were stunned. Where had the 40 percent figure come from?

All of the attention on Autopilot, and Musk's dream for fully autonomous cars, was stirring new excitement on Wall Street. The same day as the NHTSA's findings, Adam Jonas at Morgan Stanley published a research report predicting Tesla stock could rise 25 percent to $305 a share, a startlingly high level that, if achieved, would mark an unbelievable milestone: Tesla would be valued higher than Ford or General Motors. To punctuate the fact that Musk saw Tesla becoming more than just cars, he dropped the "Motors" from its official name, much as Apple had dropped "Computers" years earlier.

That February, Musk celebrated the launch of one of his rockets from the Kennedy Space Center in Florida, where NASA had once sent the first astronauts to the moon. The launch marked the first time a commercial rocket had launched from the facility. After years of struggling with SpaceX, Musk's vision for his rocket company was beginning to show promise and attracting new attention. In 2015, he was able land a rocket, a key step as he worked to develop reusable rockets for space, then went further in 2016 by doing so on a barge at sea. He captured imaginations with his talk about living on Mars while continuing to raise more and more money for SpaceX, which remained privately held and generated significantly less public drama than Tesla. He'd become the living embodiment of **Iron Man**'s Tony Stark.

Investors agreed with the rosy sentiment around Tesla. The stock began a run. By spring of 2017, Tesla had done the once unthinkable: It had overtaken Ford in market value to become the second-most valuable automaker in the U.S., behind only GM. A few weeks later, it overtook GM. Tesla, which sold a fraction of the cars and had never turned an annual profit, was now seen as more valuable than the hundred-year icons of American industry, titans who were at their most profitable in history. Investors were making a bet that Musk's vision for the new century had a greater chance of playing out than what GM or Ford had on offer.

Publicly, Musk was gleeful. On Twitter he cheerfully attacked short sellers. On Instagram, actress Amber Heard posted a photo of her and Musk, announcing a relationship that had only been whispered about and winked at in tabloids. He could be seen at a dinner in Australia, beaming with a lipstick kiss on his check.

With the excitement for the Model 3 and the famous girlfriend, Musk's own public profile rose to new levels. Tesla had propelled him into the ranks of the celebrity CEO. For the first time in early 2017, he ranked with a Q score, which measures the appeal of celebrities.

The heightened attention changed his life. On the upside, he was dating a fellow celebrity; he seemed to delight in taking her to the office to impress his senior executive team. On occasion, Heard would sit around as Musk conducted meetings; she brought a cake to celebrate this birthday. Newly elected U.S. president Donald Trump was seeking out Musk's opinions on things, asking him to join high-profile advisory councils. (During one call to Musk, according to a person familiar with the conversation, Trump sought advice on NASA: "I want to make NASA great again.") CEOs that winter were landing in hot water with customers for their association with Trump; Musk took grief from Heard, who was unhappy with his budding relationship with the Republican president. It was just one early sign to Tesla employees that his latest relationship would bring with it new personal drama for their CEO.

There were downsides to Musk's new fame. Every step he took was scrutinized. It was hard now to go out in public without being hounded by strangers. After couch surfing his way through Silicon Valley for years, he bought a forty-seven-acre estate in Hillsborough on the San Francisco peninsula. It held a century-old mansion known as De Guigne Court. The $23 million purchase gave him stunning views of the San Francisco Bay and a place to hold parties and private dinners.

Though investors showed new enthusiasm, inside Tesla there was cause for concern. Both the Fremont factory and Gigafactory were behind schedule. As Musk looked for reasons why, some were laying the blame—fairly or unfairly—on Panasonic. Almost from the beginning, the Tesla team had been pushing aggressive timelines. Accelerating production plans didn't make things easier. Just as the Model S had been worked out as production began, the Gigafactory was still being designed as Tesla approached Model 3 production.

Fearing what the delays could turn into, Dan Dees at Goldman Sachs, Musk's longtime banker, urged him to raise money in case things didn't work out smoothly. The lessons of 2008 and 2013 should've been obvious: Any sort of delay could knock over the fragile financial house that Tesla had built and eat into what limited cash it had.

Concurrent with all this, a massive new venture capital fund was being created by SoftBank, the Japanese tech conglomerate with stakes in Alibaba, Sprint, and others. CEO Masayoshi Son was on the verge of controlling a nearly $100 billion fund that was intended to rewrite the rules of investing in Silicon Valley and pick world-changing winners, infusing them with the kinds of cash that a generation ago would've seemed impossible in the private market.

Goldman thought a meeting between Musk and Son might be fruitful. A matchmaker was found for the two: Larry Ellison, co-founder of Oracle. He lived near Son's Silicon Valley home and was friendly with both men. He'd been an excited customer of the Roadster in the early days and had quietly amassed a sizable chunk of Tesla shares.

In March, a conference room on the second floor of the Fremont factory overlooking the assembly line was converted into a dining room. Caterers prepared a meal of steaks for a small group. Musk was there, as was Ellison. Son was joined by Yasir Al-Rumayyan, the managing director of the deep-pocketed Saudi Arabia sovereign wealth fund, who would soon join SoftBank's board.

Musk, Ellison, Son, and Al-Rumayyan: Collectively they controlled hundreds of billions of dollars (though unlike those of his fellow diners, Musk's fortune was largely illiquid). They all shared ambitions to make larger-than-life bets that, if successful, would change the course of humanity. If they failed, they'd burn mountains of money. As the group sat around the table and food and wine was served, they discussed a whole host of possibilities, including taking Tesla private.

As Musk and his brother, Kimbal, had noted in the past, things were so much easier at privately held SpaceX. But taking a company of Tesla's scale private was a daunting prospect.

For about a year's time, he had been quietly contemplating the idea. In his deliberations, he sought out advice from Michael Dell, who had taken his namesake computer company private in 2013, as well as from the lawyer that helped him do it. He worried about Tesla's rising value. At the current share price of around $250, they'd have to raise $60 billion—the company's current valuation with a 20 percent premium applied on top of it, a standard practice in acquisition intended to sweeten a deal and ward off bidding competition.

After a few glasses of wine, Son, wearing his usual cable-knit sweater, began to question why Musk allowed himself to be distracted with so many companies. Why not just focus on making cars, rather than getting entangled with solar—to say nothing of his occasional talk of digging tunnels for high-speed transit and developing computers that could be controlled with your mind? The implied criticism didn't sit well with Musk. Son also insisted that Musk needed to take Tesla into India, to which Musk agreed that, yes, at some point that made sense. But at the moment, **we've got a lot of other shit we're really trying to do right now.** Musk came to suspect Son's motivations were tied to business SoftBank had in the country.

It became clear that Musk and Son were no love match. Each had respect for the other, but both were alpha males who envisioned their own futures and reveled in proving doubters wrong. Any money from SoftBank would likely come with strings. Son had a reputation for being a hands-on investor—the sort of influence that Musk naturally bucked against.

The dinner came to an end with promises of further discussions.* But it seemed for now that Tesla would remain a public company. With all of the stock market excitement, Musk could again return to the market to raise more money (which he did that spring). Those dollars would provide fewer strings than Son's, and Musk's control wouldn't be challenged.

Much more pressing for Tesla's future was what was taking shape just a floor below their dinner, where workers were rushing to prepare the Model 3 assembly. Official production was slated to begin July 1, a scant few months away.

---

* The two would end up meeting a few times, including one time when Son appeared to nod off, apparently jet-legged.

CHAPTER 24

ELON'S INFERNO

I wonder how you're feeling right now—you seem a little gray about the challenge ahead," a reporter at the back of a conference room at the Tesla Fremont factory asked Elon Musk. For twenty-five minutes, Musk had sat on a stool in front of a room of journalists, gathered to discuss the start of production of the Model 3. Musk had already marked the beginning of official production weeks earlier on Twitter, when he noted that the first saleable vehicle would be ready on July 6, 2017. He tweeted that he expected 100 Model 3s to be built in the following month, then more than 1,500 in September. By December, Tesla would reach 20,000 Model 3s a month, he claimed.

That was all to come. Today, the last Friday of July, was about celebration, both for employees and for loyal customers. That night Tesla would hold an event to hand over the first thirty Model 3 cars to customers—a milestone that had seemed like a moonshot when it was laid out in Musk's vision a decade earlier. It had taken great sacrifice, grit, and no small amount of luck to get to this moment.

Despite the milestone, as he fielded questions from reporters that afternoon, Musk didn't seem particularly buoyant. He warned the assembled reporters that Tesla was about to enter at least six months of "manufacturing hell," as the company worked the kinks out of its production line to reach a weekly rate of 5,000 vehicles by the end of the year. He cautioned that it would be hard to predict when exactly it would occur but, in a monotone, he said he was confident Tesla would then reach 10,000 a week by the end of the following year. He noted that a flood, tornado, fire, or a ship sinking, "anywhere on earth," could disrupt those plans.

Many in the room needed no reminding: Tesla had struggled with all of its new cars. Tesla made a huge jump from making about 600 Roadsters a year, to 20,000 Model S sedans a year, to 50,000 Model X SUVs a year. Now, Tesla aimed for 500,000 Model 3s. "It's going to be quite a challenge," Musk said: a huge understatement.

Asked about his mood, Musk paused. "I'll be more pepped up this evening," he said. "I apologize. I've got a lot on my mind."

His staff watched in disbelief. They had tried to knock him out of his funk backstage, but here

he was before the world's media and he looked—
defeated. Something in his head seemed to acknowledge his dour mood, but his heart couldn't
muster a change in tone. Instead, he continued in a
drone about how it was a great day for Tesla, "something we've been working toward since the beginning of the company."

Hours later, Musk was back to his normal self,
onstage as hundreds of employees celebrated the
handover of keys. Timed to the event, Tesla had
worked with **Motor Trend,** which had lavished
praise a year earlier on the Chevy Bolt (the automaker's new all-electric car, not to be confused with
its hybrid Volt), to give its reviewer a chance to take
the vehicle for a first drive. The article appeared
with their first impressions, and it was gushing: "The
Tesla Model 3 is here, and it is the most important
vehicle of the century." During the reviewer's drive,
the Model 3 compared favorably in many ways
to the performance and experience of the Model S—
just as Doug Field and his team had hoped when
they began their work two years earlier.

The review concluded: "Recently I've been
spending some time in Motor Trend's long-term
Chevrolet Bolt EV and with every mile edging
closer to calling it The Automobile 2.0. With its
affordability, stress-free range, and delightful driving qualities, I'm thinking that maybe this is where
the second era of the car commences. Pause that
thought. With the Tesla Model 3's performance,

slinky style, fascinating creativity, and, critically, its Supercharger safety net, I think this is truly where it begins."

The event had given the Twitter world new pieces of meat to fight over. Almost one in four shares was tied up by short sellers, betting the company was overvalued. Larry Fossi, the man tasked with managing eccentric billionaire Stewart Rahr's private office in Manhattan, weighed in with a blistering blog post under the pseudonym Montana Skeptic. He questioned why Tesla hadn't shown its new assembly line at the event to celebrate the start of production. He predicted the company would need to raise more money by year's end.

"Bottom line: I believe the Model 3 shares the same genetic defect as the Model S and Model X: it is destined to be a chronic money loser."

Fossi was right to wonder about the assembly line. As Doug Field's team rushed to finish up work on the Model 3 line, workers had stepped up their efforts to unionize. Richard Ortiz, the hourly worker who had returned to the Fremont factory after a twenty-year stint at NUMMI, had been quietly making his way through the workforce to see who might stand up and publicly support the UAW. The Detroit-based union had sent organizers to help his effort. They stayed in a nearby hotel, opening an office in a small building down the road from the factory. On a whiteboard, they kept a list of the workers who had signed a card indicating their willingness to help organize. The list was short: Ortiz and another worker who had also worked at NUMMI, Jose Moran, a quiet man who left his home in Manteca, California, at 3:25 a.m. to drive sixty miles so that he could find parking and arrive for his shift at 5:25 a.m. He'd reached out to the UAW about helping them organize back in the summer of 2016. They decided Moran would be the perfect face for the movement.

In February, UAW sent waves through the tech media with a post under Moran's name on Medium, a website favored by startups for posting announcements masquerading as news. It described life at the Tesla factory. The headline read "Time for Tesla to Listen," and what followed was 750 words describing the tough life the 5,000 workers at the factory were enduring, including excessive mandatory overtime and preventable injuries. "Six out of eight people in my team were out on medical leave at the same time due to various work-related injuries," Moran wrote. "I hear coworkers quietly say that they are hurting but they are too afraid to report it for fear of being labeled as a complainer or bad worker by management." He wrote that he wanted to make Tesla better, and that he believed that could be achieved by forming a union. "Many of us have been talking about unionizing, and have reached out to the United Auto Workers for support."

Musk's response was quick. He attempted to discredit Moran, telling the website Gizmodo: "Our understanding is that this guy was paid by the UAW to join Tesla and agitate for a union. He doesn't really work for us, he works for the UAW." Musk added, "Frankly, I find this attack to be morally outrageous. Tesla is the last car company left in California, because costs are so high. The UAW killed NUMMI and abandoned the workers at our Fremont plant in 2010. They have no leg to stand on." In private, Musk sent workers an email deriding the UAW and promising perks ahead, including a party to celebrate the Model 3, frozen yogurt, and eventually an electric roller coaster at the Fremont facility. "It's going to get crazy good," he wrote with a smiley face.

Moran did, in fact, work at Tesla, and both he and the union denied that he was being paid by the UAW. He and Ortiz stayed busy handing out leaflets to workers in their spare moments.

There was an unfortunate truth about those thirty Model 3 sedans celebrated in July: They didn't come from a fancy new assembly line inside the Fremont factory. They were handmade by Tesla's workers. The bodies were welded together in a prototype shop, set up near the paint shop. (It was such a tight space that the different parts of an individual car had to be rolled in and out on a cart.) The body was then transferred to an area where it went through final assembly. The process took days and was physically exhausting. It would need to continue for weeks to come, too, because work crews hadn't yet finished preparing the space for the body shop, the early part of the assembly process, where giant robots welded a car's frame together before it went to the paint shop, then to final assembly.

Engineering chief Doug Field's warnings to Musk in the spring of 2016 had come true. Celebration or not, the company was still months away from having all of the equipment in place to begin regular production.

That was a relief for the folks in Sparks, Nevada, because the battery teams weren't ready either. Musk pointed the finger at Kurt Kelty, the longtime Tesla manager who had helped forge a relationship with the Japanese supplier years earlier. Panasonic wasn't any happier with Kelty; he had the unenviable job of liaising with them over Tesla's rush deadlines and its troubles at the Gigafactory. Stuck between a Musk and a hard place, Kelty decided he'd reached the end of his time with Tesla. He watched the Model 3 being celebrated at the event in July—his final day. Like Martin Eberhard, Peter Rawlinson, George Blankenship, and so many others, he'd given all he had. It was time to pull away.

Though Musk blamed Panasonic for the delay,

the blame lay with Tesla. The challenges of building a factory for the first time and starting a new kind of production system were overwhelming. Kevin Kassekert, a key Straubel deputy who was overseeing the construction project for Tesla, struggled to keep the power on for the production lines as new parts of the factory were being added at the same time. By October, it had become clear to both Musk and outsiders the extent to which Tesla was struggling to produce. The company reported making just 260 Model 3 sedans in the third quarter, including the thirty it delivered for the late July event. Tesla attributed the results to undefined "production bottlenecks."

"Although the vast majority of the manufacturing subsystems at both our California car plant and our Nevada Gigafactory are able to operate at a high rate, a handful have taken longer to activate than expected," the company said in a statement.

A few days later, **The Wall Street Journal** revealed that the factory's body shop hadn't been operational, and that cars were being handmade. Tesla responded sharply. "For over a decade, the WSJ has relentlessly attacked Tesla with misleading articles that, with few exceptions, push or exceed the boundaries of journalistic integrity. While it is possible that this article could be an exception, that is extremely unlikely." The **Journal** stood by its story and other outlets soon began writing similar ones. A shareholder lawsuit was quickly filed, claiming fraud.* Tesla's production claims attracted the attention of the Department of Justice.

It was in many ways classic Tesla: Musk making aspirational statements as part of a broader effort to motivate his teams to do the impossible, and in turn to excite investors. Except this time, Musk had said the factory would churn out as many as 200,000 cars in the second half of that year, and after three months it had done less than 1 percent of that figure. The goal had been fanciful to begin with, and Musk often conflated making 5,000 cars a week with a total-year figure. But even 5,000 cars in a single week was seeming impossible to achieve that October, as the team was rushing to get the body shop operational. When Musk had made similarly aspirational claims for the Model X in 2016 and failed to meet them, the gap between his goal and his result wasn't as great. This one you could drive a fleet of electric trucks through.

Part of the challenge was that Musk was pushing for greater and greater automation. The shop teemed with robots that needed to be set up and programmed to do specific tasks, such as welding at just a certain point on a car's frame. The lack of space at Fremont, and Musk's belief that the

* A judge would eventually toss the lawsuit on the grounds that Tesla's forward-looking statements were protected because of hedging Musk had made in announcing his targets. "Federal securities laws do not punish companies for failing to achieve their targets," U.S. District Judge Charles Breyer wrote in August 2018.

robots could work faster if put closer together, had led him to push his teams to think about density as an important design element. In response, they had about a thousand robots onsite for the Model 3 body shop, including hundreds hung upside down from the ceiling to pack more into the space.

The same was true at Gigafactory. But while the body shop was making some admittedly halting progress, Straubel's team was in a knotted mess. In some places, engineers struggled simply to get into the physical space to fix a robot, so tightly were they packed in. Battery packs needed to be made by hand in some cases; Panasonic's workers were borrowed to help out.

As the bottleneck grew, the supply of cells from Panasonic began to stack up. At one point, a Tesla executive stood among crates and crates of cells ready to be assembled. He estimated there were 100 million cells just waiting to be put into battery packs. Hundreds of millions of dollars of inventory just sitting around, eating away cash.

Gigafactory's problems drove Musk into a rage. He began flying to the Gigafactory more often to personally address the matter. He'd always been quick to fire people, but it had historically been through managers, not in person. Now, it might be whomever he came across on the factory floor. There was no reasoning with him; he blamed everyone but himself—even when they tried to explain that robots were malfunctioning because of his demands.

On the day Tesla was to announce horrendous third quarter results, he felt ill, lying for hours in the dark on the floor of a conference room. One senior executive was sent to pull him into a chair to address Wall Street. Once on the call, he sounded awful. Adam Jonas, the Morgan Stanley analyst who had long touted Tesla's potential, asked Musk: "How hot is it in hell right now?"

"Let's say level nine is the worst, okay?" said Musk. "We were in level nine. We're now in level eight."

On one occasion, Jon McNeill, the sales chief and one of his top deputies, could be heard trying to calm Musk. "I think we can fix this," he said, before repeating a proverb: **No man comes up with a good idea when being chased by a tiger.**

Perhaps sensing that he needed to bolster his team, Musk ordered up a party for some of the managers on the roof of the Gigafactory one night. He wanted a campfire and s'mores. Kassekert, director of infrastructure development, was aghast at the directive: **Musk wanted to light a fire on the roof of a factory of highly flammable batteries?** Dutifully, he figured out a way, laying a protective cloth over the roof.

That night Musk drank whisky and sang songs. He posted a short video on Instagram after 2 a.m. That video, and a **Rolling Stone** cover story that dropped in November of him waxing on about his breakup with the actress Amber Heard, spooked some observers, who wondered about his stability.

He continued to resent the fact that Tesla was publicly owned, with so many short sellers betting on its demise. "I wish we could be private with Tesla," he said. "It actually makes us less efficient to be a public company."

By late summer, efforts by Ortiz, the worker who dreamed of becoming a union president, and Moran, the face of the organizing drive, were gaining support from lawmakers at the California statehouse. Ortiz, along with three other Tesla hourly workers and UAW organizers, traveled to the state capital in Sacramento in late August to meet with legislators. They were there to lobby for language to be included in legislation for an electric vehicle tax rebate, specifying that Tesla needed to be "fair and responsible," and ensure the safety of its workplace. In response, Tesla organized some workers of its own to testify the following month at the capitol, including Travis Pratt, a lead equipment maintenance technician. During the hearing, Pratt praised the company and told lawmakers that he earned an annual income from Tesla of $130,000.

A video of the appearance made its way to Ortiz. He forwarded the video to Moran to ask him to identify the workers. "Let me know who they are . . . I want to walk up to them AND say see you in Sacramento suck ass," he texted his friend.

Moran logged in to the company's directory of

employees to check their names. He found Pratt
and took a screenshot of his picture, along with his
job title, then forwarded it to Ortiz. Ortiz posted
the photos on a private Facebook page that the
workers trying to organize had started. "These guys
been in Sacramento saying we are lying about how
things are at Tesla. Management has been taking
them . . . one of them sez he made $130,000 last
year . . . This just proves how much kissing ass and
ratting on people get you at Tesla and the ones that
do the real work get passed over."

After posting, he had second thoughts, delet-
ing the message after Pratt complained to him privately. "Say what you like about me behind closed
doors . . . I made what I did last year almost entirely as a level two maintenance technician, which
is where several of your colleagues from production now find themselves. I wish you luck but know
there will be a lot of us on the other side. Starting
with name calling may not be the approach you
want to take."

Pratt sent a screenshot of the original posting to
Tesla human resources—a gift. "Looks like we got
under some people's skin," he texted, along with a
blushing smiley face. It was the excuse Tesla needed
to open an investigation into Ortiz and, it hoped, to
squash the union drive. The company had already
been trying. Months earlier, Gabrielle Toledano,
head of HR, plotted in emails with Musk to offer
Moran and Ortiz spots on a safety team that would

convert them to salary workers, making them ineligible for the union. Now, they had a digital paper trail of Moran accessing company records in a way, they said, that violated company policy.

In late September, Ortiz was hauled into a meeting with an investigator that had been enlisted by HR. Ortiz arrived wearing a union T-shirt and pin. He was confronted with a copy of the Facebook post. He confessed to posting it and said he had pulled it down after Pratt complained. He apologized.

But who gave you the photos from the company directory? Ortiz wouldn't say. The investigator asked to see his personal cell phone. It didn't provide any clues. It was brand-new, acquired earlier that day.

Ortiz's loyalty wasn't enough to protect Moran. The company quickly traced his use of the digital records. By early October, they had a case against the men: Moran would get a slap on the wrist but Ortiz was fired.*

The effects on unionization were dramatic. The whiteboard at the UAW office listing workers who supported unionization soon went blank. For Ortiz, his dream of becoming a union president evaporated. A UAW official offered to help relocate

---

* In 2021, the National Labor Relations Board ruled that Tesla violated labor laws in handling the organizing activity, including improperly firing Ortiz. Tesla denied wrongdoing and has appealed the decision.

him to Detroit and find him a job, he said, but he wouldn't go. The East Bay was his home.

As 2018 began, it would be easy for onlookers to ignore the disarray in manufacturing. Musk was up to his old marketing misdirection. First, Straubel's battery-powered semitruck, which Jerome Guillen had taken ownership of, proved the perfect reveal for customers in November, with claims of five hundred miles on a single charge. What was more exciting, however, was the final moments of the demonstration, held alongside the Hawthorne airport runway. The semi's trailer opened to reveal snake-like headlights emerging from the dark. Then into focus came a new Roadster, a more muscular version of the original. It was longer, wider, sexier—a true supercar that zipped out of the hangar as loudspeakers blared the Beastie Boys' "Sabotage." The crowd responded to it like never before.

Tesla had spent years reaching toward the Model 3, and it was enmeshed in production hell four hundred miles to the north. But on this night, Musk unleashed the full force of his attention-generating machine. The super car's specs were enough to make car enthusiasts sweat. It would be capable of going from 0 to 60 miles per hour in 1.9 seconds, a claim that, if true, would make it the quickest production car ever made. It came as a not-so-subtle reminder that if Tesla could just survive the current storm, Musk would reward its faithful. On cue, Musk announced that Tesla was taking $50,000 preorders for the $200,000 Roadster, due out in 2020; a limited number of the so-called Founder Edition would go for $250,000 up front.

Musk reveled in this kind of marketing, of course. He strongly resisted the notion of **paid** advertising. As sales of the Model S had slowed in the past year, McNeill, the sales chief, had raised a near heresy when he explored launching a Facebook ad campaign to juice sales, an idea that was scuttled in part because of Musk's distaste, but also because the company could roll out a two-year lease deal instead, interesting buyers who might not want (or be able) to pay cash. They stoked further interest in the car in 2017 with another software upgrade, allowing certain Model Ss to accelerate even quicker (zero to 60 in 2.4 seconds), with so-called Ludicrous Mode.

Tesla had another marketing lever at Musk's disposal. SpaceX in early 2018 was preparing to test a giant rocket called Falcon Heavy. The world's most powerful rocket since the U.S. took astronauts to the moon almost fifty years earlier, it was designed to take heavy cargo into orbit around earth. To demonstrate that ability, Musk put a cherry red Roadster in its cargo hold. His team rigged video cameras on the car to capture shots of it in space. Sitting behind the driver's wheel was an empty space suit dubbed Starman. It was a stunning shot, one that put Tesla's cars in the same conversation as space travel. Tesla wasn't just an electric car company, it implied; Musk was offering you the future. He made the connection explicit to analysts after reporting a horrendous fourth-quarter loss, one of Tesla's worst ever. "I'm hopeful that people think that if we can send a Roadster to the asteroid belt, we can probably solve Model 3 production," he said.

Investors seemed to agree, remaining largely patient and betting that Musk could pull off another impossible feat. That patience was evident in Tesla's stock price, which ended 2017 up 46 percent for the year, giving the company a market value of $52 billion.

Musk looked at that stock price and told the board he believed shareholders would go along with even more grandiose long-term plans. For years, Musk had eschewed a paycheck. His real compensation had been stock options, worth millions and tied to certain milestones, such as bringing out the Model S and Model 3. Now, as it came time to negotiate a new compensation package, Musk demanded the board give him a plan that would make him the highest paid CEO ever and, in theory, if fully paid out, one of the richest people in the world. The ten-year plan he wanted would be worth more than $50 billion if Tesla achieved a variety of financial targets, including reaching a market value of $650 billion—an almost $600 billion increase from its present value. The ambition he was laying out suggested a reimagined Tesla: one with **millions** of car sales a year, and that was worth more than any other automaker by many multiples.

It was a tough plan to swallow, especially since Musk was having a hard enough time making **hundreds of thousands** of vehicles. Given the size of the potential payout, the board was divided on the incentives, according to people familiar with the deliberation; but with so many friendly members, including Musk's brother, Kimbal, it eventually sailed through. The package was announced in early 2018.

The proposal caused rancor among some of Musk's senior leadership team, who felt they too should be seeing a bigger reward for the company's success.

According to one former executive, among those hurt by Musk's actions was JB Straubel, who had been there from the beginning and was considered a co-founder. "That was one of the breaking points," the executive said. "It broke their relationship."

Straubel denied that the effect was so dramatic, noting that he understood Musk's package came with "crazy high risk" as well as a "high reward." Said Straubel: "There are days of frustration and ups and downs in any relationship but Elon and I went through a lot tougher shit than this without breaking the relationship."

The payouts were byzantine in their complexity,

staggered through twelve tranches. To receive the first tranche of 1.69 million shares, for example, Tesla would need to either boost its annual revenue to $20 billion (from $11.8 billion in 2017) or achieve an adjusted profit of $1.5 billion. Plus, the market value would need to rise to $100 billion, a level it needed to keep on average for 6 months and for 30 days. To exercise the options, he would need to pay $350.02 a share. Missing from any requirement: net profit, something the company had still failed to achieve on a total-year basis. It hadn't even been able to eke out two consecutive quarters of profitability. In the fashion of its tech contemporaries, metrics favored growth and share price over old-fashioned moneymaking.

The details of the deal required him to hold his shares for five years. Most important, Musk needed to remain employed at Tesla as either CEO or chairman. While he may have talked in recent years about stepping back as CEO, the new pay package sent a powerful message to his senior leadership team—and investors—that he wouldn't be letting go of the reins anytime soon. Some inside Tesla thought McNeill or maybe Field, the engineer overseeing the Model 3 whose authority had grown to include oversight of manufacturing, might be a future Tesla CEO. The next month, McNeill would depart Tesla to become chief operating officer of ride-hailing startup Lyft Inc.

As the new year began, Musk wasn't acting like

he wanted to give up control. It was clear that Fremont, not Sparks, was the highest hurdle. Musk's focus again shifted. The new Roadster reveal, along with the semi, helped Tesla's cash situation, though money was running low again and, for a few reasons, Musk was reluctant to seek more capital. First, the Department of Justice had begun probing Tesla's production statements amid questions of whether Musk had misled investors. Tesla hadn't yet disclosed the investigation, and it would need to do so as part of any new fundraising effort, according to people familiar with the situation. This would only add to Tesla's appearance of desperation. Second, Musk had talked publicly about the importance of raising money when it's perceived as not being needed. When it's needed, the terms are tougher—and thus more costly.

In effect, Musk was betting everything on getting the factory working. He'd blown through his deadline for 5,000 cars a week twice now. His new target: the final week of June.

As they struggled with the automated assembly line, Antoin Abou-Haydar, an engineer hired several months earlier from Audi to work on production quality, raised an observation to Field and other executives. The engineering team had done such a good job designing the Model 3 for assembly that in the previous July and August they had had a comparatively easy time making the car by hand. Instead of the complex automated assembly line that they were building, maybe they should just start over again—without robots?

The suggestion didn't go far. Musk was invested in making his line (which he had taken to calling the Alien Dreadnought) work. He was selling investors on the idea that eventually the factory would need only a few people, similar to the way Tim Watkins had programmed machines to work an overnight shift alone. Musk had envisioned a three-story assembly line, with parts of the car moving overhead on the top level to workstations on the second level, where workers would add parts delivered from the third level beneath their feet by a conveyor belt system. It seemed like an eloquent system to save space and manpower. In practice, it was a mess. Engineers couldn't get the timing right. The lack of space at the factory led to a crowded environment, one that felt like walking through a battleship.

By spring, however, it was clear that **something** needed to be done to break the logjam. The company was burning money; Abou-Haydar's idea took on new urgency. To accommodate it, a second, less robotic general assembly line was erected inside the factory, immediately resulting in a pickup of production. Musk on Twitter admitted his error: "Excessive automation at Tesla was a mistake. To be precise, my mistake. Humans are underrated."

To meet the goal of 5,000 a week, however, they'd need to do even more. They needed to add a third assembly line for the Model 3. But by this point, the factory was chock-full. They were already having to do quality inspections outside under a giant tent. The team began to wonder if they could do the same with an assembly line.

Adding additional assembly lines, especially un-automated ones, would require many more workers on the car than originally planned. In meetings, Musk lost track of the numbers. He kept talking about having around 30,000 employees, when in fact the head count with outside contracts had grown to more than 40,000. (Musk's attention was said to sometimes wander after the start of budget presentations.) Deepak Ahuja, the returned CFO, finally had to gently confront Musk with the fact that the company now had many more people.

Musk didn't take it well. It was the kind of detail that he latched on to, as if it crystallized a bigger issue, namely that Tesla's cost structure wasn't working. The plan had been for Tesla to break even on a rate of 2,500 Model 3s a week. But the added manual labor was jacking up the cost. Ahuja's new calculations that spring didn't have the company breaking even at 5,000 a week. Musk began to freeze spending, including a halt to plans to dramatically increase the size of service and delivery centers across North America. Musk wanted to slash jobs as quickly as they could.

The failures at Fremont led Musk to a predictable place. He became unhappy with Field. Though

Field had done what he'd originally been charged with doing, bringing the excitement of the Model S to the mainstream with the Model 3, the factory was in disarray. It could be argued that it was Musk's own fault for hastening production, despite the admonitions from those around him. Still, he turned to a familiar solution: Musk took over, stripping Field of oversight of manufacturing.

It was the beginning of an inglorious end for Field, who had elevated Tesla's product development operations, competing for some of Silicon Valley's best engineering talent and creating a car that by many measures was **superior** to the Model S. He wasn't without internal critics. Some longtimers said he'd instituted a more corporate outlook in the operations, that the place had become more political under his watch. Still, he was perhaps better poised to become Tesla's next CEO than any of his colleagues, if Musk ever made good on his promise to step back from day-to-day command and focus on product development.

That spring Musk in effect took control of the factory himself. Scaling back Field's duties didn't stop Musk, one evening at the factory, from calling the demoted executive up, demanding to know where he was, according to an ally who'd been told the story.

Another evening, Musk summoned to a conference room a group of engineers who were focused on making the assembly line work. He stormed in and proceeded to tell them all that their work was "complete shit." He ordered each person to go around the room and tell him "who the fuck you are and what the fuck you're doing to fix my goddamn line." As he berated the team, one of the engineers had had enough and quit to Musk's face. Musk screamed as the young engineer walked out. In another meeting, Musk walked in to find a manager that he'd grown unhappy with and said, "I thought I fired you yesterday."

It was around this time, during a tour of the factory, that Musk saw the line had stopped. He was told the automatic safety sensor halted the line whenever people got in its way. This angered him. As he sputtered about the lack of danger from a slow-speed line, he began head-butting the front end of a car on the assembly line. "I don't see how this could hurt me," he said. "I want the cars to just keep moving." A senior engineering manager tried to interject that it was designed as a safety measure. Musk screamed at him: "Get out!"

For those who had known Musk longest, these sorts of incidents were a painful mutation of what they had seen from the early days. Everyone knew he had a short-temper and didn't suffer fools. But in the early years, some felt like those who had been cut maybe deserved it. Tesla was about being the best and the toughest. Managers would trade Musk barbs like war stories: the time he fumed at a manager, for instance, that he would split his skull open and brand his brain with an "F" for failure.

But now, longtimers said, his anger seemed to lack any predictability. And instead of behind closed doors, his rage was on full display, no matter the employee's rank. It was as if the company had grown so large, he didn't know who specifically to blame anymore, so he was just yelling into the darkness.

Whatever the case, Field had reached his limit. It was a frustrating position: He had been recruited to oversee development of the Model 3 (as well as forthcoming crossover vehicle, the Model Y), but had been handed the reins of the manufacturing operations after Greg Reichow departed. Now, after his effective demotion, he was watching Musk dismantle his team person by person, and in humiliating ways. He was reluctant to leave, of course, feeling that he'd be abandoning those who remained. But it was time. His mother had died and his father was sick. Plus, his kid was about to graduate from college—all life events he was sure to miss on the factory floor.

The team was told Field was taking a leave of absence, but it was clear to many that he was gone for good. By the time Tesla officially announced his departure, the biting nature of his exit was evident. He would be among at least fifty vice presidents or higher-ranking executives to depart in the previous twenty-four months (partly fueled by the large number of big-title managers at SolarCity who had left after the acquisition). The news of his exit generated a flurry of headlines, as would be expected of the departure of a prominent car company's engineering chief. But Musk took umbrage at the press coverage. Tesla's public relations team pushed media outlets to downplay the significance. The automotive website Jalopnik ran a tongue-in-cheek correction: "A Tesla spokesperson reached out to clarify that Field was NOT the top engineer at Tesla, but rather that he was the top vehicle engineer. Much as there can only be one God, there can only be one Top Engineer at Tesla, who is Elon Musk. Second engineer at Tesla is JB Straubel."

In Sparks, Straubel's team was seeing some improvements, but they came at a cost. They had reached a rate of building three thousand packs a week and at one point had an hourly rate of building that if extrapolated would mean hitting the five thousand mark. But hitting that peak and then keeping at it, day after day—that was another thing. Amid the push, they were wasting a lot of materials; cells were being damaged by some of the automation as they rushed to figure out fixes. A report cataloging issues, compiled by quality operations supervisor Brian Nutter, underscored the latest game of whack-a-mole that Tesla was fighting: Battery cells were being found dented because of an issue at a flipper station, while elsewhere some parts were being rejected for excess adhesive in between cells. One production line was going down because of insufficient cooling tubes. An automation mistake wasn't properly moving a cure rack forward, causing a pileup of battery modules.

That spring, during a public conference call with analysts, Musk tried to explain away the first quarter's horrid results. About thirty minutes into the call, one Wall Street analyst asked when the company would reach its gross margin target for the Model 3, which, he said, the company had seemingly pushed back by six to nine months. Musk interrupted as his CFO tried to explain, saying it would be resolved in a few months. "Don't make a federal case out of it," Musk said snarkily. The analyst turned to Tesla's cash needs. Musk interrupted again as the analyst elaborated. "Boring, bonehead questions are not cool," he said. "Next." An analyst for RBC Capital Markets wanted to know what kind of effect moves to open up reservations to more customers had had on them actually configuring their Model 3s for purchase. Musk was dismissive. "We're going to go to YouTube, sorry," Musk responded, referring to a retail investor with a YouTube show who had been allowed to call in with questions. "These questions are so dry. They're killing me."

Musk's imperiousness didn't play well with

investors. In a span of about twenty minutes, the stock fell more than 5 percent. The mutterings around the conference call were ominous. Some said the outburst reminded them of the final days of Enron.

CHAPTER 25

SABOTAGE

At 1:24 a.m. on Sunday, May 27, 2018, Tesla technician Martin Tripp pushed send on an email to CNN, Reuters, Fox News, and **Business Insider.** "I currently work for Tesla, so I am requesting to remain anonymous," he began. "I am in a capacity to see daily production, production numbers, cost of failure/scrap, etc. (across every department and level). On several occasions Elon has flat out lied to the public/investors. I joined Tesla to follow it's [**sic**] Mission Statement, and it is disheartening to see how totally opposite we really are." What followed was the suggestion by Tripp that Tesla was still short of being able to make five thousand Model 3s a week and that Musk was taking risky shortcuts to speed up production.

The email to **Business Insider**'s news tips in-box

quickly found its way to Linette Lopez, a senior finance correspondent. A star at **Business Insider,** Lopez had written a few pieces about Tesla in the past year but none that pierced the corporate walls. She'd conducted an interview with short seller Jim Chanos months earlier in which she asked about his bet against Tesla. He called Musk's ability to sell his greatest quality. "He's always pitching the next great idea," Chanos told her. "The problem is, is that execution of the current ideas is falling short. And that's where I think it's problematic. And on top of that, I think—increasingly—he's making promises that he knows he cannot keep. And I think that's a much more ominous turn."

Lopez responded later that morning to Tripp: "I'm definitely interested in what you're saying here." She began digging into his claims of troubles at the Gigafactory, as workers began to erect a giant tent next to the Fremont factory more than 250 miles away in California. Many inside Tesla leadership thought that just hitting the five-thousand-a-week milestone would alleviate the pressure—taking away some of the oxygen from short sellers, steadying the company's financial footing. If they took time off before a final burst of building, they could have all of the parts ready and, hopefully, the hurdles removed to meet their number on the final day of the quarter, Saturday, June 30. The all-hands-on-deck approach was familiar to anyone who had been at Tesla from the beginning or during the trials of the Model S and Model X production ramps. Musk was simply doing the same for the Model 3: Creating a self-made deadline that they'd limp across, then regroup after the attention had drifted, at which point they could figure out a way to improve the process and cut costs.

What Tripp shared with Lopez, however, was the ugly cost of that approach. About a week after first connecting, she published a June 4 story with a headline that read "Internal documents reveal Tesla is blowing through an insane amount of raw material and cash to make Model 3s, and production is still a nightmare." Citing internal Tesla documents from Tripp (whom she didn't name), the article said the company was generating an incredible amount of waste at the Gigafactory: as much as 40 percent of the raw materials used to make the battery packs and drive units was being scrapped or needed reworking. It had already cost Tesla $150 million, the article said, though Tesla strongly denied that level of waste.

The automaker issued a statement that aimed to downplay the matter. "As is expected with any new manufacturing process, we had high scrap rates earlier in the Model 3 ramp. This is something we planned for and is a normal part of a production ramp," the company said.

In fact, the company had been watching the

issue closely. Tesla was on the cutting edge when it came to battery production, and it was experiencing something new to the auto industry. Other entrants into the market would similarly struggle with scrappage. Straubel was concerned enough about the issue that it helped convince him that he should fund a startup aimed at exploring ways to improve recycling for electric car batteries.

A few days later, Lopez followed up with another story from Tripp that said Musk's new robots for the Gigafactory weren't fully operational yet. Musk had said on Twitter just a few weeks earlier that those machines were key to getting production boosted past 5,000 a week.

Unlike Peng Zhou's leak in 2008 that exposed that Tesla was then nearly out of money, these leaks contributed to a narrative that Tesla's manufacturing effort remained flawed, adding more concern that the five-thousand-a-week goal would be missed. What else might emerge from this anonymous company insider? Musk's team didn't wait to find out. They began ferreting out the leak.

Tripp was relatively new to Tesla, having gotten an offer to work at the company the previous September, one of many new recruits as the company boosted its Gigafactory employment to more than 6,000 workers. He was hired at $28 an hour. Earlier in 2018, he'd been reprimanded for not playing well with his colleagues, company records reflected. He showed a kind of naivete that would come back to bite him.

Even as Tesla security was hunting for the leaker in its midst, Tripp included Musk and Straubel in an email about his concerns with the factory. "I would like to state that there are a LOT of concerned Tesla Employees," he told them. "A rough calculation of scrap cost for the remainder of the year is likely over $200,000,000." He added, "Even more staggering to everyone is that we physically will not have any place to put the fallout."

At 3:22 a.m. on Sunday, June 10, Musk responded that getting scrap to less than 1 percent "needs to be a hardcore goal." But the response didn't impress Tripp. He thought it was the same kind of empty promise he'd heard before. Within days, Tesla investigators had narrowed in on him after figuring out who had access to the data cited in the stories, and who had recently looked at it. He was fired.

Musk didn't take the discovery well. Late Sunday night on June 17, he sent a company-wide email warning everyone to be on the lookout for more traitors. "I was dismayed to learn this weekend about a Tesla employee who had conducted quite extensive and damaging sabotage to our operations."

Without naming Tripp, he said the leaker was motivated by being passed over for a promotion, and that the full extent of what he'd done was still being investigated. "We need to figure out if he was acting alone or with others at Tesla and if he was working with any outside organizations," Musk continued. "As you know, there are a long list of organizations that want Tesla to die. These include Wall Street short sellers, who have already lost billions of dollars and stand to lose a lot more."

A private battle followed between Musk and Tripp, who was full of bravado but lacked the financial resources to battle an angry billionaire. By Wednesday, Tesla filed a lawsuit against Tripp for allegedly stealing a gigabyte of data and for making false statements about its business. The two were exchanging blows via email. "Don't worry," Tripp told Musk, "you have what's coming to you for the lies you have told to the public and investors."

"Threatening me only makes it worse for you," Musk replied.

"I never made a threat," Tripp wrote. "I simply told you that you have what's coming."

"You should [be] ashamed of yourself for framing other people. You're a horrible human being."

"I NEVER 'framed' anyone else or even insinuated anyone else as being involved in my production of documents of your MILLIONS OF DOLLARS OF WASTE, Safety concerns, lying to investors/the WORLD. Putting cars on the road with safety issues is being a horrible human being!"

All of this bad press threatened to overshadow what Musk and Tesla saw as a long-sought milestone. To better tell the story of what was going on in the Fremont factory, Tesla began inviting reporters to see for themselves. In the middle of the factory, he spoke of how confident he was (even if had driven him to sleeping on the floor). "I'm feeling good about things," he said. "I think there's a good vibe—I think the energy is good; go to Ford, it looks like a morgue."

Asked why he had pulled Model 3 production plans ahead when members of his team had warned him of the pitfalls, he replied, "People have said that my entire life; what else is new?" He added, "They also said we couldn't land rockets."

Head of public relations Sarah O'Brien accompanied Musk on the visit. A former Apple communications manager, the thirty-seven-year-old executive oversaw a staff of more than forty in the PR department. She had worked with Musk directly for almost two years—and was beginning to feel its effects. She was exhausted, having experienced two fainting spells outside work. She kept a schedule that began at 5 a.m. and routinely ended around 9 p.m., plus weekends. She wore an Apple Watch set to alert her to Musk's tweets.

Since 2014, Musk's use of Twitter had accelerated, such that by midsummer 2018, he had posted more than 1,250 messages that year alone, or about

6 per day, on a wide variety of topics: fielding customer complaints, criticizing the media, jostling with short sellers. He'd become even more voluble in May and June, posting seven times as many messages as in January. At times, he could turn dark on Twitter, as he did when asked about his life. "The reality is great highs, terrible lows and unrelenting stress. Don't think people want to hear about the last two," he wrote. Never one for much sleep, he was struggling to get any, turning to sleeping medication for relief. "A little red wine, vintage record, some Ambien . . . and magic!"

As dark as things had been, there were some signs of hope. Researching an obscure joke about AI that he wanted to make on Twitter, he stumbled upon someone else who had made a similar one, a budding pop star named Claire Elise Boucher, better known as Grimes, then thirty years old. He struck up a conversation with her and the two soon began dating.

Meanwhile his bet that he could meet the five-thousand-a-week goal was getting closer to fruition. By the early morning of July 1, workers at the factory began their celebration, after working the entirety of the previous night. The added lines and staff had done what automation couldn't. They signed a Model 3 hood to commemorate the 5,000th car that week. Musk was nowhere to be seen, though. He had already departed for his brother's wedding in Lisbon. He sent out a company-wide email praising the effort. "I think we just became a real car company," he wrote.

What should have been a time of celebration quickly turned into the ugliest public spectacle in Elon Musk and Tesla history.

CHAPTER 26

TWITTER HURRICANES

As Tesla workers wearily marked their production milestone at Fremont, Musk landed in Portugal just in time for his brother Kimbal's wedding and a needed respite. For months, his friends had grown increasingly worried about him as he eschewed their invitations to unwind, telling them he was needed at the factory. His public comments about his now ex-girlfriend Amber Heard seemed unhinged. His newest flame, Claire Boucher, didn't fit neatly into Musk's type.

There was no denying the gravity of what he had achieved—nor was there time to bask in it. Now, with the bottleneck at the factory seemingly opened, his attention needed to turn to an equally pressing challenge: delivering cars to customers. An inability to sell the Model S five years earlier, once the factory had fixed its problems, had nearly wrecked the company. This time around, Tesla needed to deliver far more than 4,750 cars, and the goal was more than simply breaking even. This wasn't about making a point, it was about making **cash**—to pay suppliers whose bills were mounting. More than just deliver cars, Tesla needed to move beyond its lone assembly plant outside San Francisco; it had to prepare to take the company global, to give it the kind of sales volume and scale it needed to compete against the likes of GM.

And yet despite the conflicting needs—rest, refocusing on sales—Musk's mind was drawn elsewhere. He was on the verge of a public meltdown that might not only tarnish his reputation and distract Tesla from completing its goal of ushering in a mainstream electric car, but might do the one thing that Musk had fought so hard over the years to avoid: cause him to lose control of the company.

Musk's Twitter habit seemed harmless enough. He obsessively checked the social media platform throughout his day, but then, who didn't check social media obsessively? Some of his earliest public blunders had occurred on the site. He'd unnerved onlookers months earlier with an April Fool prank suggesting the company was broke. He'd taken to the platform to gloat when Tesla's market value overtook Ford's more than a year earlier, jabbing at short sellers who were feeling the pinch. His Twitter nemeses hadn't forgotten; they were busy poking holes in Tesla's latest good news. With the Martin Tripp leaks, Musk was seeing conspiracies against him at every turn. Without any proof, he privately mused that short-seller Jim Chanos was somehow behind it.

This time, an unfolding drama on the other side of the world caught his attention: a boys' soccer team was trapped in a flooded cave in Thailand. As the world watched rescuers try to save them, someone on Twitter urged Musk to intervene. At first he demurred, but within days he was proclaiming that his engineers would design a mini-submarine to rescue the kids—even if it wasn't clear that the rescuers in Thailand wanted such help. He documented his efforts on Twitter.

Tesla's team was preparing for Musk to meet with Chinese government leaders, to celebrate the automaker's deal to open a factory in China—a hugely consequential deal that could redefine the company and propel it beyond a niche car company. Instead of drawing attention to that triumph, Musk had other plans. En route to China, Musk had his jet stop in Thailand, where he rushed to the cave site. He posted pictures on Twitter. "Just returned from Cave 3. Mini-sub is ready if needed. It is made of rocket parts & named Wild Boar after kids' soccer team. Leaving here in case it may be used in the future. Thailand is so beautiful," he wrote on July 9, even as a daring (and ultimately successful) rescue attempt was under way.

Musk's offer had become part of the narrative and spectacle of the rescue, which turned into a mostly heartwarming story of survival for the boys when all of the players were finally pulled out on July 10. "Great news that they made it out safely," Musk tweeted. "Congratulations to an outstanding rescue team!"

Musk's sub was never used, and Narongsak Osottanakorn, head of the operation coordinating the rescue, told reporters the submarine wouldn't have been practical for the mission. By then, Musk was in China. He received a text from Boucher, better known as Grimes, alerting him to the statement and warning that the media was turning negative against him. He reached out to his staff: "I just woke up in Shanghai. What's happening?" As the team tried to figure out who Osottanakorn was, Sam Teller, Musk's chief of staff, weighed in with an email: "He's the fucking regional governor who has ignored our calls."

Musk couldn't let the perceived slight go. He wrote back: "We need to go all out and make this guy retract his comment."

The sentiment only grew worse. A couple days later, a British man named Vernon Unsworth, a spelunker who helped rescuers with his knowledge of the caves, was interviewed by CNN. In a passing question, he was asked about Musk's submarine. He called it a PR stunt and said that "it had absolutely no chance of working" and that Musk had "no conception of what the cave passage was like." He said Musk could "stick his submarine where it hurts."

The video clip quickly began making the rounds on Twitter. By July 15, Musk was furious, attacking Unsworth in a series of messages on Twitter that included this one: "Sorry pedo guy. You really did ask for it." When another Twitter user noted that Musk was calling Unsworth a pedophile, Musk responded: "Bet ya signed dollar it's true."

This didn't create a Twitter storm. It caused a category 5 hurricane. Shares plunged 3.5 percent, wiping out $2 billion of valuation. James Anderson of Baillie Gifford, one of Tesla's largest investors, weighed in during an interview, calling the event "a regrettable instance" and saying Tesla needs "peace and execution." Major news outlets began reaching out to Tesla's communications department to ask if Musk was, in fact, calling Unsworth a pedophile. The team closely monitored the coverage, following more than two dozen headlines from the BBC to Gizmodo. One aide wrote a memo analyzing the situation: "Media continue to cover E's tweet, with some stories mentioning that the 'outburst' comes 'just a week after he said in a Bloomberg interview that he would try to be less combative on Twitter.'" It went on to say that a number of investors and analysts "believe his comments are adding to their concerns that he's distracted from Tesla's main business."

A Reuters opinion piece summarized the dilemma that Tesla's board of directors faced: "Firing him for his 'pedo guy' tweet could precipitate a crisis of confidence among investors. Unlike the eventual ouster of [CEO] Travis Kalanick at then–privately held Uber, for a capital-hungry public company that might prove terminal. Directors should consider stripping him of either the chairman or chief executive title."

Early the next morning, July 17, Teller, the thirty-two-year-old chief of staff, tried to reason with Musk, saying it was time to apologize. He told him that he had talked with all of the people Musk held in deep regard—board member Antonio Gracias, CFO Deepak Ahuja, general counsel Todd Maron (his former divorce lawyer), and others—and they all agreed that an apology and a break from Twitter "sets you back on the right path internally and externally." Teller had even taken the liberty of writing out what an apology letter could say. He told his boss: "Everyone will love and respect you more for openly admitting the mistake and showing how much you care about your employees and the company mission."

An hour later, Musk responded. "After sleeping on this, I'm not happy about the suggested approach." Musk worried an apology offered so quickly after Tesla's shares had dropped would be dismissed as disingenuous and cowardly. "We need to stop panicking," Musk said.

Later that night, however, Musk relented. In another tweet, he said: "My words were spoken in anger after Mr. Unsworth said several untruths & suggested I engage in a sexual act with the mini-sub, which had been built as an act of kindness & according to specifications from the dive team leader."

Around the same time, Montana Skeptic's real-life identity—as investment manager Larry Fossi—began circulating on Twitter among Tesla supporters, who posted personal information about him. Bonnie Norman, the early Roadster owner turned investor and Tesla evangelist, was relaying to Musk and Maron intel she had helped gather. In an email, she noted that an anonymous group of Tesla investors had managed to crack the case after Fossi posted a photo of his Montana home, which they scraped for metadata. It gave them the photo's location, allowing them to ID Fossi. "I just burst out laughing when I heard," she wrote in an email with the subject heading "they're not so smart." It was disclosed that Fossi worked for billionaire Stewart Rahr, who was renowned for his midlife playboy antics.

"Wow, really interesting," Musk wrote Norman at 1:22 a.m. on July 6. Musk knew Rahr. "He bought some early Model S cars, got my direct number somehow, started stalking me, left long drunken messages on my vm and then got really [mad] that I didn't want to hang out with him."

He reached out to Rahr. According to Fossi, Musk had a message for his boss: If Fossi continues to write about Tesla, he would sue him and drag Rahr into it. The next day, Fossi announced he was retiring his blog: "Elon Musk has won this round. He has silenced a critic."

As the Twitter winds swirled, Tesla's executives kept their heads down. They had met the goal of churning out five thousand Model 3s in a single week at the end of June, but re-creating that and more, as Musk promised, was proving an uphill battle. The delays meant Tesla hadn't been generating the kind of sales it had hoped for to keep the business afloat. Cash on hand had fallen to $2.24 billion at the end of June. Now Tesla not only needed to increase sales, it needed to cut costs—and quickly.

In an unusual move that signaled how dire things were getting, Tesla began asking some of its suppliers to refund a portion of what the car company had already paid them. The request, sent the same week that Musk was grappling with Unsworth, was couched in a bid to help the company turn a profit in 2018. A manager charged with sending out the request said an immediate discount or rebate was the most obvious way to help. "This request is essential to Tesla's continued operation," the memo read.

It set off immediate alarm bells among some of those suppliers, who were growing increasingly wary of Tesla. It wasn't the first time they had seen such a tactic from a carmaker. They had grown accustomed to it in the dark days leading up to General Motors' bankruptcy.

The gravity of Tesla's financial situation weighed on Musk. At one point, he thought aloud about how Apple and its war chest of $244 billion might help. By all accounts, the iPhone maker's efforts to develop a car had been a struggle. Musk's supposed bravado with CEO Tim Cook years earlier, when Tesla was last in trouble, may have ended a possible acquisition.

This time, with his hat in his hand, Musk reached out to Cook about meeting for a possible deal. Perhaps Apple would be interested in acquiring Tesla for about $60 billion, or more than twice its value when Cook had originally inquired? A back and forth began between the two men's camps to find a time to meet, but it quickly became clear that Cook's side was dragging its feet, seemingly uninterested in finding an actual meeting time, a person familiar with the situation said. Instead, Apple hired back Doug Field, fresh from his work on the Model 3, to help guide its own car program.

Musk awoke at one of his five Los Angeles mansions on August 7 and was greeted with a story in the **Financial Times** that revealed something that had been quietly brewing at Tesla. Saudi Arabia's sovereign wealth fund had taken a $2 billion stake in the company, instantly making it one of the car-maker's largest shareholders. Minutes later, as Musk headed to the airport to fly to the Gigafactory in Nevada, he typed out a fateful message on Twitter: "Am considering taking Tesla private at $420. Funding secured."

It was the kind of half-cocked, uncensored mes-saging that Musk was known for—precisely what made his Twitter account a must-read for tens of millions of people, fans and detractors alike. Musk wasn't remotely prepared for the onslaught that these nine words would bring.

The reaction on Wall Street was instantaneous. Shares, which had already been rising, soared. Musk arrived at the Gigafactory almost giddy, asking managers if they knew what 420 stood for? It was a marijuana reference, he told them. He laughed.

Normally, a company alerts the NASDAQ ahead of the maneuver Musk was now casually propos-ing, and trading is halted. It isn't a courtesy, it's the trading exchange's rule; companies are supposed to notify the exchange at least ten minutes before any news that might create significant volatility in the stock price, such as an intention to go pri-vate, so trading can be stopped to allow investors to digest the new information. The announcement caught them off guard—Tesla hadn't said a thing. NASDAQ officials frantically tried to reach their contacts at the company.

Little good that did. Tesla's head of investor relations was caught off guard, too. He messaged Musk's chief of staff, Sam Teller: "Was this text legit?" Reporters reached out, too. "Quite a tweet! (Is it a joke?)," one wrote. Another emailed Musk directly: "Are you just messing around?"

About thirty-five minutes after the tweet, CFO Deepak Ahuja texted Musk: "Elon, am sure you have thought about a broader communication on your rationale and structure to employees and potential investors. Would it help if Sarah [O'Brien, head of PR], Todd [Maron, general counsel] and I draft a blog post or employee email for you?" Musk said that would be great.

He seemed to be going about his day as if it were just another one. Its highlight was a scheduled dinner with senior executives back in Silicon Valley. During moments of free time, he sent more tweets. Almost an hour after the first, he wrote: "I don't have a controlling vote now & wouldn't expect any shareholder to have one if we go private. I won't be selling in either scenario." Twenty minutes later, he added: "My hope is *all* current investors remain with Tesla even if we're private. Would create special purpose fund enabling anyone to stay with Tesla. Already do this with Fidelity's SpaceX investment." More than two hours after the first tweet, he elaborated on his reasoning: "Hope all shareholders remain. Will be way smoother & less disruptive as a private company. Ends negative propaganda from shorts."

The next day the Securities and Exchange Commission opened an investigation.

Musk and Masayoshi Son of SoftBank may not have hit it off, but after the heady dinner that Musk's friend and investor Larry Ellison had assembled at the Fremont factory in March 2017, the Saudis had kept in touch with Musk. As his war with the short sellers moved through July 2018, the fund requested a meeting. Musk met with them on Tuesday evening, July 31, a day ahead of announcing Tesla's second-quarter results, in which he promised to continually turn a profit going forward, and a week before the tweet heard around the internet. Teller and Ahuja joined him for the short meeting.

In hindsight, their exchange was open to interpretation. The Saudis informed Musk that they had bought up a stake in Tesla from shares on the open market amounting to almost 5 percent of the company, just shy of the threshold at which they'd have to publicly announce their position. As Musk and Yasir Al-Rumayyan, the managing director of the fund, had discussed at their dinner a year earlier, they raised the prospect of taking Tesla private. Musk didn't get into the details of how such a deal would be structured. The Saudis wanted a Tesla car factory built in their home country, a prize several Middle East countries had been vying for over the years. After about half an hour, according to Musk, Al-Rumayyan left the ball in Musk's court: Let us know how you want to do a going-private deal, and as long as the terms were "reasonable," it could happen.

Musk thought on it. On Thursday, the day after announcing second-quarter results, he watched the stock rise 16 percent, giving Tesla a market value of $59.6 billion. He worried that if Tesla kept growing in value, he might miss his chance to go private. After the market closed, he sent his board of directors a memo. He was tired of Tesla being targeted with "defamatory" attacks by short sellers. The constant barrage was harming the Tesla brand. He wanted his proposal for going private put to shareholders as quickly as possible and said the offer would expire in thirty days. He'd come up with a price of $420 a share, which would value Tesla at about $72 billion, based on a 20 percent premium of where the stock closed that day. (An actual 20 percent increase would've been $419, but he thought it would make his girlfriend laugh if he went with $420.)*

To make matters more complicated, he told the board in a special meeting called the next night that he wanted to keep the door open to existing

---

* Musk later admitted to the SEC that picking a price as a pot joke for his girlfriend was "not a great reason."

investors. Essentially he wanted to let investors remain with a newly private Tesla if they wanted to, and to buy out those who didn't. Small retail investors, such as Bonnie Norman, had been some of his biggest boosters over the years; he wanted to keep them in the fold.

Some on the board were skeptical, but it approved his contacting some of the larger investors about a deal. He would report back.

The Monday after the emergency board meeting, Musk called Egon Durban at the private equity firm Silver Lake about the deal. The firm, which included Larry Ellison as an investor, had carved out a high profile in Silicon Valley. In 2013, they helped Michael Dell take the company he founded, Dell, private in a $25 billion leveraged buyout. Durban cautioned that Musk's hope of allowing all current investors to stay with the company would be unprecedented. He said the number of remaining shareholders needed to be below 300. Tesla had more than 800 institutional shareholders alone. There were also countless small investors, the kinds of people who had excitedly shown up at Tesla's annual meetings.

It was the next morning, with this caution still fresh in his ears, that Musk sent out his explosive tweet.

—

His approach wasn't inconsistent with the way he had run Tesla for much of its history: Announce something, then figure out how to make it happen. The problem here was that as the CEO and chairman of a publicly traded company, his public statements about Tesla took on greater weight. It's a prosecutable crime to say something about your business that you know to be false. Given that his announcement on Twitter seemed so off-the-cuff, and the scant details that emerged seemed not fully thought through, suspicion was immediately ignited. Companies normally introduce such deals only after extensive vetting, with statements well parsed by lawyers. Tesla was racing to put together a team to evaluate the deal, **post**-announcement.

A week after the tweet, the board said it would set up a committee to consider a deal. It was helmed by Brad Buss, who had joined in 2009 and spent a brief stint as SolarCity CFO; as well as Robyn Denholm, who had helped steer Tesla through its SolarCity acquisition. They would be accompanied by a new board member, Linda Johnson Rice, a Chicago media executive who had been added the previous year following complaints that the company had too few directors without deep ties to Musk. They began looking to hire lawyers and advisers.

Musk tried to put out the fire he'd set; instead he fanned the flames. He wrote a blog post indicating that the details of the deal were far from complete, that all would come in due time. "It would be premature to do so now," Musk wrote. "I continue to have discussions with the Saudi fund, and I also am having discussions with a number of other investors, which is something that I always planned to do so since I would like for Tesla to continue to have a broad investor base." He tried to explain why he'd broadcast an idea that was only half-baked. "The only way I could have meaningful discussions with our largest shareholders was to be completely forthcoming with them about my desire to take the company private," he wrote. "However, it wouldn't be right to share information about going private with just our largest investors without sharing the same information with all investors at the same time." As for the "funding secured" part of his tweet, he explained that he'd met with the Saudi fund at the end of July, when he "expressed his support for funding a going private transaction for Tesla at this time."

He concluded that "if and when a final proposal is presented," the company's board would consider it and, if approved, shareholders would get a chance to vote.

The post only threw Wall Street into greater confusion, sending shares plummeting. Rumors began circulating that Musk might not be able to survive this latest misstep. Influential **New York Times** business columnist James Stewart heard that Jeffrey Epstein, the disgraced financier who had pled guilty to a sex crime involving a teenage girl, was compiling a list of candidates for Tesla chairman, at Musk's behest. It was a wild claim, in the midst of an unbelievable period. Stewart reached out to Epstein about the rumor, then found himself at the financier's Manhattan home on August 16 for an interview on the condition it would be "on background," meaning the information could be reported on but couldn't be attributed directly to Epstein. Stewart found Epstein evasive.

The newspaper reached out to Musk as well, and Musk went ballistic when he heard. "Epstein, one of the worst people on Earth, actually told NYT that he was working with Tesla and me on the take-private," Musk fumed to Juleanna Glover, a high-powered PR consultant in Washington, D.C., brought in to help him navigate the media, "and, under that guise, confided in them 'concerns' that he had about me. That was incredibly creepy and diabolical." He got on the phone with the **Times** to deny Epstein's claim, but once they were connected, Musk proceeded to self-implode for an hour, recounting all of the hardships he'd faced in recent months trying to get the Model 3 out—nearly missing his brother's wedding, spending his birthday on the factory floor.

The resulting headline read "Elon Musk Details 'Excruciating' Personal Toll of Tesla Turmoil." It described him as emotional, having "choked up multiple times" during the interview. He talked of his use of Ambien to combat sleep troubles; some board members were said to be worried about this. They claimed it contributed to his late-night Twitter sessions.

As if all of this noise weren't enough, Musk's celebrity profile and his budding relationship with Boucher added to the cacophony. Around this time, rapper Azealia Banks, who had a reputation for public beefs, took to Instagram to complain about a soured deal between her and Boucher to collaborate on music. First, she claimed to be holed up at one of Musk's LA homes waiting for days for Boucher, then she seemed to suggest Musk had published his take-private tweet while on acid.* When asked by a reporter to elaborate, Banks said she'd been at Musk's home the weekend after he posted the infamous tweet, as he tried to undo the damage. "I saw him in the kitchen tucking his tail in between his legs scrounging for investors to cover his ass," she said. "He was stressed and red in the face."

Musk just wanted it all to end. "Don't they have something else to write about?" he asked his PR consultant. "It is so tiresome to see myself in the news!"

Musk's behavior over the past year had troubled close observers, but his latest missteps catapulted him and his woes into the mainstream. Investors freaked out; shares plunged almost 9 percent the

* On occasion, employees over the years had suspected Musk might be high.

day after the **New York Times** interview was published. Wall Street analysts began pulling back their expectations for the company, telling investors they thought the stock was overvalued. Tesla's board, close allies of Musk, was put in a tough position. They could be held liable if they turned a blind eye to his latest episode.

The following Saturday, they held a conference call. From LA, Musk and his brother, Kimbal, who had been working in the background to try to help with PR hits, dialed in. Musk conveyed news that his hope of keeping small investors might not pan out. Musk's team of advisers at Silver Lake, as well as others from Goldman, wrestled with how to make the numbers work. One of Musk's chief assumptions had been that large shareholders would stick with Tesla, even as a private company. It was a naive belief. He learned that mutual funds, due to regulatory requirements, would be forced to trim their stake. His plan had assumed two-thirds of shareholders would follow him; if the likes of Fidelity and T. Rowe, which owned a combined 20 million shares, couldn't go along, they'd have to be bought out at $420 a share. Or put another way, Musk would need to muster yet another $8 billion to bring his plan to fruition.

It wasn't just a matter of finding the money. Musk had gotten pushback from inside Tesla. There were those who questioned what it meant for an electric car company, one that had set out to make the gas-guzzling car extinct, to be taking money from a large foreign oil provider. Meanwhile, the Saudis were unhappy with how the proposed idea was rolled out. They'd never made a formal proposal. (The fund's leader, Al-Rumayyan, would later tell government lawyers that he hadn't agreed to a deal with Musk.) The idea of going private was dividing their senior leadership, especially as Musk's behavior raised questions about both his stability and his central role in the company.

This left Musk's advisers looking for other deep pockets to take the Saudis' places, including possibly Volkswagen AG. The team presented him with a plan to raise as much as $30 billion. But new investors of that scale would inevitably want a say in how the company operated. This worried Musk. After all, wasn't the point of going private to limit outside influence? Musk was unhappy with some of the proposed investors, Volkswagen among them.

On Thursday, sixteen days after Musk had sent his going-private tweet, Tesla's board of directors flew to the Fremont factory to go through their options with a small cadre of advisers and lawyers. They presented their case, then left the room. With the board remaining, attention to turned to Musk. What did he think?

He said that based on the information he had gathered, he was withdrawing the proposal. Tesla would remain public. "In my opinion, the value of Tesla will rise considerably in the coming months

and years, possibly putting any take-private beyond the reach of any investors," Musk said in an
email after the decision was made. "It was now or
perhaps never."

It marked the end of two of the most unsettled
weeks in Tesla's history. But for as much as Musk
might have liked to simply retract his tweet and
move on, he had painted himself into a corner. Now
he had to convince the SEC that he had in fact had
funding secured for the deal when he tweeted it, that
he hadn't misled investors. Musk and other board
members were slated to give depositions under oath
to investigators in coming days. The SEC, led by
the San Francisco office, was moving quickly.

A less brash executive might have been chastened.
But amid all of this, Musk returned to Twitter. His
emotional outburst in **The New York Times** earlier
in the month had led to a debate about whether
female founders could get away with crying on the
job. "For the record, my voice cracked once during the NY Times article. That's it. There were no
tears," he tweeted at 8:11 a.m. on August 28. That
drew scorn from some. "Elon, your dedication to
facts and truth would have been wonderful if applied to that time you called someone a pedo," a
Twitter user wrote. Musk responded: "You don't
think it's strange he hasn't sue me? He was offered
free legal services."

By then, Unsworth had a lawyer, who chimed in:
Check your mail.

The tweets set off another round of media, piquing the interest of BuzzFeed reporter Ryan Mac. Mac
emailed Musk on August 29. Emails went back and
forth until Musk, a day later, began an email "Off the
record," then suggested Mac call people in Thailand
"and stop defending child rapists, you fucking asshole." Musk dug himself a deeper hole. "[Unsworth
is] an old, single white guy from England who's been
traveling to or living in Thailand for 30 to 40 years,
mostly Pattaya Beach, until moving to Chiang Rai
for a child bride who was about 12 years old at the
time.* There's only one reason people go to Pattaya
Beach. It isn't where you'd go for caves, but it is where
you'd go for something else. Chiang Rai is renowned
for child sex-trafficking. He may claim to know how
to cave dive, but he wasn't on the cave dive rescue
team and most of the actual dive team refused to
hang out with him. I wonder why . . ."

He then added: "I fucking hope he sues me."

Mac had never agreed to go off the record, abid-
ing by a long-standing tradition in journalism that
said a reporter and interviewee could agree to go
off the record only **before** an exchange. No similar
promise had been made. BuzzFeed published Mac's
story on September 4.

---

* Unsworth, in fact, didn't take a child bride. His longtime Thai
girlfriend was then 40.

Musk knew immediately that he was in trouble. Glover, the PR consultant, moved in D.C. political circles; she forwarded Musk an email from Jeff Nesbit, a politically savvy environmentalist, who offered assistance and voiced concerns about what Musk's Twitter rants might mean for the company: "One or two more of these and I can guarantee that there will be a no confidence vote on the BOD [board of directors]."

Musk wrote back that he knew it was "extremely bad." He had intended only for BuzzFeed to investigate the guy. "I'm a fucking idiot," he concluded.

Glover suggested he do an interview, on the record, "to kill this nonsense, speculation around your mental state." She wanted to get him out in the public again, presented in a way that made him look decisive, droll, and self-aware. Musk suggested comedian Joe Rogan's podcast, **The Joe Rogan Experience.** Rogan, a standup comic, Ultimate Fighting Championship (UFC) commentator, and the former host of TV's **Fear Factor,** had carved out a wildly popular corner of the media landscape interviewing thought leaders, academics, and celebrities, as well as strident voices whose extreme positions most media refused to touch.

Two days later it was arranged. Glover advised Musk that Rogan's interviews can run several hours. "Joe doesn't interrupt much so he will let you roll (he is funny and curses on air as no FCC rules for podcasts)," she prepped Musk. He needed to figure out with his lawyers what to say if Rogan asked about the ongoing SEC investigation. Also, if he asked about Unsworth, she pleaded, avoid answering. "Pls, pls, pls, pls if the Thai diver comes up pls just say you think you have gotten into enough trouble on that already and are not going to say more," she wrote.

The live interview, streamed on YouTube, began late on the West Coast. Musk, dressed in a black T-shirt that read "Occupy Mars," seemed in good spirits. In many ways, Rogan was the perfect interviewer for Musk, allowing him to talk at length about his interests, from space travel to tunnel digging. As the night wore on, Rogan and Musk began drinking whiskey. Near the end of the nearly three-hour interview, Rogan lit what he said was a marijuana-tobacco blunt and asked if Musk had ever smoked marijuana before. "I think I tried one once," Musk said laughing. "You probably can't because of stockholders, right?" Rogan asked.

"I mean, it's legal right?" Musk said from the California studio.

"Totally legal," Rogan replied. He then handed the spliff to Musk, who took a puff. The conversation turned heady. Rogan wondered about the role of inventors in furthering society. What if there were a million Nikola Teslas? he asked. Musk said things would have advanced very quickly. Right, Rogan added, but there's not a million Elon Musks. "There's one motherfucker," Rogan said. "Do you think about that?"

Musk checked his phone.

"You getting text messages from chicks?" Rogan asked.

No, Musk said. "I'm getting text messages from friends saying, 'What the hell are you doing smoking weed?'"

When the Saturday edition of **The Wall Street Journal** landed the next day, its front cover featured a picture of Elon Musk in a cloud of smoke, holding the blunt. To those watching, it wasn't clear that Musk or the company he ran would ever exit the haze.

CHAPTER 27

THE BIG WAVE

I t was a jarring image, omnipresent in the media in the days that followed. But Musk had little time to dwell on his latest PR disaster. With three weeks before the third quarter of 2018 ended, time was ticking. Twenty-three days to save Tesla.

Three months earlier, Musk had raced to pull off a dramatic achievement, reaching his goal of five thousand Model 3s in a week, but none of it mattered if the company didn't keep building vehicles at that rate, not to mention selling those vehicles to customers. Musk had promised a profit, and after so many broken promises, he was obsessed with delivering.

By August, Tesla's extra cash had fallen to a bare minimum of $1.69 billion—putting the company right on the brink of having enough money to do business. Internally, Musk was pushing the team

to deliver 100,000 vehicles in the third quarter—
roughly as many as the company had sold in all of
2017. It wasn't clear if the Fremont factory could
even make such an amount, especially as it struggled to pump out vehicles free of defects. One of
the latest pinch points was the paint shop, which
had been stung by several fires earlier in the year.
Board member Antonio Gracias was trying to find
a fix while the sales team raised the price of red-
paint models—the most troubling variety for them
to create.

Musk's plan counted on the company delivering
almost 60 percent of its vehicles in the final weeks
of September. Cars were being carefully timed for
their arrival with customers, at which point they
could be marked as "sold." Those destined for shipment to the East Coast would be manufactured
earlier in the quarter, to accommodate their longer
delivery times. West Coast cars would be made only
after those bound for faraway markets. Both would
be coordinated to land just before quarter's end,
so they could be billed toward that period's earnings. The process was known internally by some
as "the wave," for how it spread cars out to customers all at once. But this time, its scale had grown so
large, so quickly that a big wave threatened to crush
the company.

Another consequence of this approach was that
the company's success or failure—whether they

would meet or miss their quarterly targets—could only really be known in the final days of September.

Adding to the pressure Musk was feeling, the Securities and Exchange Commission, which was investigating his claims of having secured money to take Tesla private, was eyeing the end of the quarter for its own purposes. That was the end of the SEC's fiscal year. Musk's lawyers had gone to see government lawyers just hours after his appearance on the Rogan show to inquire about a potential deal in hopes of avoiding litigation. If they could reach an agreement, the government would be predisposed to finalize it by month's end, to count any fine levied on Tesla in its own year-end total.

But on the Saturday after Rogan, Musk's attention was on deliveries. He relocated his roving desk from the Fremont factory to a Tesla delivery center about two miles down the road, where he held nightly calls with Tesla managers across the U.S. In an office park of startups and other businesses, the front entrance of the Fremont center looked like any of the welcoming stores first created by George Blankenship seven years earlier. But the expansive back end housed what was essentially an assembly line, where customers on one side were matched with vehicles on the other. Delivery centers just like it around the U.S. were at the center of Tesla's current woes. Similar to when it had attempted to turn a one-dollar profit in 2013, building the cars was only half the problem. They needed to deliver at a scale that they were unaccustomed to. For Tesla 2013's near collapse had come as a surprise when deposits weren't converting into sales; this time around the team had fought in advance to ward off Delivery Hell 2.0.

This time, though, they would face a nightmare of Musk's own making, one that resulted from a slate of impossible choices. Before Jon McNeill quit as president in February, his team had been fashioning a plan to handle the massive growth of deliveries that they targeted for year's end. It was expensive and extensive, calling for a consolidation of deliveries to large regional centers, maybe twenty-five or thirty globally, at the cost of hundreds of millions of dollars. As Doug Field and JB Straubel struggled to ramp up production in the first half of 2018, however, Musk looked at the books and concluded Tesla simply couldn't afford the sales and delivery teams' plans. He told them to come up with another solution. Entrepreneur Dan Kim, whom McNeill had recruited to oversee global sales, began working on improving the company's online sales process, in hopes of encouraging more sales through the Web and the company's smartphone app (and diminishing their reliance on stores).

Tesla wasn't the same brand it had been in 2013. Back then, the Model S represented a challenging sell to buyers uncertain about an upstart carmaker offering an untested technology. After the success of the Model S, however, those concerns weren't as prominent for the Model 3. Those buyers still needed some handholding, though, especially when it came to financing and trading in current vehicles. Kim worked to beef up call centers, where in-house sales teams would take the lead in closing deals. Tesla wasn't going to make the mistake it had made with the Roadster and the Model S, assuming reservations would automatically convert into sales.

Cayle Hunter had been hired in January 2018 to oversee the new in-house sales and delivery team based in Las Vegas, in an old SolarCity office not far from the Strip.* Other, smaller teams were based in Fremont and New York. Their job was clear: close deals. They worked their way through a list of 500,000 people who had put down refundable $1,000 deposits for the Model 3.

In Hunter's first eight months on the job, he had successfully grown his group of 35 to 225. At first, they had had no problem finding buyers for the cars inching their way through the Fremont factory. Customers of previous Tesla vehicles who had put money down were placed at the front of the line. Hunter's team found it took little effort to convert them. The question they asked was not **why** buy,

---

* Despite Musk's public claims that he'd boost SolarCity's business, after the company's acquisition in 2016, he'd had little time or resources to carry them through. Instead, most of what remained of the solar business was now being directed to help either build or deliver the Model 3.

but **when**: When could they get their hands on the Model 3? This wasn't sales, Hunter thought to himself in those early months. The sales goals he'd been set for the first two quarters of 2018 had been daunting, but the targets dwindled—as it became increasingly clear that the factory couldn't churn out enough vehicles to meet those quotas—and the numbers seemed less foreboding.

By summer, however, that had changed. Musk had effectively untangled the snares at the Fremont factory, and production was steadily increasing. Hunter's team was shifting from more or less effortlessly slotting a car with an eager customer to true, hard salesmanship. As they made their calls, more and more often they found resistance on the other end of the line.

Part of the success of the 2016 reveal had stemmed from Musk's promise that the Model 3 would start at $35,000. The actual sales price in August 2018 wasn't anywhere near that. The cheapest version started at $49,000 with a top-end, performance version going for around $64,000. As the team went through its backlog of reservations, it was becoming clear that many had put down money with the intent of buying **only** a $35,000 car. For some it was a stretch to get into a car even at that price.

There were others who could be upsold; Hunter's team would talk about how they could take possession of the car now rather than wait—maybe another year?—for the $35,000 version. Plus, the buyer could take advantage of the $7,500 federal tax credit that would begin phasing out the following January. The sales force would hit talking points about how the cost of Tesla ownership was less than for a traditional car because it didn't require trips to the gas station. It often worked, but it wasn't proving as easy a sell as in those early days.

During this period, Musk told the team that it was time to capitalize on those buyers willing to spend up to $65,000 for a car. There were only so many people willing to do so, and early next year it would just be harder, as the most eager customers had already been served a car. Tesla currently couldn't make a profit on the $35,000 Model 3, Musk said; each delivery at that cost would lose the company more than $1,000. This was how Tesla was going to be profitable in the third quarter: flood the market with high-end versions of their "consumer" car, with their higher profit margins.

For all its growth, it still wasn't clear the factory could churn out its targeted 100,000 vehicles in that time. And even if they did, Tesla lacked the space to deal with that much inventory. To help handle the wave of deliveries they knew was coming, the team kept a list of four thousand employees who had volunteered to be dispatched to overloaded delivery centers around the country. But they found themselves hamstrung, waiting for the go-ahead.

Early that quarter, as Musk focused on the Thai soccer team and taking Tesla private, it seemed to some senior managers that CFO Deepak Ahuja was holding back the resources needed to prepare for deliveries, raising questions in their minds about whether this would actually be the quarter that Tesla again mustered a profit, or if the c-suite was playing some kind of game, planning to stack the sales in the fourth quarter instead. Musk's presence in the Fremont delivery center that September, however, signaled that deliveries were now the company's top priority.

But even with the extra hands, the delivery centers each had a certain number of slots for cars on-site. Each was booked for an hour. Across the U.S, Tesla might have a total of 100,000 slots available in the third quarter, if calculated using every working hour and day for a three-month period. The problem was that the vehicles would only arrive in the second half of the quarter—thousands and thousands of deliveries needed to occur before the final day of the reporting period. Model 3s were being parked anywhere they could find room: parking garages, rail yards, shopping centers. Short sellers began taking notice, posting photos on social media and postulating theories of defective inventory being hidden away.

They weren't entirely wrong: many cars needed fixes before going to customers. In Marina del Rey, in Southern California, one of Tesla's most important delivery centers came under pressure during Musk's nightly call. Customers there had begun complaining about their cars on social media. Musk was displeased. He threatened to begin firing people at the center if he heard any more complaints about defective vehicles. Hunter's boss, Kim, stopped sending cars out until they could fix paint defects (even as Jerome Guillen asked him if he knew how much money he was costing the company). Kim deployed crews to Marina del Rey to redo body panels and fix paint jobs, hiring outside contractors to help.

Unlike the factory teams, the sales and delivery teams had been relatively sheltered from Musk that year. No longer. Musk's (mostly) nightly conference calls with sales leaders around the country, scheduled around his calendar and often late at night for East Coast managers, were high pressure, with commands often coming with implicit—if not explicit—threats of being terminated for failing. One such call occurred for Hunter that summer as Musk continued to speed up the hand-off process.

Is Vegas on the line? Musk asked, directing his question at Hunter. **How many people did you sign up for pickup today?** This was Hunter's big moment: His team had just that day scheduled 1,700 people to pick up their Model 3s in coming days—a record—and he was proud to announce the achievement.

Musk wasn't pleased—he ordered him to more than double the result the next day or else he'd take over. On top of that, Musk said, he heard that his team had been relying on calling people directly to schedule their appointment to pick up their car. That stopped now. Nobody likes phone calls, they take up too much time, Musk said. Text customers instead. That would be faster. If he heard about any calls being made tomorrow, Hunter was fired.

A sense of panic hit Hunter. His wife and kids had only finally joined him in Las Vegas; they had just finished unpacking their boxes. Now Musk was threatening to fire him if he didn't do the impossible in twenty-four hours? The sales organization didn't have hundreds of company cell phones for his team to use for texting, and they didn't want their employees using their own personal phones. The company had built a system to track customer interactions, avoid miscommunications, and ensure sales leads were followed up on. He would have to work around it.

Overnight, Hunter and other managers pieced together a solution, employing software that allowed his team to text from their computers. They stopped the practice of walking customers through the reams of sales paperwork that would eventually need to be completed and signed. If Musk's goal was to have people in a queue to pick up their cars, then that's what they would do. They'd just start assigning pickup times for customers. For example: **Can you come in at 4 p.m. on Friday to get your new Model 3?** Often, Hunter didn't even wait for a response before putting a customer on the list for pickup. If a customer couldn't make it, they might be told they would lose their spot in line for a car that quarter. Customers became more motivated to pony up the personal information needed to finalize a sale when there was a Model 3 dangled in front of them. Hunter's team began telling customers to have it all completed forty-eight hours prior to delivery.

The team began racing through their list of customers, assigning random times at pickup centers around the U.S. By 6 p.m. the next day, they had reached 5,000 appointments. Hunter gathered the team to thank them for their work. He fought back tears. He hadn't told them that his job was on the line; all they knew was that it was super important to schedule a bunch of deliveries. That night on the call, Hunter reported the results to Musk.

"Wow," Musk said.

It was a major breakthrough, one that would be looked back on by some senior managers as a defining moment for the quarter. But there was little time for celebration. They moved on to the next fire. With delivery centers packed, Musk wanted to begin delivering Model 3s directly to customers' homes. The company had already developed a system for finalizing sales remotely; it was how they had worked around laws in Texas and other states, where franchise dealers had been victorious in keeping Tesla stores out. Customers at "galleries" were directed to computers to contact the company about buying a car. Salespeople in Las Vegas or elsewhere would follow up to close the deal. The company had come up with a packet of documents that needed "wet signatures," which would get sent to a customer's house along with an overnight envelope and instructions to return it with a check within two days. The cars would then be shipped from southern California to Texas for delivery. As the pace grew more frantic, Hunter began shipping as-yet-unsold cars to Texas on the bet that he'd have the check in hand from the buyers of those specific vehicles by the time the truck crossed the Texas state line. If the timing was off, Hunter had to pay for the cars to be shipped back for redelivery.

A small percentage of sales had so far gone through this direct process. Now Musk wanted 20,000 cars in the third quarter delivered directly. In theory, it would save money on having to expand delivery centers; in practice it would require an army of people to physically take the vehicles to customers. There was no way they were ready yet to do 20,000 home deliveries.

Kim, who had worked to improve the online buying process, turned to members of his sales staff who had worked at Amazon and Uber, for their expertise in tracking packages and hiring gig workers. Musk wanted cars delivered in covered trucks. Kim and chief designer Franz von Holzhausen huddled to develop the look of the carriers until it became clear it would be too costly and take too much time. Instead, Kim proposed to Musk that employees simply drive cars to buyers' homes and hand over the keys. Tesla's drivers would return to the office by calling an Uber or Lyft. Home delivery was unusual in the auto industry, and an uncertain proposition for some buyers.

There were other strategies for hastening delivery. Instead of spending an hour with each customer introducing the new Tesla owner to their brand-new car's features, Kim wanted his deliverers to aim for five minutes. They would tell customers to watch a training video instead. Some of the drivers were so eager to return that they would summon an Uber or Lyft ahead of their arrival at a customer's door—a clever time-saving ploy unless the ride-hailing car showed up before they did and knocked on the eager Model 3 buyer's door.

As they raced toward the end of the quarter, it became clear that the team had failed to anticipate how many trucks it would require to deliver an ever-increasing volume of cars to delivery centers. Third-party car carriers didn't have enough space for them. Managers had just assumed they could keep upping and upping their shipments as cars rolled off the line.

During one nightly call, Kate Pearson, who had recently been hired as head of customer experience and operations, spoke up. She'd spent thirteen years overseeing supply chains for the Army National Guard and joined Tesla from Walmart, where as a vice president she worked on the retailer's e-commerce business. She was deeply experienced in operations, and looking at the numbers, she had bad news for Musk. The company couldn't meet his goal of 100,000 deliveries this quarter. They were on pace for around 80,000.

Musk didn't accept that. He said it needed to occur. Within days Pearson was ousted. Musk told the nightly conference call of sales leaders that it wasn't because she didn't kiss enough ass, but was rather about her "fundamental inability to perform." In reality, she had given him an answer he didn't want to hear. He wanted to hear: **We'll do our best.** The managers had been conditioned against telling him the unvarnished truth.

On another occasion, one senior sales manager had had enough, after almost two years with the company. He put in notice that he was quitting. News of the decision found its way to Ahuja, the CFO, who didn't want to lose the sales leader and began trying to keep him in the fold. Musk, however, had the opposite reaction: rage. At the Fremont delivery center, he approached the manager, screaming profanities as he towered over him, telling him to leave. "I don't want anyone here who is going to quit on me during a time as important as now," Musk yelled, according to a person who watched. Musk followed him into the parking lot. It was a scene that was ugly and public enough that the board ultimately felt a need to investigate, amid accusations that Musk had physically pushed the manager.

Manhandling an employee, then, was added to the list of accusations against him, as his lawyers were busy trying to negotiate a deal with the SEC. Musk wasn't making that task any easier. The two sides believed they had reached a deal late September 26, and the SEC planned to announce it the next day. Early the next morning, though, Musk's lawyer called the SEC. Musk had changed his mind: the deal was off. He was worried about how a settlement might affect his ability to tap debt markets for SpaceX.

Stunned, the SEC rushed to the courthouse. After the markets closed, they filed a lawsuit formalizing their claim that Musk had misled investors when he announced that he had lined up funding to take Tesla private. The SEC's lawyers were asking a judge to ban Musk from ever running a publicly traded company again—banishing him from the leadership ranks of Tesla for life. The filings marked a dramatic turn, surprising investors and surely pleasing financial detractors. After the lawsuit was announced, Tesla's shares fell 12 percent, making short sellers, on paper, an estimated $1.4 billion profit.

Wall Street analysts had begun contemplating a Tesla without Musk, wondering aloud if there was a Musk premium built into the trading price. Others wondered if lenders would be as excited to continue loaning Tesla money without Musk's vision behind it.

There was, however, a dynamic working in Musk's favor through the SEC squabbling. He knew that Tesla's demise would hurt the SEC as much as it would hurt him. When the commission levies punishments on a company, it can hurt shareholders, which ultimately hurts the SEC. They are loath to use the full extent of their power. For that reason, many close observers doubted the SEC would follow through on a ban. What they wanted was to get Musk under control, to implement new safeguards that would keep him from pulling future stunts.

Musk's lawyers spent that night trying to change Musk's mind about his refusal of the settlement, even asking celebrity investor Mark Cuban to prod him into a deal. Cuban, the billionaire owner of the Dallas Mavericks basketball team, had had his own public battle with the SEC that dragged on for five years, after he was charged with insider trading. It was like a scene out of Showtimes' **Billions,** as Cuban counseled the beleaguered CEO, warning him that he would have a bruising, years-long fight ahead of him if he persisted. A settlement wouldn't be as damaging as a court battle.

Musk was conflicted. He believed he had had a verbal agreement with the Saudis and that the SEC was flawed in thinking a written agreement and fixed price in writing were required for a deal. Middle Eastern business deals routinely operate on verbal agreements in principle. Plus, Musk believed he could have led the deal to go private using his stake in SpaceX, which was now worth billions itself.

But at the end of the day, Musk could be pragmatic, especially when left with no other choice. His lawyers reached out to the SEC Friday morning asking if they could reconsider the previous deal.

The SEC had the upper hand now, and they were going to use it.

As the clock ticked down to the end of the third quarter and Tesla's outrageous sales goal seemed out of reach, Musk put out an unusual cry for help on Twitter. He asked his loyal customers: **Help us deliver vehicles.**

Longtime Tesla owner Bonnie Norman, now retired and living in Oregon, rose to the challenge. She wanted to see Tesla succeed; she turned up at the Portland delivery center. Others showed up at other centers, too. They focused on showing customers how to operate their new cars, and explained life with an electric vehicle, freeing up the paid staff to handle the overflow of paperwork. Musk and his new girlfriend Claire Boucher, the musician known as Grimes, worked at the Fremont delivery center, joined by board member Antonio Gracias. Kimbal Musk showed up at the store in Boulder, Colorado. It was truly an all-hands-on-deck moment.

Surrounded by friends and kin, Musk seemed at his happiest, one manager recalled. "It was like a big family event . . . He likes that—he likes loyalty."

That support was needed. After leaving the SEC at the altar, Musk's lawyers returned now seeking a final deal. They ultimately agreed to the SEC's new terms: Musk could keep his role as CEO but he was to give up his title of chairman for three years rather than two, as originally proposed. Musk personally would have to pay a $20 million fine, $10 million more than in the first deal. Tesla would have to pay a $20 million fine and agree to placing two new directors on its board. The company would also have to put in place a plan to monitor Musk's public comments. He wouldn't be allowed to tweet material information without prior approval. No more "funding secured" messages without a lawyer looking at it first.

The parties agreed. They announced it Saturday, September 29, and investors breathed a sigh of relief that could be heard throughout Wall Street on the first day of trading following the deal's announcement. Shares soared 17 percent that day, the biggest one-day change in a year of staggering volatility for the carmaker. (Happily for Musk, short sellers lost an estimated $1.5 billion on paper in that time.)

The company was ready to tabulate the quarter's final delivery results. It was close. Deliveries reached 83,500—a record that exceeded Wall

Street's expectations but that was 15 percent shy of the internal goal of 100,000. (It was also uncannily close to the estimate of Pearson, head of customer experience, who had seemingly been fired for suggesting it.) Almost 12,000 vehicles had been caught en route to customers, missing the deadline for the third quarter.

While short of Musk's goal, it was still an enormous achievement. It was also enough to push the company into profit—largely because many of those cars were the high-priced ones that Hunter's team had been told to upsell, but also because Ahuja, the CFO, had worked to delay payments to suppliers. The company's accounts payable—the money it owed to suppliers and others—rose 20 percent from the second quarter and 50 percent from a year earlier.

Effectively, Tesla was making its numbers work on the backs of its suppliers, a trick that big automakers had done for years and a sign of its new might. It wasn't pretty but it would register as a win to investors when the final figures were released in October: a $312 million profit. It was the largest the company had ever made, and it came as a surprise to Wall Street analysts who had been predicting a loss. The momentum continued into the fourth quarter, allowing the company to report in January its first back-to-back periods of profitability. During a call for investors and analysts, Musk sounded confident about the year ahead, saying he expected to make a small profit in the first three months of 2019 and then "for all quarters going forward." More than eight years since Tesla had gone public, investors could enjoy the blue skies long promised—or so they were being told.

For months, Musk had obsessed with the pricing options for the Model 3, tweaking things online as if he were crafting the perfect cocktail. The buyer profile for the compact car was different from that of the Model S. First, there was the price, which, depending on the options you picked, could be tens of thousands of dollars lower than for the luxury sedan. Then there were other facts of life: Model 3 owners typically depended upon the vehicle for daily use. They often required financing and needed to trade in their old cars. To avoid hiring a larger sales force, Musk had directed Dan Kim to make their online configurator, a tool for customizing each purchase, easier to use. It was supposed to be as close to one-stop shopping as you could get with a car.

Musk pushed Kim to build out a home-delivery team, too. While it didn't get them to Musk's goal of 20 percent of deliveries through direct customer delivery in the third quarter, it did in the fourth quarter, according to a person familiar with the figures. All of these steps were taken to reduce spending.

Musk had become obsessed with making Tesla profitable. He began wondering aloud about whether they could shut down all of the company's stores.

Internally, Musk warned his managers of a "dark winter," pushing them to slash cost and focus on boosting production by as much as possible. Tesla needed greater scale, he told them. His new attention to spending came as the sales team could see a slowdown coming in orders for the Model 3 in the first quarter of 2019. Largely thanks to Tesla's success, the U.S. federal tax credit for buying a fully electric car would begin to phase out on January 1, dropping to $3,750 from $7,500. Midyear it would go to $1,875. By year's end it would be gone. Effectively, the Model 3, which was already costly, was getting a price increase every six months at the very time when Tesla needed it to be cheaper.

Hunter's team in Las Vegas had spent the final three months of 2018 converting remaining deposit holders into buyers to achieve another record quarter of deliveries. It grew harder as the year dwindled, with the ranks of buyers with an appetite for the higher-priced version growing thinner and more and more customers holding out for the $35,000 model. His team was looking forward to taking some time off after closing the year. Some would quit altogether. They'd reached their goal—but also the end of their rope. Many, like Hunter himself, would be laid off, as Musk focused both on cutting costs and taking the Model 3 to Europe and China, where there were still early adopters presumably willing to pay for the more expensive versions of the Model 3.

It was a time of change for Tesla. Todd Maron, Musk's lawyer since his first divorce, was ready to venture over to a startup. Ahuja, the CFO, was ready to depart—again. JB Straubel hadn't given up the fight yet, but he was exhausted and in need of a vacation.

As part of the SEC deal, the board added two new members, including Larry Ellison, the longtime investor and Oracle co-founder. Robyn Denholm, who had helped guide Tesla through the SolarCity acquisition and the abortive attempt to go private, would take the reins from Musk as chairman, though it was clear to anyone watching that, title or not, Musk was still in charge. He couldn't help himself, noting on Twitter that he had deleted his titles from the company website. "I'm now the Nothing of Tesla. Seems fine so far." In a CBS News **60 Minutes** interview, he showed total contempt for the SEC, saying he didn't have anyone routinely reviewing his social-media messages. "To be clear," he said, "I do not respect the SEC"—which on Twitter he mocked as the "Shortseller Enrichment Commission."

Any question as to whether Denholm might rein Musk in quickly vanished. The board's investigation into his alleged physical behavior at the delivery center went nowhere. As he revealed the

Model Y compact SUV at the design center in late 2018, a vehicle he thought would ultimately outsell the Model 3, Denholm could be seen standing in the front row pressed against fans and customers, cheering Musk on. Asked later about his use of Twitter, she told a reporter: "From my perspective, he uses it wisely."

To offset the price increase that came with the federal tax credit's expiration, the carmaker decided it would lower the prices of all its vehicles. The Model 3 would now start at $44,000, down from $46,000 (though still a long way from $35,000). If Tesla thought the move would appease investors, they guessed wrong. Investors interpreted the move as an indication that demand was slowing, an ominous sign for a company whose narrative depended on bottomless growth, and which seemingly hadn't figured out a way to curtail costs. Shares fell almost 7 percent the day the price cut was announced.

Musk struck a reassuring tone, telling investors in January that interest in the car was high. "The inhibitor is affordability. It's just like people literally don't have the money to buy the car. It's got nothing to do with desire. They just don't have enough money in their bank account. If the car can be made more affordable, the demand is extraordinary." To further counter concerns, Musk took to Twitter to celebrate that Model 3s were being loaded on a boat and headed to Europe for the first time. He noted that Tesla "will make around 500k [cars] in 2019," a statement he followed up hours later with another tweet to say he meant an **annualized production rate** of 500,000, but that total deliveries for the year were still expected to be about 400,000.

It was the kind of boastful, unguarded message that his agreement with the SEC was meant to prevent. Regulators had already suspected he wasn't taking the deal seriously, especially after his performance on **60 Minutes.** The day after his latest tweet, they asked Tesla if anyone had approved it. Not surprisingly, the answer was no. Only after the tweet, Tesla's team said, did a lawyer help craft Musk's clarifying tweet. He argued that he didn't think he needed pre-approval, because he was simply repeating past statements. The SEC wasn't buying it. By the end of February, they asked a judge to hold him in contempt for violating his deal.

It was beginning to feel like the summer of 2018 again. And the drama wasn't over. A few days later, Musk announced that Tesla would be closing most of its stores to slash costs, enough so that the company could finally come through with its long-promised $35,000 Model 3. The move to online-(almost)-only sales was a dream he'd long held, but which his team had bucked at because of the challenges inherent in selling an electric car to first-time buyers.

In theory, it may have seemed like an easy way to slash costs; in reality, Tesla had hundreds of leases for stores around the world. It had a total of $1.6 billion in lease obligations, with most of that money due in the next few years. It couldn't just shut the lights off and save a bundle. Tesla "is a company with a viable balance sheet that is going to owe a lot of landlords a lot of money," Robert Taubman, chief executive of Taubman Centers Inc., said at a conference days later. His properties had eight stores, including the Denver location.

Investors had largely forgiven Musk's head-spinning antics over the years, largely because the company had continued to show impressive growth (even if it was devouring cash in the process). By April, their forbearance had nearly reached its limit. Then the bottom fell out of the Tesla fantasy.

Tesla reported a collapse of sales; deliveries fell 31 percent in the first three months of the year compared to the previous quarter, as the company struggled to find U.S. buyers and to get Model 3s into Europe and China fast enough to compensate. By mid-April, Tesla quietly pulled back on its tentative promise of a $35,000 starting price, now listing $39,500 as the lowest price, claiming that an off-the-menu $35,000 price **was** available—but only if a customer visited or called stores, stores that may or may not be open. It would finish the month posting one of its largest quarterly losses. It cautioned that the second quarter would be in the red, too. So much for blue skies.

Musk just needed the company to hang on a little longer. The arrival of the Model 3 in Europe would help bolster Tesla's results in a familiar way. Just as Tesla's first quarterly profit in 2013 was aided by the sale of regulatory credits in California to rivals who failed to meet vehicle emission goals, Tesla managers in Europe were quietly negotiating a new deal to pool sales there with Fiat Chrysler Automobiles so that the rival could avoid fines for violating tough new European Union emissions rules. The deal, which would be announced in the spring of 2019, would be valued at more than $2 billion spread out over several years, sprinkling pure profit onto Tesla's books when it would be needed most. And there was still China, Musk soothed, where the company was on track to begin production later that year.

Still, he acknowledged during an analyst call what was obvious to many: Tesla needed to raise more money.

The company was careening. In a few short weeks, it had swung from making money, with claims of profits into the distant future, to a carmaker that grappled with unnecessary, self-inflicted crises and that was running dangerously low on cash yet again. Finally, Wall Street lost its patience. Shares began to free-fall, reaching a low of

$178.97 a share in June, fully half its value at the start of the year. Short-seller bets were finally paying off, with paper profits estimated at more than $5 billion in the first half of the year (almost as much as they were estimated to have lost in 2016 through 2018).

Even longtime Tesla bull Adam Jonas, the Morgan Stanley analyst who was among the first to see its potential, seemed fed up with Musk. In a private call with investors, Jonas cautioned that Tesla wasn't a growth story anymore but rather a "distressed-credit story and restructuring story." In other words, a possible bankruptcy in the making. The company's debt had ballooned to about $10 billion, in part because of the SolarCity deal years earlier. That would have been palatable if Tesla had continued to grow and generate cash, and if it could maintain access to investors, something that had now been called into question. It would need to raise a massive amount of money, he warned, or "to be seeking strategic alternatives"— banker talk for a sale or merger. Three-fourths of the company's stock that week went to short sellers. Its debt had fallen to 85.75 cents on the dollar— a sign that debt holders worried about getting their money back from the beleaguered carmaker.

Worst of all, it seemed, Musk's plan to bank Tesla's salvation on China might come too late, as new auto sales in that nation fell for the first time since 1990 and relations between the countries grew frostier. Could Tesla have staggered its way through a marathon, fought off rivals and detractors, only now to be collapsing in sight of the finish line?

Or as Jonas put it, "Could there be a worse time to depend on China?"

CHAPTER 28

RED TIDINGS

On a cold January day in 2019, Musk stood in a muddy field on the outskirts of Shanghai. Dressed in a suit and overcoat, he was accompanied by his college friend Robin Ren, as well as Shanghai mayor Ying Yong, for a ceremonial groundbreaking. Tesla would be establishing its second assembly plant at this site, the company's first foray into building cars outside of Fremont. The three smiled for photos that would circulate around the world. The moment represented a win for China and a win for Tesla.

It ran the risk, however, of being short-lived: Tesla was on the precipice. The Model 3 was here now—it was on the road, no longer just a fantasy bandied about by a few Silicon Valley dreamers. And despite the setbacks that had plagued Musk in his summer of discontent—many of them self-imposed—the company was flirting with long-term profitability. But Musk hadn't carried the mission to its conclusion. He had as yet failed to make the Model 3 a truly viable car. To do that, he needed scale, which would drive down costs. And to achieve scale, he would need money—a lot of it. It all pointed him to one place: China.

On Twitter, Musk promised his factory's initial construction would be completed by summer, with Model 3 production rolling by the end of the year. To Tesla observers, it prompted the customary eyerolls: yet another totally unrealistic timeline for a seemingly impossible goal. It was only the latest in a series of bullish announcements, as Tesla tried to regain its footing following its roubled 2018.

But investors that fled the company that spring failed to fully understand (or else refused to believe) what had been brewing for years behind the scenes at Tesla.

It had begun with leadership. Musk knew that if he was going to carry out his global ambitions, he needed the right help. It had to be someone he could trust, someone who knew how he thought and who could represent him on the other side of the world, where Musk couldn't just drop in for a night of roll-up-your-sleeves engineering, or else camp out on a factory floor for a few days until an assembly line was unsnagged. To find the right person, Musk reached back—way back to his college days.

As a student at the University of Pennsylvania, Musk had grown frustrated when he wasn't setting the grading curve in his physics class (or so goes the story he told executives). He approached the professor to complain: Who could possibly be better than him? The answer: Robin Ren, a student from Shanghai. Musk confronted his academic rival only to quickly realize Ren hadn't just set the curve in his class—he had been among the best physics students in all of China, earning the then-rare opportunity to study in the U.S. The two outsiders became fast friends; so much so that they traveled together to California after graduation, when Musk thought he would attend Stanford but instead veered toward startup life. While Musk followed his path, Ren studied for a master's in electrical engineering at Stanford and went on to a career at Yahoo! and Dell, eventually rising to the rank of chief technology officer of Dell's flash drive subsidiary, XtremeIO. He was trustworthy and had the right experience. If Musk was brash, Ren was reserved. When Musk needed help restarting his China effort, after the company's disastrous turn in 2015, Ren took the helm.

Behind the scenes, Ren and Jon McNeill (head of sales before he departed in 2018) began with one of Tesla's most pressing missions: winning approval for an assembly plant in China. They were warmly received by Shanghai, which was excited to have another automaker in its fold; but the nation's law required Tesla to have a local business partner and, characteristically, Musk wouldn't budge.

Fortunately, Ren inherited two managers who had experience in navigating both construction sites and government hallways. Tom Zhu, who had been hired to build out Tesla's charging network in China, had a Duke University MBA and had run large construction projects in Africa; he was put in charge of overseeing the production of the new assembly plant. Grace Tao, meanwhile, handled government relations. A former China Central Television correspondent, she was widely known in Beijing power circles and helped the company navigate the bureaucracy. Tao's family ties went way back in the Communist Party, according to a person familiar with the dynamics, and she was showing an acute understanding of how to work the levers of government power. At Tesla, she hung a large organizational chart tracking the upper reaches of central government and senior leaders in key provinces. **Bloomberg Businessweek** cited unnamed employees who said that she once claimed she could get a message to President Xi Jinping by going through just one intermediary, "which would be an astronomical level of access in China" (a statement Tesla has denied). Nevertheless, her colleagues marveled at her sway. "Grace is really smooth," one said. "She knows exactly how to play the politics." Musk's growing public profile didn't hurt, either.

During a trip to Beijing in 2016, as he and colleagues sat stuck in a snarl of cars, he began talking about the need to fix the city's traffic congestion. He proposed an idea for digging tunnels beneath the city. It was the kind of thinking he was known for: His head would cock upward, his eyes seeming to access a download from the cloud. "What if we . . . ," he'd begin.* What followed might be a crazy whim, soon forgotten. Or it might be a massive undertaking, one that would eat up years and years of some manager's life. It was that kind of no-limits thinking that had gotten him to China to begin with, and that came again as he entertained serious conversations about entering the country without a partner—an idea that a team of lawyers had told him over the years was impossible.

By 2017, though, the outlook for such a venture was looking brighter. Late that summer, a deal was reached to allow Tesla to build a factory, and the Chinese asked if they could announce it soon, presumably in the fall, when President Donald Trump was expected to visit the country amid trade tensions, said people familiar with the talks and plans.

---

* By year's end, he had tweeted out a similar idea, which foreshadowed the creation of another Musk venture, the Boring Co. "Traffic is driving me nuts. Am going to build a tunnel boring machine and just start digging . . . ," he wrote, followed with another message: "I am actually going to do this."

But Musk said the deal couldn't go through yet. With Model 3 production snared, Tesla didn't have the money for the factory. Musk wanted the company to punt the announcement down the road, one of the people said.

As things dragged into 2018, however, questions began to arise about whether Tesla was ever going to achieve its ambition of building a Chinese factory without a local partnership. Those concerns coincided with intensifying trade talks between the United States and China. Ultimately, when the deal was announced, it allowed the Middle Kingdom to appear more open to working with U.S. business—at least when it suited them.

The terms of the final agreement were generous for Tesla in other respects—in some ways even more generous than what the company had gotten in Nevada. The China Gigafactory would sit on 214 acres of land granted by Shanghai, under a deal in which Tesla agreed to invest about $2 billion into the project. Separately, it received a loan of $1.26 billion on favorable terms from politically connected Chinese banks, money it would use to build the factory, plus another $315 million to pay for labor and parts. Tesla, in other words, had been given the chance to build a factory in China with the country's own money.

If the welcome was generous, it's because China needed Tesla as much as Tesla needed China. The country was eager to spark an electric vehicle market, and what better way to motivate rivals than to let Tesla in. The Chinese car market had been booming. Almost 40 percent of GM's vehicle sales in 2018 had come from the country, its largest market; Volkswagen AG depended heavily upon the country's buyers. This fact, paired with the country's tightening emissions regulations, was forcing global auto giants to hasten their preparations for an electric future.

The rush came even as their past offerings continued to falter, especially compared to Tesla's. GM announced it was killing off its Chevrolet Volt plug-in hybrid. Sales of the vehicle had never amounted to much, dwindling to less than 19,000 in the U.S. in 2018. GM's all-electric Bolt did even worse. Despite its advancements, it paled in comparison to the Model 3.

But the automakers had learned from their mistakes, and they were doubling down. Both GM and Volkswagen planned to concentrate their investment dollars on fully electric, not hybrid, cars, a tacit acknowledgment that Tesla's strategy had been correct all along. VW was planning for two-fifths of its sales to be electric by 2030. GM planned to introduce at least twenty electric models by 2023. Both companies were rushing to secure lithium-ion battery supplies. GM would partner with LG Chem to invest a combined $2.3 billion in a giant battery factory in Ohio, similar to the Gigafactory. Volkswagen had committed to investing

$1 billion in a European startup, which was building its own plant in Sweden. The startup was made up of executives who had helped create Tesla's Sparks facility.

More and more, the collective vision for the future of the car was Musk's. VW chief executive Herbert Diess couldn't be goaded by reporters in Germany into disparaging his rival. "Tesla is not niche," the CEO said. "We have a lot of respect for Tesla. It's a competitor we take very seriously."

Musk had always been fixated on the notion of **fairness.** While he might not act in a way that others thought was equitable, few things set him on the warpath more than his perception that he was somehow being treated unfairly. His list of such grievances was lengthy, involving Martin Eberhard, the designer Henrik Fisker, the media, and, lately, the SEC. His September 2018 deal with the commission to avoid a lengthy court battle gnawed at him, and his latest spat, the SEC threatening to hold him in contempt, proved a major distraction.

After growing up in South Africa, Musk had an unusual place in his heart for the U.S. court system, according to those who dealt with him on legal matters. He was if anything overconfident in a judge's ability to see things his way. In this latest

fight, he was arguing that the letter of his agreement with the SEC allowed for his tweet about
production numbers, and furthermore, the U.S.
Constitution protected his ability to speak publicly. He argued the SEC was trying to muzzle his
freedom of speech—an argument that seemed to
overlook that he had inked an agreement to avoid
precisely the kind of confusing tweets that had now
gotten him into this predicament. Musk had per-
fected the ability to put things in a way that some
investors took them as **plans,** while leaving himself
enough wiggle room to later say that he had only
been offering vague **ambitions.**

In a Manhattan courtroom that spring, Musk's
belief in the court system was rewarded. U.S.
District Judge Alison Nathan chastised government lawyers for rushing to hold him in contempt,
suggesting that the language of the agreement was
imprecise. She told the two sides to "put their rea-
sonableness pants on." As he exited the courthouse
to a throng of reporters, Musk couldn't contain
his glee, telling them he "was very happy with the
results."* By the end of the month, they would

---

* Musk's luck in court continued later in 2019 when a jury cleared
him of a defamation claim that came after he suggested Vernon
Unsworth was a pedophile during the ugly Twitter episode in 2018.
Unsworth's team, led by L. Lin Wood, had tried to frame their case as
one of a billionaire using his power to hurt their client while Musk's
lawyer had painted his comments as joking taunts.

settle their disagreement with a more detailed list of things Musk needed permission for before tweeting. More important, the cloud was removed; he could now focus on the real battle.

If the future seemed to more and more belong to Tesla, the present was less certain. After the difficult early months of 2019, it was understandable that Musk was having cold feet as he stood backstage with JB Straubel, the chief technology officer, that June as a crowd of investors awaited the company's annual shareholder meeting. Normally the event, held at the Computer History Museum in Mountain View, was a boisterous affair—practically a family reunion. While Ford might hold its annual meeting virtually and hope for as little attention as possible, Tesla's served as a rah-rah session for small investors and longtime owners, who had since the early days held on to hope for what Tesla could become. Musk had used the occasion over the years to hint at future products.

This time would be something else. He planned to announce that Straubel—who had not only been instrumental in many of Tesla's successes but who had spent essentially his entire adult life at Tesla, now with a wife and young twins of his own—would be departing the company.

With fifteen years of shared history, their

relationship had grown strained, especially as Tesla floundered in getting the Gigafactory up and running.* Musk was as demanding as ever, requiring nothing less than perfection. In late 2018, Straubel had opted to step away from work for some needed rest. A vacation turned into second thoughts about what was ahead. His vision for cars powered with lithium-ion batteries had, in his time working with Tesla, gone from a dangerous novelty to the future of a global industry. His plan for cutting battery costs in Nevada seemed to be working too, even if it had started off messy. Analysts estimated the Model 3's battery costs had fallen below $100 a kilowatt hour, a magic number long sought after by the automotive industry, at which point the cost of making an electric car was seen as equivalent to making its gas-powered twin. He'd accomplished what he set out to do. And as he looked ahead, he believed Tesla needed something he couldn't give it: manufacturing expertise, large operational experience. The company was

---

* Both men stressed that they remain on good terms. "If I could wind the clock back, I should just have started Tesla with JB as originally intended and not teamed up with Eberhard, Tarpenning and Wright. It ended up with just JB and me anyway, but only after a ton of painful drama that almost killed the company," Musk said. "The mistake I made was trying to have my cake and eat it too. I like creating products, but I don't like being CEO, so I tried to have someone else run the company while I developed the car. Unfortunately, that did not work."

no longer a startup, and startups, he knew, were where his passions lay. For Straubel, it was time to leave.

But as they were about to go onstage to make it official, Musk backed away from their plan. Today wasn't the right day for such an announcement, he decided. Straubel needed to stay just a little longer.

Musk walked onstage to cheers. "It's been a hell of a year, but a lot of good things are happening," he said. "And I think it's worth going over those things." After talking about how the Model 3 was outselling all of its gas-powered luxury competitors in the U.S.,* he called for Straubel to join him to discuss the success of the Gigafactory. They stood on the stage answering questions for almost an hour. As their time wound down, Musk grew reflective. Straubel's still-secret departure was likely on his mind. He talked about the first time he had met Straubel years earlier, at that fateful lunch with Harold Rosen. "That was like a good conversation," Musk said awkwardly.

"I think we've made pretty amazing progress since then," Straubel told him. "We didn't exactly envision how this would unfold."

"I thought we just for sure would fail—"

---

* In 2018, Tesla sold an estimated 117,000 Model 3 cars compared to 111,000 of the Lexus RX, to make it the top-selling luxury vehicle for the year, according to Edmunds.

"But it needed to be done," Straubel interrupted. "I mean it was clearly something that was worth doing, even if the odds of becoming where we are today—or even 10 percent of that, or 1 percent of that—were slim. It was still worth doing. But it's pretty awesome to see EVs driving around on every road all over the place. It's incredible."

Musk had always shown himself capable, when the spirit struck him, of obsessing over details of all sizes. His unwillingness to relinquish control of facets of Tesla, big and small, was perhaps the defining feature of his leadership, going back to the company's earliest days. He may not have been able to sleep on a Chinese factory floor with the same ease as in Fremont, but he still wanted to be close to the ground. Every day construction was under way, with Zhu now leading the China factory project. With Zhu, it seemed Musk had found his Jerome Guillen of China, an executive who could drive the hammer. Zhu's aggressive management style found its way into local media reports, including how he often called and messaged employees after midnight, sometimes venting anger, only to be in the office early the next morning, snapping photos of empty seats and posting them to company chats asking where the workers were. He seemed to understand the importance of a close relationship with Musk, sending him pictures daily of the factory's progression and traveling every few weeks to California during the height of the work to deliver updates in person. Zhu learned the ultimate lesson in dealing with Musk and keeping him from painfully stepping into the day to day: Deliver results. He was helped, of course, by the fact that Musk, separated by an ocean, couldn't as easily meddle.

The picture so far was encouraging. Tesla's new factory was seemingly benefitting from the company's previous experience—it was avoiding the sins of Fremont and Sparks in important ways. For one, thanks largely to Doug Field's initiative, the Model 3 was much easier to build than previous cars. The company still bore the scars of the impossibly complex Model X. This relative ease of assembly meant a line that Tesla knew how to manage.

Also, the new plant wouldn't be bogged down with automation—in part because Chinese labor would be dramatically cheaper than its American counterpart. In essence Tesla was going to replicate the assembly line they had built under the tent in Fremont, only six thousand miles away (and under a proper roof). Tesla had learned to make cars in one of the most expensive places on earth; now it was turning to a country known for its cheap labor. Analysts expected huge savings, making the Model 3 as much as 10 percent to 15 percent more profitable.

Blessings from the Chinese government meant that the wheels of bureaucracy moved quicker on the ground for Tesla in China than they had in Nevada. The state-owned company that ran the Shanghai Free-Trade Zone, where the factory was being built, worked closely to expedite construction. The automaker was allowed to submit just a portion of the blueprints normally required for building permits, for example. The electric grid extended electricity to the site in half the time it would normally take.

Lessons learned from the first Gigafactory were also being applied to how the team designed and built the new factory. As Kevin Kassekert, Straubel's deputy, worked on the final touches of the Nevada factory, he began building a team of construction experts that could be deployed around the world to open factories—taking the lessons they had learned and applying them globally. Much as the Model 3 had in many ways improved upon the Model S, they were trying to think of the factory as a product that could be enhanced with each iteration. The team in China convinced Musk not to do a total carbon copy of previous factories, but instead to use more typical Chinese industrial construction, which was less expensive and faster to raise, according to a construction manager briefed on the progress. The plants themselves were becoming a scalable product. Meeting Musk's growth goals would call for making millions of vehicles annually; Tesla would need more than just two assembly plants. They scouted land in Germany for a European factory.

Setting up assembly in China, however, required more than just a factory. Tesla needed suppliers to feed it parts, none so important as the batteries, which would need to eventually come from local factories in order for the cars to qualify for tax breaks. Tesla's longtime partner, Panasonic, was balking at going into China.

The two companies' relationship had soured with the problems at the Gigafactory. Musk's erratic behavior in 2018 didn't help either, especially appearing to smoke marijuana during a live interview. One Panasonic executive watched the interview on the way to the office with alarm: "What will our investors think?" Some in the Japanese company had already been lobbying to limit further exposure to Tesla as its stock fell almost 50 percent.

When it had been announced, Tesla said the Gigafactory in Nevada would span more than one hundred football fields, but it hadn't reached that size in 2019. With Kurt Kelty, who had helped forge the supplier relationship for Tesla, already gone and Straubel on his way out, Tesla's tie to the partner was weakened. Yoshi Yamada, the senior Panasonic executive who helped bring the companies together for the Gigafactory, had reached mandatory retirement age. Musk quickly hired him in hopes he could help preserve the relationship. But increasingly, it was Musk himself calling the Panasonic president Kazuhiro Tsuga directly, not only to ask for price cuts for the batteries coming from the Nevada factory but also to agree to build another factory in China. Put more bluntly, to double down on what appeared to be a losing hand.

In view of all that, Tesla prepared for a Chinese factory without Panasonic. Drew Baglino, who had taken over battery leadership from Straubel, began looking for an alternate partner. Initially, the preferred choice was LG Chem, the South Korean supplier behind GM's EV program. But some worried about being beholden yet again to a singular supplier. Another name was proposed, Contemporary Amperex Technology Co., a Chinese battery maker more commonly known as CATL that had grown from supplying parts to Apple to becoming the world's largest EV battery maker. Musk seemed initially to chafe at the idea. He apparently worried about partnering with a company that supplied so many rivals with cells.

In what had become a common refrain for Musk—just like in 2010 and 2013—he called up his executives on a Saturday demanding that Tesla begin developing its own battery cells. Baglino set up a skunkworks shop for a program dubbed Roadrunner.

In the meantime, Musk would still need a supplier for China, where production was supposed to begin by year's end. His team rushed to lock down an agreement with LG, while gently working to put CALT in position for a deal, too. In August, as Musk went to China for meetings, an introduction to CATL founder Robin Zeng was arranged on the sidelines of an AI conference, where Musk was speaking with Alibaba co-founder Jack Ma. In private, Musk and Zeng quickly hit it off, the Tesla CEO identifying a fellow engineer at heart. "Robin is hardcore," Musk told his team. The meeting cleared the way for an eventual deal. No longer was Tesla's fate tied solely to Panasonic.

That summer trip to China left Musk seeing plenty of reason to be excited about the potential for a Chinese-made, locally sold Model 3. Even with hefty tariffs, Model 3 sales in China were already helping boost Tesla's bottom line. Revenue rose 64 percent in the third quarter of 2019. Those numbers were buoyed by Tesla's friends in Europe, too. Norway was gobbling up Model 3s as soon as they arrived. Revenue from the country swelled 56 percent in the first nine months of the year. Deliveries of vehicles were on pace to make the Model 3 the country's top-selling vehicle for 2019. Better still, as in the U.S. the previous year, initial Model 3 sales were of the more expensive variety. The international expansion had been a little messy, but as summer turned into fall, it was clear that Tesla's overall plan for growth was working.

By the time the company reported its robust third-quarter profit in November 2019, the company was already on its way to a profitable fourth. More importantly, it had done what Musk had promised in January: Tesla was, unbelievably, ready to begin production in China. The speed of it all put the carmaker's efforts in Nevada to shame. Social media captured the progress, showing first videos of a muddy field in January, cranes seen among shipping containers and steel. By August, when Musk was in China, the structure of a giant factory had emerged. By October, the government gave the green light for car production to begin.

First came trial production. Tesla released photos of pristine machinery. Workers in blue Tesla caps and nifty new uniforms appeared to toil over a blue Model 3, which sat near the end of an assembly line. After the Model 3's overblown start of production in 2017, it was easy to dismiss the photos as mere hype. Still, investors were reassured by what they saw. Short sellers, meanwhile, were feeling the pain. They were estimated to have lost more than $3 billion on Tesla since June. The pain wasn't over for them—not even close.

In the final days of December, Tesla began delivering Chinese-made Model 3s, first to its employees. One worker took the opportunity to propose marriage to his sweetheart on the factory floor. The biggest celebration was still to come, as shares continued to climb, cementing Tesla as the world's third-most-valuable automaker behind Volkswagen and Toyota—a stunning ascent.

A few days after that milestone—during a time in China that would go on to have world-shaking repercussions—Musk flew to the Shanghai factory to celebrate the official start of production, with vehicles going to non-employee customers. Onstage at the factory in front of hundreds of customers and employees stood a jubilant Musk. Tesla had crossed the finish line. It not only had a car that consumers wanted, and that rival carmakers coveted, but the means to produce it at previously unimaginable scale, all while turning a profit. Musk was celebrating what a year earlier would have sounded like a fantasy—if not a delusion.

But then, at every turn since Martin Eberhard showed up at Musk's door in 2004, Tesla's development had always sounded like delusion. Eleven years earlier, almost to the day, Musk had nearly lost Tesla for the first time, betting everything, including his personal fortune, on his vision for what the Roadster and Model S could do. Each small success had given him the confidence to go a step further.

Of course, Tesla's success into the distant future was still far from guaranteed. What automaker's was? Given the whims of the industry—the strong headwinds that established carmakers face, even more than a century into their lives; the husks of broken-down rivals that dotted the road to carmaking fortune—its continued success would always remain something of a question mark.

On this day, however, as he stood onstage as

CEO of a legitimately global electric automaker, Musk had won the day. He had made a mainstream electric car, the envy of the industry he had sought to overturn. His competitors could only shake off the dust—and hope to catch up.

As beats blared from speakers, Musk flung off his suit jacket. A giant grin crossed his face, and he began an awkward strut—a dance of victory.

"If we could get this done," he said a moment later, catching his breath, "what else can we do?"

EPILOGUE

The coronavirus panic is dumb," Elon Musk tweeted on March 6. It was the same day Apple began encouraging employees to stay home, one of many tech giants taking efforts to slow the spread of the novel virus.

The global pandemic of early 2020 was threatening to ruin Tesla's moment. Just weeks earlier, Musk had been onstage in Shanghai celebrating the start of Model 3 production in China, defying skeptics who thought he couldn't pull off such a feat in less than a year. Two days after his performance, the World Health Organization announced the discovery of a mysterious pneumonia-like illness in Wuhan, a large Chinese city more than five hundred miles to the west. It was still early days for what would become known to the world as COVID-19. For many around the globe, it was easy to dismiss the potential threat as something a world away—if they were even paying attention.

This was especially true if you were a Tesla shareholder joining Musk in celebrating, as the company's stock continued to surpass record highs. His unexpected victory in China, not to mention racking up two quarters of profitability at the end of 2019, had given him renewed credibility. Adding to the excitement in late January, Musk announced he was pulling ahead plans to begin production of the Model Y compact SUV in coming weeks, from the previously announced deadline of fall 2020. With two assembly plants online, and a third in the works in Germany, Tesla said it would comfortably exceed delivering 500,000 vehicles in 2020—a figure that, if achieved, would amount to a 36 percent jump from 2019. The Tesla growth story was alive again.

In the following days, the stock would continue to soar, valuing Tesla at more than $100 billion and overtaking Volkswagen AG as the second most valuable automaker in the world, behind only Toyota Motor Corp. It also put Musk closer to receiving the first of twelve payouts from his compensation plan, the ambitious program that ultimately aimed to raise the company's market value to $650 billion. It was a valuation that many thought unlikely anytime soon. Despite a few profitable quarters here and there, Tesla had never, since its 2003 founding, had a full year of profitability; 2020, however, held new promise.

"If Tesla proves to be profitable . . . we think this removes one of the biggest impediments for why legacy [automakers] were hesitant to go 'all in' on EVs," noted Adam Jonas, the Wall Street analyst whose longtime optimism had been tested a year earlier. During a conference call with Jim Hackett, Ford's CEO (its second since Tesla overtook the automaker in market value in 2017), Jonas pressed the executive on Tesla's ascent. "It's kind of a historic day because Tesla's now worth over 5x the market cap of Ford," he said. "Does it make sense to you? What's the message the market's sending Ford?" Hackett's answer would ultimately matter little; within a few weeks, he would announce his retirement. Ford, General Motors, and other automakers were falling even further behind in the minds of investors betting on the future of the automobile. Musk had done what he set out to do: Convince the world that the car should be electric—even if actual car buyers hadn't turned out en masse to buy them just yet. The forces of change seemed to be on his side.

Each piece of good news for Tesla was like another log on a flame that was turning into a bonfire. In the three months since posting the surprise third quarter profit, shares had doubled in value. In the month after Musk had danced onstage in

Shanghai, the shares doubled again. In the days after posting the fourth quarter profit, shares kept rising. Investors weren't just betting that the future of the automobile was electric, they were wagering that Tesla would be the dominant player in that new world.

Perhaps overly so. Brian Johnson, an analyst for Barclays, cautioned that the pricing signaled that the market believed Tesla would be "the sole winner," adding that it was reminiscent of the tech boom of the 1990s—which came crashing down on Martin Eberhard and Marc Tarpenning, among so many others.

Despite two strong quarters, Tesla still toiled under a fundamental weakness that dogged it from the beginning, a need for cash to fuel Musk's ambitions. He had shown an ability time and time again to tap investor enthusiasm when he had to. But the question always remained: What would happen when the music stopped? The boom-and-bust nature of the automotive industry had tripped up many car companies over the years. The Great Recession had nearly killed Tesla in its early days, when cash suddenly dried up. Musk had found a way to survive—just barely. In the coming weeks, he'd face another test, and this one would determine if Tesla was truly a car company or a house of cards.

In early 2020, Tesla was positioned to pull ahead production of the Model Y compact SUV thanks in large part to Doug Field, the executive who had overseen new vehicle development before departing in 2018 amid the Fremont factory troubles. Much in the way that Tesla had planned to build the Model S sedan and Model X SUV off the same vehicle underpinnings, the Model Y was to share engineering with the Model 3. But to avoid the kind of tinkering and expense that had turned the Model X into a disaster, Field had kept the compact SUV's development largely away from Musk, who had been consumed with production hell anyway. On the occasions that the vehicle had come up, Musk had shown signs of the development creep he was known for, such as arguing the Model Y didn't need a steering wheel because it was going to be fully self-driving.

By the time Musk got around to focusing on the Model Y, though, Field's team (or what remained of it) could present the vehicle with a bow on it. It was truly a top hat to the Model 3, carrying over about 70 percent of the parts. To purists, the vehicle was more of a fastback version of the Model 3 than a true SUV. It was maybe 10 percent larger. The driver sat a few inches higher. The exterior looked somewhat different from the Model 3, sporting a more bulbous rear to make room for a hatchback. At its core, though, it was a Tesla, with a sleek interior, large center screen, and quick acceleration.

The Model Y was also fundamental to Tesla's going mainstream, aimed at buyers of so-called crossovers, an SUV built on a car platform instead of a truck chassis—a change that allowed for a smoother ride while keeping the benefits of a traditional SUV's high-seated driving position and roomy interior. It was among one of the fastest-growing vehicle segments, especially in China, where about one in five vehicles sold were compact SUVs. Musk believed the Model Y could outshine the Model 3.

Then, as February began, in what seemed like the blink of an eye, Tesla's bright future seemed very dim. Shares plunged 17 percent on February 5, one of its worst days ever, on news out of China that the company's locally made Model 3s would be further delayed because of COVID-19. Excitement about Tesla's potential for growth in China had fueled a rocket-like rise in value, but now there was growing concern that this new virus, which scientists cautioned was deadly and easily spreadable, might bring the country's economy to a standstill.

As investors began to digest that news, it became increasingly clear, too, that COVID-19 wasn't a threat to just China or Asia, as the SARS outbreak had largely been almost twenty years earlier. In Tesla's home state of California, local government officials were growing increasingly concerned. Already a few cases had been identified in Santa Clara County, where Tesla headquarters was located.

On February 13, the company unexpectedly

announced it was raising $2 billion in a stock sale to bolster its balance sheet. Had Musk taken a lesson from his 2007 and 2008 travails, that a rainy day fund was a good thing to have?

Roughly a month later, just a few days after President Donald Trump banned European travel and the NBA suspended its season in efforts to stop the virus's spread, local governments in the Bay Area issued a "shelter-in-place" order, instructing residents to stay home and shutting down businesses it deemed "nonessential"—a term that would become open to debate. Mostly lost in the fear: Tesla had that day begun delivery of the first Model Ys coming out of its Fremont factory.

Musk was hell-bent on continuing, no matter what local officials said. Late that night, he sent a defiant email to Tesla employees: "I will personally be at work, but that's just me," he said. "Totally ok if you want to stay home for any reason." The Fremont factory kept humming, even as the county sheriff's office the next day said publicly that Tesla should stop production. By week's end, as similar shutdowns spread across the United States, Musk relented, announcing a temporary halt to production and assuring the public it had enough cash on hand to weather the storm. By then, California governor Gavin Newsom warned that 56 percent of people in the nation's most populous state could be infected within eight weeks without aggressive action.

All things considered, the timing for Tesla could have been worse. The China factory turned out to be offline only for a short while. Fremont closed toward the end of the quarter, at a moment when much of the period's inventory was ready for delivery, during the typical end-of-the-quarter push.

Better still, some of the wild decisions Musk had made in the dark periods of 2018 were suddenly seeming prescient, even if they had only been chanced upon. His push to build out a home-delivery sales team payed unexpected dividends in an era when showrooms were being closed. Tesla's global deliveries rose 40 percent in the quarter, short of analysts' expectations but still way better than the rest of the industry. In China, where overall the industry's sales plummeted 42 percent in the first quarter of 2020, Tesla's **rose** 63 percent, and it was poised to continue on that course if Tesla could keep its Chinese factory humming. The government seemed eager to ensure it would, arranging for dormitories and transportation for hundreds of workers and helping to secure ten thousand masks, thermometers, and cases of disinfectant, all of which allowed Tesla to resume work in Shanghai the first working day after the nation's extended Lunar New Year break. The effort was hailed in local media, with footage of locally made Model 3s being assembled.

Musk's decision to push ahead with the Shanghai factory also proved wise. It meant Tesla had a lifeline if its only U.S. factory was kept offline for months to come. Accustomed to moving quickly, Musk slashed salaried workers' pay and furloughed without pay employees unable to work from home. Tesla also began asking landlords of its stores for rent reductions.

Even taking those steps, a prolonged shutdown wouldn't be pretty. After its latest capital raise, Tesla ended March with $8.1 billion in cash. Jonas, the Wall Street analyst, estimated the company would burn about $800 million a month if it was knocked mostly offline. Musk hoped to resume work on May 4—the day after the local shutdown order was scheduled to lift. But as that time came, local authorities, still worried about the threat of the virus, extended the shutdown.

All of which helps explain why Musk, despite the rather rosy sales results, couldn't contain his anger for the well-meaning public officials keeping his Fremont factory closed. "If somebody wants to stay in their house, that's great. They should be allowed to stay in their house and they should not be compelled to leave," Musk said on a public call with analysts in late April. "But to say that they cannot leave their house, and they will be arrested if they do, this is fascist. This is not democratic. This is not freedom. Give people back their goddamn freedom."

Behind the scenes, he was pushing his people to get ready for a reopening—with or without local authorities' blessing. Roughly a week later, he told his workers they should return to the factory—even as the government continued to push for a later reopening. That would've been after Michigan was letting its auto factories resume work. Tesla might find itself disadvantaged compared to its U.S. rivals thanks to California's caution. "Over the past week, I have received several complaints that Tesla was in fact violating the Alameda County Health Order by ordering their employees back to work to re-open the production line," local police said in an email to Tesla on May 8.

The next morning, a Saturday, Musk opened a multifront war through Twitter and the court. He announced he was suing the local government and called the county health officer "unelected & ignorant" on Twitter. "Frankly, this is the final straw," he continued. "Tesla will now move its HQ and future programs to Texas/Nevada immediately. If we even retain Fremont manufacturing activity at all, it will be dependen [**sic**] on how Tesla is treated in the future. Tesla is the last carmaker left in CA."

Workers watched with a mixture of feelings; a desire to see the company they had sacrificed so much for survive, a need to earn a living, but also a real concern about their safety, especially in a factory that hadn't always seemed to prioritize well-being. "I'm concerned about what Tesla can realistically do to keep us safe," one production associate in her twenties said. Others were worried about the pressure to make up for more than a month of lost production in order to meet Musk's ambitious production goals.

Some decided it was time to quit. Fresh in their minds was the grueling pace of the past few years, as the company struggled to increase production of the Model 3. **What new hell awaited?** They didn't want to stick around to find out.

Not lost on many was the way Tesla's meteoric rise earlier that year had put Musk closer to receiving the first part of his massive $50 billion-plus payday. He needed Tesla to remain at a market value of $100 billion on average for a set period of time to unlock 1.69 million shares, which would allow him to nominally net more than $700 million, were he to sell immediately (which he wasn't going to do). The ten-year plan wouldn't fully pay out unless Tesla reached a market value of $650 billion. Given the sky-high valuation requirement, this first goal might be one of the only tranches Musk could reasonably reach, some thought at the time.

Still, in the wake of his "fascist" comment, investors seemed to reward his defiance. Shares were rebounding from a low in February. With his further outbursts downplaying the danger of COVID-19, Musk waded right into the growing political unrest that roiled the nation. Debate hinged on whether to prioritize containing the pandemic or fueling the economy, and it was split largely along party lines. Trump weighed in, supporting Musk in his battle to reopen.

To raise the stakes, Musk publicly dared local officials to stop him. He announced on Twitter that work was resuming, with him at the factory. "I will be on the line with everyone else," he tweeted on May 11. "If anyone is arrested, I ask that it only be me."

Faced with enormous pressure, local authorities flinched. Within a few days, they declared an agreement had been reached for the factory to resume work, citing safety protocols proposed by Tesla.

By then, a race was already underway to turn out as many cars as possible to try to make up for lost time. It was familiar ground for Tesla—another make-or-break end of quarter, coming down to the final days. "It is very important that we go all out through end of June 30 to ensure a good outcome," Musk told his workers. "Wouldn't bring this up if not very important."*

The push helped. Wall Street had expected sales to fall about 25 percent; instead, Tesla managed just a 4.9 percent decline from the same quarter a year

* Musk's predictions that new cases of COVID-19 in the U.S. would fall likely to zero by the end of April 2020 proved wrong. Months later, he announced on Twitter that he had tested positive.

earlier. In the upside-down world of the COVID era, such results were tantamount to wild growth. Tesla's rivals had taken a much harder fall globally. The strong results allowed Musk to post a $104 million quarterly profit—a fourth consecutive quarter in the black, the longest streak in its history. (The $428 million sale of emissions credits again played a big role in the victory.)

To top things off: Musk announced he had picked a site outside of Austin, Texas, for Tesla's next assembly factory. The company's center of gravity was slightly moving.

That summer, a decade after going public, it was Tesla's stock that went into Ludicrous mode. If Musk had found renewed credibility six months earlier by delivering on his promise to open a factory in China, he had now cemented his place in automotive industry lore. His results were made even more impressive by the fact that the rest of the industry was still reeling from factory closures and sales falloffs. Fueled in part by young stock buyers stuck at home during the pandemic, the stock rose past $1,000; the company's valuation overtook Toyota.

Little Tesla was now the world's most valuable automaker.

And the stock kept rising. In just weeks, the

company was worth as much as Toyota and Volkswagen combined. Musk hit his $100 billion market value payout target, then went on to clear thresholds for three more payouts (then two more by spring 2021). The stock passed the $700 billion market value milestone that Musk had thrown down years earlier like Babe Ruth calling his shot. It soared from a value of $100 billion to more than $800 billion in 244 days, accomplishing something it took Apple almost a decade to do. With the stock he already owned, his wealth was surging from an estimated $30 billion at the start of 2020 to around $200 billion at the start of 2021, overtaking Amazon founder Jeff Bezos's spot as the world's richest person, according to **Bloomberg**'s Billionaire Index.

The excitement—some would say mania—spread to other related companies. In the following months, several startups went public. Among those headed in that direction was Lucid Motors, whose CEO was Peter Rawlinson. After the Model S engineering chief left Tesla in 2012, he'd been working on what could be considered the next generation of the luxury sedan. The company had been founded by Bernie Tse, also tied to Tesla—he was the former board member who left after Musk squashed his efforts to create a battery division.

The electric car boom was also adding renewed excitement for a car venture at Apple, the company

Doug Field had returned to after leaving Tesla in 2018. The former engineering chief was helping guide the iPhone maker's secret car project.

JB Straubel, the battery wiz, was getting lots of attention, too, including investments from Amazon in his new startup: a company called Redwood Materials that aimed to recycle electric car battery waste for use in future cars. It was an idea that came to him as he struggled to rein in waste during the spring of 2018 at the Gigafactory. His wealth from Tesla afforded him the ability to pursue many of his ideas. The Tesla shares that he held when he left the company would be worth more than $1 billion at the start of 2021—if he kept them all.

Even Martin Eberhard could watch Tesla's success with some sense of ownership. While he had once told **The New York Times** he planned to sell all of his shares after his fight with Musk, Eberhard confided that in fact he had held on to a few.

"I'm the longest owning Tesla shareholder in the world," he said with a sense of pride. He still owns his original Roadster, too, and his Mr. Tesla license plate.

Musk has said many times that his money is mostly tied up in Tesla and SpaceX. Even as those investments have soared in value, court records again shed light on his personal finances late in 2019, indicating he was at the time cash poor.

He continued to fund his life off loans tied to his shares. Eventually, Musk has said, he plans to sell his Tesla stock to pay for a Mars colonization effort, not to mention Earth-bound charitable acts, a process he imagines might begin in earnest as he reaches retirement age in his late sixties. Though it's hard to imagine him ever truly giving up control of Tesla.

As 2020 came to an end, Musk's faith in the judicial system was further renewed when Martin Tripp, the former Gigafactory worker who in 2018 claimed to be a whistleblower, agreed to settle the legal battle with Tesla. As part of the deal, Tripp wouldn't contest Tesla's claims that he released the company's private data and would pay $400,000. Perhaps even more satisfying to Musk, however, was the revelation during the litigation that suggested a short seller had been funding part of Tripp's legal battle, which stirred up many negative headlines. It wasn't famed short Jim Chanos, who Musk had long thought behind Tesla's woes. But the discovery did, in some ways, confirm what Musk had long suspected: shadowy forces were out to get him.

Amid criticism that he was a billionaire forcing workers to return to dangerous jobs during the pandemic, he announced he was selling all of his homes. "This is very much a personal choice and not something I would advocate broadly. If someone wants to have or build a great house and that makes them happy, I think that is totally cool," Musk said. "I'm just trying to make my life as simple as possible right now, so will only keep things that have sentimental value." After that, he was mostly living near SpaceX's Boca Chica launch facility in southern Texas and jetting between Berlin and Austin to inspect progress on factories he aimed to open later that year.

His plans for Tesla were as ambitious as ever. His next vehicle, a pickup dubbed the Cybertruck, had many of the signatures of a car developed by Musk—features that appealed to him directly (including a dystopian appearance, supposedly bulletproof steel and shatter-proof windows) and that are likely hard to industrialize.

Battery costs remained his biggest challenge. He conceded that his plans for manufacturing cells might cause delays. As ever, he promised investors in 2020 that Tesla's Roadrunner project was going to drive costs down by half from improvements to cell manufacturing and battery chemistry.

His next insane goal: becoming the world's largest automaker by deliveries. He wanted to reach 20 million vehicles a year by 2030—about twice what the sales leader Volkswagen sold in 2019. "One of the things that troubles me the most is that we don't yet have a truly affordable car, and that is something that we will make in the future," he said. "In order to do that, we've got to get the cost of batteries down." He was still chasing the goal of a truly affordable electric car. Now, instead of $35,000, he was targeting $25,000—someday.

For Tesla, the practical effect of the euphoria around its stock price was that the company could easily tap money from investors yet again. Tesla executed a 5-for-1 stock split to make it easier for small shareholders to buy stock, and it issued millions of new shares to amass $19.4 billion—roughly three times as much money as it had lost in the sixteen years it worked to bring out the Roadster, Model S, and Model 3. It was a war chest that meant that for years to come, Musk could continue to fuel his pricey ambitions. All of which created a virtuous cycle: Investors believed in the potential for growth, allowing Tesla to cheaply raise money to fuel that growth, which in turn further stoked excitement for **more** growth.

It was a logic that still vexed Musk's critics, especially as they pointed out the flaws they continued to see in the business—profit thanks to the sale of emission credits, the inconsistent quality of vehicles, the question of ultimate demand for its cars, the missed goals, the overpromises of Autopilot, the challenges of handling an ever increasing customer base. As the world's most valuable automaker, Tesla faced enormous scrutiny. It was chasing a sales crown that has tripped up many automakers in recent years. It could be one nasty government recall away from a devastating blow, a poor product introduction away from being overtaken by an older rival. It was always a heartbeat away from finding itself without a strong leader. The bubble would have to burst at some point, critics argued.

Short sellers such as Chanos conceded that their bet against Tesla has been painful, especially in 2020, when on paper they collectively lost more than $38 billion. Still, they couldn't shake the feeling that they might be right someday. Tesla shares remained among the most shorted on the market. "I've never met Elon Musk. I've never had a conversation with him," Chanos said in late 2020 as Tesla shares approached an 800 percent rise for the year. If they did cross paths, Chanos figured, "I'd say, 'Job well done—so far.'"

As 2021 began, it was easy to argue that Tesla was overvalued—even as the company notched its sixth straight profitable quarter (and its first full-calendar year of profitability), targeting a greater than 50 percent sales gain for the year ahead. Musk himself had even conceded months earlier that Tesla was overvalued. As analysts stumbled over themselves to justify its stock price, Jonas, the long-time analyst, attributed the run up, in part, to the "power of hope."

The idea of the Roadster had been a beacon of hope for Musk, an idea he gave just a 10 percent chance of working out. The Model S that followed had been a gamble that he could make an electric car as good as, if not better than, any other car on the road. And the Model 3 was the product of his belief that everyday people, given the chance, would want a car run not on gasoline but on something more sustainable.

Musk was selling a vision of the future. He had bet everything on the idea that others would share his vision, if only they were given the chance. Now, with their money, their words, and their faith, they were telling him they too wanted a piece of the action.

Popular myth is that Elon Musk, sleeping on the factory floor, willed Tesla into being. His determination and stubbornness surely played a big part in the company's rise and there wouldn't be a Tesla Inc. without him. But how the company went from a rather unlikely idea in the summer of 2003 to the world's most valuable automaker in 2020 is much more complex than one man's moxie. This book aims to the tell the story of how Tesla came to be. It relies on hundreds of interviews conducted with Tesla insiders—past and present. Many of those interviews were conducted on the condition of anonymity, in part because some had signed non-disclosure agreements while others said they feared Musk's reprisal. The motivations for these people were mixed. Some felt slighted by Musk while many felt proud of what they had accomplished, and they wanted the full story of the company to finally be told.

While this book is based on thousands of company records, court filings, and video recordings, it also relies on the memories of people who were there over the course of almost two decades. Memories, of course, can be fallible. Dialogue and scenes have been re-created from firsthand witnesses and every effort was made to confirm through additional sources. Some of the characters in the book may have participated while others may only seem to have done so based on the depth of reporting around them.

As for Musk, he was given numerous opportunities to comment on the stories, facts, and characterizations presented in these pages. Without pointing to any specific inaccuracies, he offered simply this: "Most, but not all, of what you read in this book is nonsense."

ACKNOWLEDGMENTS

This book was made possible by the people who trusted me with their stories. I'm grateful to them. It was also built on the work of those who went before me—dogged reporters who have covered Tesla over the years and blazed a trail for me to follow. Ashlee Vance wrote the definitive biography of Elon Musk. Dana Hull, Lora Kolodny, Kirsten Korosec, Edward Niedermeyer, Alan Ohnsman, Susan Pulliam, Mike Ramsey, and Owen Thomas are among those journalists who broke some of the biggest stories about the company. In particular, I'm indebted to Ramsey for his advice.

Personally, I have benefited from the help of many people who have looked out for me over the years. It has been a great honor to report and write for **The Wall Street Journal.** Without the support of Matt Murray, Jamie Heller, Jason Dean, Scott

Austin, Christina Rogers, John Stoll, and many others at the **Journal,** I wouldn't have been able to do this book.

Early in my career, editors Paul Anger and Randy Essex convinced me to give up covering Iowa politics to report on cars for the **Detroit Free Press.** Editor Jamie Butters taught me everything I know about the auto industry—first at the **Free Press,** then at Bloomberg News. Tom Giles, Pui-Wing Tam, and Reed Stevenson, at Bloomberg, then introduced me to the world of tech reporting in Silicon Valley. A combination of experiences in Detroit and San Francisco prepared me well for diving deeply into Tesla.

I'd like to thank my agent, Eric Lupfer, for guidance and support, my Doubleday editor, Yaniv Soha, for patience and thoughtfulness, and my fact-checker, Sean Lavery, for eagle eyes. I'd also like to thank my longtime writing coach/editor John Brecher, for his wisdom and encouragement. Fellow authors Sarah Frier, Alex Davies, and Tripp Mickle formed a unique fraternity as we all tried our hands at writing our first books. Finally, a project like this is supported by the enduring love of family. Thank you, Karin.

NOTES

PROLOGUE

xvi It also became: Scott Corwin, Eamonn Kelly, and Joe Vitale, "The Future of Mobility," Deloitte, September 24, 2015, https://www2 .deloitte.com/us/en/insights/focus/future-of -mobility/transportation-technology.html. Kim Hill et al., "Contribution of the Automotive Industry to the Economies of All Fifty States and the United States," Center for Automotive Research (January 2015).

xvii The average car: Average North America operating profit for U.S. automakers in 2018, according to research from Brian Johnson of Barclays.

xix "Either they become": Stephen Lacey, "Tesla Motors Raises More Than $1 Billion from Debt Equity," Reuters, May 17, 2013.

xx The world's largest automakers: William Boston, "Start Your Engines: The Second Wave of Luxury

Electric Cars," **Wall Street Journal,** June 22, 2018, https://www.wsj.com/articles/start-your -engines-the-second-wave-of-luxury-electric -cars-1529675976.

xxi If Tesla had been valued: Philip van Doorn, "Tesla's Success Underscores the Tremendous Bargain of GM's shares," MarketWatch (Oct. 28, 2018), https://www.marketwatch.com/story/ teslas-success-underscores-the-tremendous -bargain-of-gms-shares-2018-10-25.

CHAPTER I

5 Lithium-ion cells: Sam Jaffe, "The Lithium Ion Inflection Point," Battery Power Online (2013), http://www.batterypoweronline.com./articles/ the-lithium-ion-inflection-point/.

8 Rosen Motors had: Larry Armstrong, "An Electric Car That Hardly Needs Batteries," Bloomberg News, Sept. 23, 1996, https://www .bloomberg.com/articles/1996-09-22/an-electric -car-that-hardly-needs-batteries.

9 "There are not": Karen Kaplan, "Rosen Motors Folds After Engine's '50%' Success," **Los Angeles Times,** Nov. 19, 1997.

10 The result: Chris Dixon, "Lots of Zoom, with Batteries," **New York Times,** Sept. 19, 2003.

12 "If you like space": Video posted by Stanford University from Entrepreneurial Thought Leader series (Oct. 8, 2003), https://ecorner.stanford .edu/videos/career-development/.

14 It didn't align: YouTube video posted by

shazmosushi on July 12, 2013: https://youtu.be/
afZTrfvB2AQ.

CHAPTER 2

17 In 2000, ahead: Michael Kozlowski, "The Tale
of Rocketbook—the Very First E-Reader," Good
E-Reader (Dec. 2, 2018), https://goodereader
.com/blog/electronic-readers/the-tale-of
-rocketbook-the-very-first-e-reader.

18 "It's kind of foolish": Author interview with
Martin Eberhard.

19 The EV1 battery pack: Data for average sedan
weight pulled from U.S. Environmental
Protection Agency's Automotive Trends Data.
https://www.epa.gov/automotive-trends/explore
-automotive-trends-data.

22 He agreed to pay $100,000: Details included in
California court records reviewed by author.

22 Henry Ford's wife had: Douglas Brinkley, **Wheels
for the World** (New York: Viking Adult, 2003).

23 An electric car that might cost: Michael
Shnayerson, **The Car That Could** (New York:
Random House, 1996).

28 "It felt like a racecar": Ian Wright, "Useable
Performance: A Driver's Reflections on Driving
an Electric Sportscar," business document created by Tesla Motors (Feb. 11, 2004).

29 "Elon has money": Author interview.

31 What about converting the sports car: Emails reviewed by author.

32 They had run the numbers: Review of Tesla

Motor Inc.'s "Confidential Business Plan," dated Feb. 19, 2004.

33   "Convince me you know": Author interviews with people familiar with the talks.

35   Musk had been kicked: Jeffrey M. O'Brien, "The PayPal Mafia," **Fortune** (Nov. 13, 2007), https://fortune.com/2007/11/13/paypal-mafia/.

CHAPTER 3

41   While the EV1 motors: Author interviews with early Tesla employees.

45   He'd later learn: Author interviews with multiple former Tesla employees familiar with the matter.

48   In 2004 and 2005: Damon Darlin, "Apple Recalls 1.8 Million Laptop Batteries," **New York Times** (Aug. 24, 2006), https://www.nytimes.com/2006/08/24/technology/23cnd-apple.html.

48   When LG Chem realized: Author interviews with multiple former Tesla employees familiar with the matter.

50   "Guys, that's like between one": Author interviews with multiple former Tesla employees.

CHAPTER 4

54   "He's not a man": Justine Musk, "I Was a Starter Wife," **Marie Claire** (Sept. 10, 2010), https://www.marieclaire.com/sex-love/a5380/millionaire-starter-wife/.

55   He discussed his dreams: Video of CNN

interview posted on YouTube by misc.video on Nov. 17, 2017, https://youtu.be/x3tlVE_QXm4.

55 "I'm the alpha": Justine Musk, "I Was a Starter Wife."

59 Eberhard had sought out: Author interviews.

61 In the post–World War II era: Stewart Macaulay, **Law and the Balance of Power: The Automobile Manufacturers and Their Dealers** (Russell Sage Foundation, Dec. 1966).

63 Musk pushed for selling: Author interviews with people involved in the discussions.

65 "Martin was getting antagonistic": Author interview with a person involved in the due diligence.

65 Musk told them they: Details of negotiations from author interviews and Musk's interview with Pando Daily posted on YouTube on July 16, 2012, https://youtu.be/NIsYT1rqW5w.

CHAPTER 5

68 "Elon is the perfect investor": Author interview and color about relationship taken from emails between the men reviewed by the author.

70 "Why the fuck": Michael V. Copeland, "Tesla's Wild Ride," **Fortune** (July 21, 2008), https://fortune.com/2008/07/21/tesla-elon-musk-electric-car-motors/.

73 He suggested being vague: Details taken from emails between the men reviewed by the author.

75 Musk fired them: Michael V. Copeland, "Tesla's Wild Ride."

76 Guests included: Sebastian Blanco, "Tesla

Roadster Unveiling in Santa Monica," Autoblog (July 20, 2006), https://www.autoblog.com/2006/07/20/tesla-roadster-unveiling-in-santa -monica/.

77 A crowd lined up: Description of event taken from video posted on YouTube by AP Archives, https://youtu.be/40pZmDdKqt0.

79 Joe Francis, creator of: Author interviews with early Tesla employees.

80 It was the first sign: Anecdote comes from interviews and records, including emails between the parties reviewed by the author.

82 "What I want to hear": Quotes and details taken from emails between the men, reviewed by the author.

83 Musk wanted special headlights: Michael V. Copeland, "Tesla's Wild Ride."

85 "I am sure you": Emails reviewed by author include the conversation and details of the presentation.

86 "There are several burning": Email reviewed by the author.

86 "It is the view": Email reviewed by the author.

CHAPTER 6

89 Born in Detroit: Lynne Marek, "Valor Equity Takes SpaceX Approach to Investing," **Crain's Chicago Business** (May 14, 2016), https://www.chicagobusiness.com/article/20160514/ISSUE01/305149992/valor-equity-takes-spacex -approach-to-visionary-investments.

90 The firm raised $270,000, plus $130,000: Antonio Gracias, Hispanic Scholarship Fund bio, https://www.hsf.net/stories-detail?storyId=101721718.

92 "I had never seen": Author interview.

92 Their relationship would contribute: Antonio Gracias, Hispanic Scholarship Fund bio.

94 During Musk's first trip: Author interviews with Tesla executives at the time.

96 "Many times, I have": Email exchange reviewed by author.

103 "If this is true": Michael V. Copeland, "Tesla's Wild Ride," **Fortune,** July 21, 2008, https://fortune.com/2008/07/21/tesla-elon-musk-electric-car-motors.

103 He calculated that the cost: Tim Watkins's declaration filed with California court on June 29, 2009.

104 "Martin seems to be focused": Email reviewed by the author.

104 "Lots of issues at this company": Email reviewed by the author.

CHAPTER 7

106 "I've noticed a few things": Interviews with Tesla employees at the time.

108 "we've been having": Author interviews with people at the table that day.

109 "That tore it for me": Keith Naughton, "Bob Lutz: The Man Who Revived the Electric Car," **Newsweek** (Dec. 22, 2007), https://www

.newsweek.com/bob-lutz-man-who-revived-electric-car-94987.

113   Some Tesla managers began calling: Author interviews with people working on the project.

128   Anonymous sources said: Josée Valcourt and Neal E. Boudette, "Star Engineer Quits Chrysler Job," **Wall Street Journal** (March 26, 2008), https://www.wsj.com/articles/SB1206475 38463363161.

129   Donoughe was granted: Donoughe Offer Letter (June 4, 2008), filed with the SEC.

130   Straubel took the car apart: Author interviews with multiple Tesla employees at the time.

131   Basically, the only parts: Details of differences between the Elise and the Roadster come from a blog posting made by Darryl Siry, "Mythbusters Part 2: The Tesla Roadster Is Not a Converted Lotus Elise," Tesla.com (March 3, 2008), https://www.tesla.com/blog/mythbusters-part-2-tesla-roadster-not-converted-lotus-elise.

132   Kelley sent an email: Author interview with Kelley; Poorinma Gupta and Keven Krolicki, "Special Report: Is Tesla the Future or the New Government Motors?" Reuters (June 28, 2010), https://www.reuters.com/article/us-tesla-special-reports-idINTRE65R5EI20100628.

134   There they personally yanked: Author interviews with people involved in the effort.

135   The Mercedes CLS large sedan: Author interviews with people who worked on the project.

CHAPTER 8

137   "His father had the Encyclopaedia": Sissi Cao, "At 71, Elon Musk's Model Mom, Maye Musk, Is at Her Peak as a Style Icon," **Observer** (Jan. 7, 2020), https://observer.com/2020/01/elon-musk -mother-maye-model-dietician-interview-book -women-self-help/.

138   Years later, Tesla executives: Author interviews with Tesla workers at the time.

139   "Unequivocally, I will support": Kim Reynolds, "2008 Tesla Roadster First Drive," **Motor Trend** (Jan. 23, 2008), https://www.motortrend.com/ cars/tesla/roadster/2008/2008-tesla-roadster/.

139   "I really wanted the car": Author interview.

139   "I want to be very clear": Jennifer Kho, "First Tesla Production Roadster Arrives," Green Tech Media (Feb. 1, 2008), https://www.greentech media.com/articles/read/first-tesla-production -roadster-arrives-546.

140   An editor from **Motor Trend**: Kim Reynolds, "2008 Tesla Roadster First Drive."

140   Michael Balzary, better known: Michael Balzary, "Handing Over the Keys IV," Tesla blog (Nov. 6, 2007), https://www.tesla.com/blog/ handing-over-keys-iv-michael-flea-balzary.

140   Leno marveled: Description taken from video posted April 19, 2020, by Jay Leno's Garage on YouTube. https://youtu.be/jjZf9sgdDKc.

140   It had been a rocky road: Author interviews with people involved in the funding plan.

141   Musk complained: Author interviews with people involved with the funding plan.

143   "We either do this": Author interviews with Tesla
workers at the time.

143   Reclaiming the narrative: Elon Musk, "Extraordinary times require focus," company blog
(Oct. 15, 2008).

144   "I actually talked a close friend": Owen Thomas,
"Tesla Motors Has $9 Million in the Bank,
May Not Deliver Cars," Valleywag (Oct. 30,
2008), https://gawker.com/5071621/tesla
-motors-has-9-million-in-the-bank-may-not
-deliver-cars.

145   "The past month has": Owen Thomas, "The
Martyr of Tesla Motors," Valleywag (Nov. 4,
2008), https://gawker.com/5075487/the-martyr
-of-tesla-motors.

146   Employees had overheard: Author interviews with
Tesla workers at the time.

147   Tesla "was building a car": Author interviews
with Tesla workers at the time.

147   "Yeah, I know no one": Author interviews with
Tesla workers who witnessed Musk's efforts.

148   Back in LA: Anecdote told by Jason Calacanis
during a podcast conducted by **Business Insider**'s
Alyson Shontell (Aug. 3, 2017), https://play
.acast.com/s/howididit/investorjasoncalacanis
-howiwasbroke-thenrich-thenbroke-andnow
have-100million.

148   There were other: Ibid.

148   "Elon, looks like": Ibid.

149   but, as Musk: Ashlee Vance, **Elon Musk: Tesla,**

SpaceX, and the Quest for a Fantastic Future (New York: HarperCollins, 2015), 157.

149 Musk suspected the delay: Ibid.

150 To stoke their competitive juices: Ibid.

151 Now Musk: Chuck Squatriglia, "Tesla Raises Prices to 'Guarantee Viability,'" **Wired** (Jan. 20, 2009).

152 Billionaire Larry Ellison: Author interview with Tesla workers at the time.

152 "It didn't seem worth": Tom Saxton's Blog (Jan. 15, 2009), https://saxton.org/tom_saxton/2009/01/.

152 "I cannot understate": transcript of filming from **Revenge of the Electric Car,** 2011.

154 "I hope you like what you see": Description of event from video posted by Sival Teokal on June 30, 2015, https://youtu.be/ZV8w OQsKV8Y.

CHAPTER 9

161 And after months: Kate Linebaugh, "Tesla Motors to Supply Batteries for Daimler's Electric Mini Car," **Wall Street Journal** (Jan. 13, 2009), https://www.wsj.com/articles/SB123187253507878007.

166 "I don't have the budget": This anecdote comes from author interviews with Peter Rawlinson with certain details corroborated by other interviews with Tesla workers at the time.

170 They argued that Tesla: Author interviews with two people who were part of the discussions.

CHAPTER 10

175 "If you were my employee": Justine Musk, "I
     Was a Starter Wife," **Marie Claire** (Sept. 10,
     2010), https://www.marieclaire.com/sex-love/
     a5380/millionaire-starter-wife.

175 The practical effect: Elon Musk, "Correcting the
     Record About My Divorce," **Business Insider**
     (July 8, 2010), https://www.businessinsider.com/
     correcting-the-record-about-my-divorce-2010-7.

176 Looking for a way out: Jeffrey McCracken,
     John D. Stoll, and Neil King Jr., "U.S. Threatens
     Bankruptcy for GM, Chrysler," **Wall Street
     Journal** (March 31, 2009), https://www.wsj
     .com/articles/SB123845591244871499.

178 "Early on it wasn't clear": Author interview with
     Yanev Suissa.

179 Herbert Kohler, head of Daimler's: Author inter-
     view with people familiar with the interactions.

181 But neither side wanted: Author interviews with
     people involved in the negotiations.

181 "It became about the press release": Author inter-
     view with Suissa.

183 Also, he ultimately: **Martin Eberhard v. Elon
     Musk,** California superior court, filed May
     2009.

184 The backup option: Elon Musk said on Twitter
     (Dec. 8, 2018), https://twitter.com/elonmusk/
     status/1071613648085311488?s=20.

185 "Probably the most difficult": Emails reviewed
     by the author.

CHAPTER 11

188 His requests for help: Leanne Star, "Alumni Profile: Deepak Ahuja," **McCormick Magazine** (Fall 2011), 42.

189 By the middle of 2009: Author interviews with Tesla workers at the time.

193 Straubel's team had equipped: Author interviews with people involved with the demonstration.

193 And with that he stormed: Details of the IPO process come from author interviews with several people involved in the effort.

200 "It's like Gutenberg saying": Jay Yarow, "Revealed: Tesla's IPO Roadshow," **Business Insider** (June 22, 2010), https://www.businessinsider.com/teslas-ipo-roadshow-2010-6.

204 "People at this point": Description of scene taken from video posted by CNBC on June 29, 2010, https://www.cnbc.com/video/2010/06/29/tesla-goes-public.html.

204 "Fuck oil": Author interviews with Tesla workers at the event.

CHAPTER 12

205 "Sales suck": Author interviews with Tesla workers in attendance.

207 Gracias and Watkins's theory: Steven N. Kaplan, Jonathan Gol, et al., "Valor and Tesla Motors," University of Chicago case study (2017), https://faculty.chicagobooth.edu/-/media/faculty/steven-kaplan/research/valortesla.pdf.

212   "Elon Musk would like": Nikki Gordon-Bloomfield, "From Gap to the Electric Car: Tesla's George Blankenship," Green Car Reports (Nov. 24, 2010), https://www.greencarreports .com/news/1051880_from-gap-to-the-electric -car-teslas-george-blankenship.

213   Car customers tend to be loyal: "R.L. Polk: Automakers Improve Brand Loyalty in 2010," **Automotive News** (April 4, 2011), https://www .autonews.com/article/20110404/RETAIL/ 110409960/r-l-polk-automakers-improve-brand -loyalty-in-2010.

216   "Is that what it should": Author interview with Blankenship.

218   "No, no, no": Author interview with Tesla worker at the time.

CHAPTER 13

223   "Culturally, we're so different": Ariel Schwartz, "The Road Ahead: A Tesla Car for the Masses?" **Fast Company** (Jan. 11, 2011), https://www .fastcompany.com/1716066/road-ahead-tesla-car -masses.

224   In a two-year period: John Voelcker, "Five Questions: Peter Rawlinson, Tesla Motors Chief Engineer," Green Car Reports (Jan. 14, 2011), https://www.greencarreports.com/news/1053555 _five-questions-peter-rawlinson-tesla-motors -chief-engineer.

225   "What if we hit 50 bucks": Author interview with Avalos.

225 One engineer left Musk: Author interview with Tesla workers at the time.

225 The plan was: John Voelcker, "Five Questions: Peter Rawlinson, Tesla Motors Chief Engineer."

225 He subscribed to: Author interviews with multiple Tesla workers who worked at the company throughout the years.

226 "Rapid decision-making may": Musk to author in an email conversation.

226 Engineers would email requests: Author interviews with Tesla workers at the time.

227 "I'm going to sell": Author interview with a passenger aboard the airplane that day.

228 "This is the stupidest": Author interview with Tesla workers familiar with the episode.

230 The Japanese company: Tesla press release (Nov. 3, 2010), https://ir.teslamotors.com/news -releases/news-release-details/panasonic-invests -30-million-tesla-companies-strengthen.

232 One of Tesla's engineers: Author interview with people in those meetings.

233 They resolved the problem: Mark Rechtin, "From an Odd Couple to a Dream Team," **Automotive News** (Aug. 13, 2012), https://www.autonews .com/article/20120813/OEM03/308139960/ from-an-odd-couple-to-a-dream-team.

233 "What the fuck is this?": Author interview with person involved in the matter.

235 Musk's personal experience weighed: Author interviews with Tesla employees who worked on the project.

239 In his rare free time: Hannah Elliott, "At Home

with Elon Musk: The (Soon-to-Be) Bachelor," **Forbes** (May 26, 2012).

243 "I still love her": Hannah Elliott, "Elon Musk to Divorce from Wife Talulah Riley," **Forbes** (Jan.18, 2012).

CHAPTER 14

249 "We learned that quickly": Author interview with a Tesla manager.

250 Passin needed: Pui-Wing Tam, "Idle Fremont Plant Gears Up for Tesla," **Wall Street Journal** (Oct. 21, 2010), https://www.wsj.com/articles/SB10001424052748704300604575554662948527140.

250 They knew the car: Philippe Chain and Frederic Filloux, "How Tesla Cracked the Code of Automobile Innovation," Monday Note (July 12, 2020), https://mondaynote.com/how-the-tesla-way-keeps-it-ahead-of-the-pack-358db5d52add.

251 "Solve it, guys": Ibid.

253 The direction came over: Mike Ramsey, "Electric-Car Pioneer Elon Musk Charges Head-On at Detroit," **Wall Street Journal** (Jan. 11, 2015), https://www.wsj.com/articles/electric-car-pioneer-elon-musk-charges-head-on-at-detroit-1421033527.

253 "Scaling Model S production": Email reviewed by the author.

254 So instead he and his team: Author interviews with Tesla workers who developed the effort.

254 The crew worked almost: Author interviews with Tesla workers.

255 A team was created: Author interviews with Tesla workers.

255 Some of Straubel's: Linette Lopez, "Leaked Tesla Emails Tell the Story of a Design Flaw . . ." **Business Insider** (June 25, 2020), https://www .businessinsider.com/tesla-leaked-emails-show -company-knew-model-s-battery-issues-2020-6.

256 called him the Hangman: Author interview with Tesla manager.

257 "What would have been deemed": Philippe Chain and Frederic Filloux, "How Tesla Cracked the Code of Automobile Innovation."

261 "They presented a better face": Elon Musk's appearance recorded by C-Span (Sept. 29, 2011), https://www.c-span.org/video/?301817-1/future -human-space-flight.

CHAPTER 15

263 If Tesla delivered: Author interview with Tesla workers at the time.

265 "I know I've asked": Author interviews with Tesla workers at the meeting.

266 "The mere fact the Tesla Model S": Angus MacKenzie, "2013 Motor Trend Car of the Year: Tesla Model S," **Motor Trend** (Dec. 10, 2012), https://www.motortrend.com/news/2013-motor -trend-car-of-the-year-tesla-model-s/.

267 "We're not doing this": Description of events taken from Tesla video recording of the event

posted by the company on YouTube on Nov. 17, 2012, https://youtu.be/qfxXmIFfV7I.

268 The company was facing: Author interviews with Tesla workers at the time.

269 He told an assistant: Author interview.

269 "This looks promising": Author interview with Blankenship.

271 Musk complained to his staff: Author interviews with Tesla workers at the time.

271 While the firm: Susan Pulliam, Rob Barry, and Scott Patterson, "Insider-Trading Probe Trains Lens on Boards," **Wall Street Journal** (April 30, 2013), https://www.wsj.com/articles/SB1000142 4127887323798104578453260765642292.

273 The review was uncharacteristically rapturous: "Tesla Model S review," **Consumer Reports** (July 2013), https://www.consumerreports.org/cro/magazine/2013/07/tesla-model-s-review/index.htm.

274 Musk's relationship with Blankenship: Ashlee Vance, **Elon Musk: Tesla, SpaceX, and the Quest for a Fantastic Future** (New York: HarperCollins, 2015), 216.

274 "For Tesla to succeed": Author interview with Blankenship.

277 He had quietly reached: Ashlee Vance, **Elon Musk,** 217.

CHAPTER 16

281 But their innovations: Author interviews with people familiar with Akerson's thinking.

281   The first fire occurred: Tom Krisher and Mike Baker, "Tesla Says Car Fire Began in Battery After Crash," **Seattle Times** (Oct. 3, 2013), https://www.seattletimes.com/business/tesla -says-car-fire-began-in-battery-after-crash/.

282   A second Model S: Ben Klayman and Bernie Woodall, "Tesla Reports Third Fire Involving Model S Electric Car," Reuters (Nov. 7, 2013), https://www.reuters.com/article/us-autos-tesla -fire/tesla-reports-third-fire-involving-model-s -electric-car-idUSBRE9A60U220131107.

282   "I had a Tesla": Tom Junod, "George Clooney's Rules for Living," **Esquire** (Nov. 11, 2013), https://www.esquire.com/news-politics/a25952/ george-clooney-interview-1213/.

282   As they studied the fires: Author interviews with engineers involved in the matter.

284   "That newly crowned leadership": Author interview with member of the task force.

285   The 2012 Mercedes-Benz: Historical pricing data was provided to author from Edmunds, an automotive industry researcher.

285   The company justified: Don Reisinger, "Tesla Kills 40 kWh Battery for Model S over 'Lack of Demand,'" CNET (April 1, 2013), https:// www.cnet.com/roadshow/news/tesla-kills -40-kwh-battery-for-model-s-over-lack-of -demand/.

287   Compared to competitors: Research first released on July 7, 2014, by Pied Piper Management Company LLC. Evaluations were conducted between July 2013 and June 2014, firm founder

Fran O'Hagan told author in a December 2019 email.

287   "When I first sat": Ronald Montoya, "Is the Third Drive Unit the Charm?," Edmunds.com (Feb. 20, 2014), https://www.edmunds.com/tesla/model-s/2013/long-term-road-test/2013-tesla-model-s-is-the-third-drive-unit-the-charm.html.

289   "If you can't get this": Author interview with person familiar with the matter.

CHAPTER 17

291   "If there is a party": Tatiana Siegel, "Elon Musk Requested to Meet Amber Heard via Email Years Ago," **Hollywood Reporter** (Aug. 24, 2016), https://www.hollywoodreporter.com/rambling-reporter/elon-musk-requested-meet-amber-922240.

291   Some said they tried: Tim Higgins, Tripp Mickle, and Rolfe Winkler, "Elon Musk Faces His Own Worst Enemy," **Wall Street Journal** (Aug. 31, 2018), https://www.wsj.com/articles/elon-musk-faces-his-own-worst-enemy-1535727324.

292   Dealers in Massachusetts: Mike Ramsey and Valerie Bauerlein, "Tesla Clashes with Car Dealers," **Wall Street Journal** (June 18, 2013), https://www.wsj.com/articles/SB10001424127887324049504578541902814606098.

292   To Wolters, it didn't: Author interview with Wolters.

293   "I really admire what you've": Author interview with Wolters.

295 "I'm going to spend": Author interview with Wolters.

297 The Texas dealers association: Texans for Public Justice, "Car-Dealer Cartel Stalled Musk's Tesla," Lobby Watch (Sept. 10, 2013), http://info.tpj.org/Lobby_Watch/pdf/AutoDealers vTesla.pdf.

297 "I love what you're doing": Author interview with a person familiar with the moment.

CHAPTER 18

304 The cells were costing: Research provided by Simon Moores of Benchmark Mineral Intelligence.

304 That meant that: Csaba Csere, "Tested: 2012 Tesla Model S Takes Electric Cars to a Higher Level," **Car and Driver** (Dec. 21, 2012), https://www.caranddriver.com/reviews/a15117388/2013 -tesla-model-s-test-review/.

305 As Straubel discussed his math: Author interview with Straubel.

307 While Musk had humored: Author interviews with several Tesla workers from that period to detail the evolving relationship with Panasonic.

312 To address: Author interviews with people who worked on the effort.

316 Back in Japan: Author interviews with people familiar with the deliberations at Panasonic.

317 He had to make it: Author interviews with people who worked on the effort.

CHAPTER 19

320   "This thing was packed": Author interview with Varadharajan.

322   Guillen had two sides: Author interviews with Tesla managers who worked with him.

323   Musk began 2014 telling: Alan Ohnsman, "Musk Says China Potential Top Market for Tesla," **Bloomberg News** (Jan. 24, 2014), https://www.bloomberg.com/news/articles/2014-01-23/tesla-to-sell-model-s-sedan-in-china-from-121-000.

326   It was a non-starter: Author interview with Tesla workers who worked in this area.

328   Sales dropped 33 percent: China registration figures provided to author by research firm JL Warren Capital.

329   Their data showed that: Author interview with Tesla workers from this period.

329   Unlike in the U.S.: Survey data of U.S. customers provided to author by Alexander Edwards of research firm Strategic Vision.

332   Musk turned to his cousins: Author interviews with people who worked on the matter.

CHAPTER 20

335   A 2013 article in **Forbes**: Caleb Melby, "Guns, Girls and Sex Tapes: The Unhinged, Hedonistic Saga of Billionaire Stewart Rahr, 'Number One King of All Fun,'" **Forbes** (Sept. 17, 2013), https://www.forbes.com/sites/calebmelby/2013/

09/17/guns-girls-and-sex-tapes-the-saga-of
-billionaire-stewart-rahr-number-one-king-of
-all-fun/#3ca48b2d3f86.

340 "It struck me that": Author interview with Fossi.

341 An estimated loss: Research provided to author by research firm S3 Partners.

343 It required more and more: Cassell Bryan-Low and Suzanne McGee, "Enron Short Seller Detected Red Flags in Regulatory Filings," **Wall Street Journal** (Nov. 5, 2001), https://www.wsj.com/articles/SB1004916006978550640.

344 Chanos called the company: Jonathan R. Laing, "The Bear That Roared," **Barron's** (Jan. 28, 2002), https://www.barrons.com/articles/SB1011910694160632340?tesla=y.

344 America Online: **Ibid.**

347 He had taken out personal loans: Tesla filings with the SEC.

347 He'd long loathed: Susan Pulliam, Mike Ramsey, and Brody Mullins, "Elon Musk Supports His Business Empire with Unusual Financial Moves," **Wall Street Journal** (April 27, 2016), https://www.wsj.com/articles/elon-musk-supports-his-business-empire-with-unusual-financial-moves-1461781962.

348 "I was nervous watching": Emails reviewed by the author.

348 "You know that I don't": Detailed in a deposition Kimbal Musk gave on April 23, 2019.

CHAPTER 21

354   More than six thousand work: Wellford W. Wilms, Alan J. Hardcastle, and Deone M. Zell, "Cultural Transformation at NUMMI," **Sloan Management Review** 36:1 (Oct. 15, 1994): 99.

355   In 1991: **Ibid.**

356   "All he does is brag": Author interview with Ortiz.

357   The average tenure: Harley Shaiken, "Commitment Is a Two-Way Street," white paper prepared for the Toyota NUMMI Blue Ribbon Commission (March 3, 2010), http://dig.abclocal.go.com/kgo/PDF/NUMMI-Blue-Ribbon-Commission-Report.pdf.

358   He was making $21: GM pay data from the Center of Automotive Research.

359   But as the requests: Author interviews with Tesla workers involved in the projects.

360   The hydraulics weren't standing: Author interviews with people who worked on the vehicle.

362   Musk was cool under: Author interviews with people who worked on the vehicle.

364   In 2015: Tim Higgins, "Tesla Faces Labor Discord as It Ramps Up Model 3 Production," **Wall Street Journal** (Oct. 31, 2017), https://www.wsj.com/articles/tesla-faces-labor-discord-as-it-ramps-up-model-3-production-1509442202.

364   Those fancy second-row seats: **Ibid.**

365   That spring while Depp: "Elon Musk Regularly Visited Amber Heard . . . ," **Deadline** (July 17, 2020), https://deadline.com/2020/07/elon

-musk-amber-heard-johnny-depps-los-angeles
-penthouse-1202988261/.

366 Musk seemed to be: Author interviews with Tesla executives.

366 He was spotted by: Lindsay Kimble, "Amber Heard and Elon Musk Party at the Same London Club Just Weeks After Hanging Out in Miami," **People** (Aug. 3, 2016), https://people .com/movies/amber-heard-and-elon-musk -party-at-same-london-club-weeks-after-miami -sighting/.

366 "Lack of sleep": Author interview with former Tesla executive.

367 Among the first: Tim Higgins and Dana Hull, "Want Elon Musk to Hire You at Tesla? Work for Apple," **Bloomberg Businessweek** (Feb. 2, 2015).

369 Many knew that Musk: Interviews with Tesla employees at the time, and Will Evans and Alyssa Jeong Perry, "Tesla Says Its Factory Is Safer. But It Left Injuries Off the Books," Revealnews.org (April 16, 2018), https://www.revealnews.org/ article/tesla-says-its-factory-is-safer-but-it-left -injuries-off-the-books/.

371 Musk found himself: Author interviews with people familiar with the matter.

372 It showed that Tesla's: Author reviewed J.D. Power presentation of "Tesla: Beyond the Hype" (March 2017).

CHAPTER 22

377 "There was not a single": Author interview with a Tesla executive from that period.

391 "You are now working": Charles Duhigg, "Dr. Elon & Mr. Musk: Life Inside Tesla's Production Hell," **Wired** (Dec. 13, 2018), https://www.wired.com/story/elon-musk-tesla-life-inside-gigafactory/.

392 If a critical workstation: Author interviews.

392 Musk told him to: Claim laid out in a federal lawsuit against Tesla filed in 2017.

392 They were still struggling to: Author interview with Tesla managers from that period.

CHAPTER 23

394 Brown died on impact: Details taken from National Highway Traffic Safety Administration report (Jan. 19, 2017), https://static.nhtsa.gov/odi/inv/2016/INCLA-PE16007-7876.PDF.

395 It made no attempt: Ibid.

396 He had forty-eight hours: Jason Wheeler deposition taken on June 4, 2019.

398 His senior executives: Author interviews with multiple people familiar with the discussions.

399 As they looked at: Antonio Gracias deposition taken April 18, 2019.

400 "I don't negotiate": Courtney McBean deposition taken June 5, 2019.

400 "Tesla's mission has always": Tesla blog posting

on June 21, 2016, https://www.tesla.com/blog/ tesla-makes-offer-to-acquire-solarcity.

402 **Fortune** magazine's Carol Loomis: Carol J. Loomis, "Elon Musk Says Autopilot Death 'Not Material' to Tesla Shareholders,'" **Fortune** (July 5, 2016), https://fortune.com/2016/07/05/ elon-musk-tesla-autopilot-stock-sale/.

402 That very question raised: Jean Eaglesham, Mike Spector, and Susan Pulliam, "SEC Investigating Tesla for Possible Securities-Law Breach," **Wall Street Journal** (July 11, 2016), https://www.wsj .com/articles/sec-investigating-tesla-for-possible -securities-law-breach-1468268385.

402 Denholm was getting an earful: Denholm de- position page in stockholder litigation against Tesla, taken on June 6, 2019, 154.

403 "Honestly, we hate being public": Kimbal Musk deposition taken on April 23, 2019.

403 "Elon was risky but": Author interview with Tesla manager.

404 "Lousy sentiment outside": Brad Buss deposi- tion taken on June 4, 2019.

404 Wheeler's team calculated: Presentation pre- sented to the Tesla board of directors, dated July 24, 2016.

405 "We are going to": Elon Musk deposition taken on Aug. 24, 2019.

408 "Latest feedback from major": Emails reviewed by the author.

409 It wasn't foolproof: Author interviews with Tesla engineers.

409   They had been monitoring: Author interviews with people familiar with Anderson's efforts.

410   Tesla's legal and PR: Author interviews with several people involved with Autopilot.

414   CEOs that winter: Author interviews with people around Musk.

416   In March, a conference room: Author interviews with several people involved with the meeting.

CHAPTER 24

423   The list was short: Author interviews and Tim Higgins, "Elon Musk has an Awkward Problem at Tesla: Employee Parking," **Wall Street Journal** (April 11, 2017), https://www.wsj.com/articles/ elon-musk-has-an-awkward-problem-at-tesla -employee-parking-1491926275.

423   "Six out of eight": Jose Moran, "Time for Tesla to Listen," Medium.com (Feb. 9, 2017), https:// medium.com/@moran2017j/time-for-tesla-to -listen-ab5c6259fc88.

424   They were handmade: Tim Higgins, "Behind Tesla's Production Delays: Parts of Model 3 Were Being Made by Hand," **Wall Street Journal** (Oct. 6, 2017), https://www.wsj.com/articles/ behind-teslas-production-delays-parts-of-model -3-were-being-made-by-hand-1507321057.

424   It was such a tight: Author interview with workers.

425   Panasonic wasn't any happier: Author interviews with Panasonic and Tesla workers at the time.

427   Tesla's production claims: Dana Cimilluca,

Susan Pulliam, and Aruna Viswanatha, "Tesla Faces Deepening Criminal Probe over Whether It Misstated Production Figures," **Wall Street Journal** (Oct. 26, 2018), https://www.wsj.com/articles/tesla-faces-deepening-criminal-probe-over-whether-it-misstated-production-figures-1540576636.

428 In response, they had: Author interviews with Tesla workers.

428 Battery packs needed: Lora Kolodny, "Tesla Employees Say to Expect More Model 3 Delays, Citing Inexperienced Workers, Manual Assembly of Batteries," CNBC.com (Jan. 25, 2018), https://www.cnbc.com/2018/01/25/tesla-employees-say-gigafactory-problems-worse-than-known.html.

428 He estimated there were 100 million cells: Author interview with Tesla workers.

429 On one occasion: Charles Duhigg, "Dr. Elon & Mr. Musk: Life Inside Tesla's Production Hell," **Wired** (Dec. 13, 2018).

429 Dutifully, he figured out: Author interviews with Tesla workers.

430 "I wish we could": Neil Strauss, "Elon Musk: The Architect of Tomorrow," **Rolling Stone** (Nov. 15, 2017), https://www.rollingstone.com/culture/culture-features/elon-musk-the-architect-of-tomorrow-120850/.

430 "Let me know who": Details taken from the findings of an administrative judge's findings on Sept. 27, 2019, in a National Labor Relations Board case against Tesla.

432   The whiteboard at: Author interview with organizer.

434   As sales of the Model S: Details about advertising plans came from author interviews with former Tesla executives.

438   Tesla hadn't yet disclosed: Author interviews with Tesla executives at the time.

438   As they struggled with: Author interview with Tesla managers at the time.

440   He kept talking: Author interviews with Tesla executives at the time.

441   according to an ally: Author interviews with people familiar with the matter.

442   "who the fuck you are": Author interview with former Tesla engineer.

442   "I don't see how": Tim Higgins, Tripp Mickle, and Rolfe Winkler, "Elon Musk Faces His Own Worst Enemy," **Wall Street Journal** (Aug. 31, 2018), https://www.wsj.com/articles/elon-musk -faces-his-own-worst-enemy-1535727324.

443   But it was time: Author interview with people familiar with Field's thinking.

445   An automation mistake: Email reviewed by the author.

445   Musk's imperiousness didn't play: Tim Higgins, "Tesla's Elon Musk Turns Conference Call into Sparring Session," **Wall Street Journal** (May 3, 2018), https://www.wsj.com/articles/teslas-elon -musk-turns-conference-call-into-sparring -session-1525339803.

CHAPTER 25

447 "I currently work for": Emails reviewed by the author after Marty Tripp released them on Twitter.

448 "He's always pitching": Video of interview posted on **Business Insider**'s website on Feb. 21, 2018: https://www.businessinsider.com/jim-chanos -tesla-elon-musk-truck-video-2018-2.

451 He thought: Details from Martin Tripp deposition taken as part of litigation between him and Musk.

453 She kept a schedule: Sarah O'Brien deposition taken on June 5, 2019.

453 Since 2014, Musk's use: Susan Pulliam and Samarth Bansal, "For Tesla's Elon Musk, Twitter Is Sword Against Short Sellers," **Wall Street Journal** (Aug. 2, 2018), https://www.wsj.com/ articles/for-teslas-elon-musk-twitter-is-sword -against-short-sellers-1533216249.

454 He struck up a conversation: Emily Smith and Mara Siegler, "Elon Musk Quietly Dating Musician Grimes," **New York Post** (May 7, 2018), https://pagesix.com/2018/05/07/elon -musk-quietly-dating-musician-grimes/.

CHAPTER 26

459 "I just woke up": Emails reviewed by the author.

460 "peace and execution": Sarah Gardner and Ed Hammond, "Tesla Needs Period of 'Peace and Execution,' Major Shareholder Says," **Bloomberg News** (July 11, 2018), https://www

.bloomberg.com/news/articles/2018-07-11/
tesla-ought-to-pipe-down-and-execute-major
-shareholder-says.

462   "We need to stop": Email exchanges reviewed by
the author.

462   "I just burst out": Emails reviewed by the author.

463   Tesla began asking some: Tim Higgins, "Tesla
Asks Suppliers for Cash Back to Help Turn a
Profit," **Wall Street Journal** (July 22, 2018),
https://www.wsj.com/articles/tesla-asks
-suppliers-for-cash-back-to-help-turn-a-profit
-1532301091.

464   Perhaps Apple: Tim Higgins, "Elon Musk Says
He Once Approached Apple CEO About Buying
Tesla," **Wall Street Journal** (Dec. 22, 2020),
https://www.wsj.com/articles/elon-musk-says
-he-once-approached-apple-ceo-about-buying
-tesla-11608671609.

464   A back and forth: Author interview with a person familiar with the effort.

466   "Was this text legit?" Emails detailed in court filings by the SEC.

469   In 2013, they helped: Miriam Gottfried, "Dell
Returns to Public Equity Markets," **Wall Street
Journal** (Dec. 28, 2018), https://www.wsj.com/
articles/dell-returns-to-public-equity-markets
-11546011748.

472   Stewart reached out: James B. Stewart, "The
Day Jeffrey Epstein Told Me He Had Dirt on
Powerful People," **New York Times** (Aug. 12,
2019), https://www.nytimes.com/2019/08/12/
business/jeffrey-epstein-interview.html.

472   "Epstein, one of the": Email exchange reviewed by the author.

472   Musk proceeded to self-implode: David Gelles, James B. Stewart, Jessica Silver-Greenberg, and Kate Kelly, "Elon Musk Details 'Excruciating' Personal Toll of Tesla Turmoil," **New York Times** (Aug. 16, 2018), https://www.nytimes.com/2018/08/16/business/elon-musk-interview-tesla.html.

473   "I saw him in the kitchen": Kate Taylor, "Rapper Azealia Banks Claims She Was at Elon Musk's House over the Weekend as He Was 'Scrounging for Investors,'" Business Insider (Aug. 13, 2018), https://www.businessinsider.com/azealia-banks-claims-to-be-at-elon-musks-house-as-he-sought-investors-2018-8.

473   "Don't they have something": Email exchange reviewed by the author.

474   From LA: Liz Hoffman and Tim Higgins, "Public Bravado, Private Doubts: Inside the Unraveling of Elon Musk's Tesla Buyout," **Wall Street Journal** (Aug. 27, 2018), https://www.wsj.com/articles/public-bravado-private-doubts-how-elon-musks-tesla-plan-unraveled-1535326249.

475   The fund's leader: Bradley Hope and Justin Scheck, **Blood and Oil: Mohammed Bin Salman's Ruthless Quest for Global Power** (New York: Hachette, 2020), 251.

475   This left Musk's advisers: Kimbal Musk deposition taken on April 23, 2019.

475   Musk was unhappy with: Liz Hoffman and Tim

Higgins, "Public Bravado, Private Doubts: Inside the Unraveling of Elon Musk's Tesla Buyout," **Wall Street Journal** (Aug. 27, 2018), https://www.wsj.com/articles/public-bravado-private-doubts-how-elon-musks-tesla-plan-unraveled-1535326249.

475   "In my opinion": Elon Musk told the author in an email on Aug. 25, 2018.

478   "One or two more": Email exchange reviewed by the author.

CHAPTER 27

481   By August, Tesla's extra cash: Tim Higgins, Marc Vartabedian, and Christina Rogers, "Some Tesla Suppliers Fret About Getting Paid," **Wall Street Journal** (Aug. 20, 2018), https://www.wsj.com/articles/some-tesla-suppliers-fret-about-getting-paid-1534793592.

481   Internally, Musk was pushing: Author interviews with Tesla managers at the time.

483   Musk's lawyers had gone: Susan Pulliam, Dave Michaels, and Tim Higgins, "Mark Cuban Prodded Tesla's Elon Musk to Settle SEC Charges," **Wall Street Journal** (Oct. 4, 2018), https://www.wsj.com/articles/mark-cuban-prodded-teslas-elon-musk-to-settle-sec-charges-1538678655.

484   Before Jon McNeill quit: Author interviews with Tesla executives who worked on the plan.

486   As they made their calls: Author interviews with Tesla sales managers.

489   Kim deployed crews: Author interviews with Tesla managers familiar with the situation.

493   As they raced toward: Author interviews with Tesla managers involved in the effort.

494   They were on pace for: Author interviews with people on the call.

494   "I don't want anyone": Author interview with Tesla worker who witnessed the episode.

494   It was a scene: Dana Hull and Eric Newcome, "Tesla Board Probed Allegation That Elon Musk Pushed Employee," **Bloomberg News** (April 5, 2019), https://www.bloomberg.com/news/articles/2019-04-05/tesla-board-probed-allegation-that-elon-musk-pushed-employee.

495   After the lawsuit was announced: Research provided to author by research firm S3 Partners.

496   Musk's lawyers spent: Susan Pulliam, Dave Michaels, and Tim Higgins, "Mark Cuban Prodded Tesla's Elon Musk to Settle SEC Charges," **Wall Street Journal** (Oct. 4, 2018), https://www.wsj.com/articles/mark-cuban-prodded-teslas-elon-musk-to-settle-sec-charges-1538678655.

496   He believed he had: Author interviews with people familiar with Musk's thinking.

498   "It was like a big": Author interview with a Tesla manager.

498   Happily for Musk: Research provided to author by research firm S3 Partners.

502   The board's investigation into: Dana Hull and Eric Newcomer, "Tesla Board Probed Allegation That Elon Musk Pushed Employee."

502   As he revealed the Model Y: Author observations of Denholm and event.

503   "From my perspective": Angus Whitley, "Tesla's New Chairman Says Elon Musk Uses Twitter 'Wisely,'" **Bloomberg News** (March 27, 2019), https://www.bloomberg.com/news/articles/2019 -03-27/tesla-chair-defends-musk-tweets-even-as -habit-lands-him-in-court.

503   Shares fell almost 7 percent: Tim Higgins, "Tesla Shares Sink on Model 3 Delivery Miss, Price Cut," **Wall Street Journal** (Jan. 2, 2019), https:// www.wsj.com/articles/tesla-plans-to-trim-prices -as-fourth-quarter-deliveries-rise-11546437526.

504   By the end of February: Dave Michaels and Tim Higgins, "SEC Asks Manhattan Federal Court to Hold Elon Musk in Contempt," **Wall Street Journal** (Feb. 25, 2019), https://www.wsj.com/ articles/sec-asks-manhattan-federal-court-to -hold-elon-musk-in-contempt-11551137500.

504   The move to online: Tim Higgins and Adrienne Roberts, "Tesla Shifts to Online Sales Model," **Wall Street Journal** (Feb. 28, 2019), https:// www.wsj.com/articles/tesla-says-it-has-started -taking-orders-for-35-000-version-of-model-3 -11551392059.

505   Tesla "is a company": Esther Fung, "Landlords to Tesla: You're Still on the Hook for Your Store Leases," **Wall Street Journal** (March 8, 2019), https://www.wsj.com/articles/landlords-to-tesla -youre-still-on-the-hook-for-your-store-leases -11552059041.

506   The deal: Peter Campbell, "Fiat Chrysler to

Spend €1.8bn on CO2 Credits," **Financial Times** (May 3, 2019), https://www.ft.com/content/ fd8d205e-6d6b-11e9-80c7-60ee53e6681d.

507 Short-seller bets were finally: Research provided to author by research firm S3 Partners.

507 Its debt had fallen: Sam Goldfarb, "Tesla Faces Steeper Costs to Raise Cash," **Wall Street Journal** (April 29, 2019), https://www.wsj.com/ articles/tesla-faces-steeper-costs-to-raise-cash -11556535600.

507 Worst of all: Trefor Moss, "Global Auto Makers Dented as China Car Sales Fall for First Time in Decades," **Wall Street Journal** (Jan. 14, 2019), https://www.wsj.com/articles/chinese -annual-car-sales-slip-for-first-time-in-decades -11547465112.

CHAPTER 28

511 As a student at the University of Pennsylvania: Author interviews with former Tesla executives.

511 They were warmly received: Author interviews with Tesla managers from that period.

512 **Bloomberg Businessweek**: Matthew Campbell et al., "Elon Musk Loves China, and China Loves Him Back—For Now," **Bloomberg Businessweek** (Jan. 13, 2021), https://www .bloomberg.com/news/features/2021-01-13/ china-loves-elon-musk-and-tesla-tsla-how-long -will-that-last?sref=PRBlrg7S.

513 He proposed an idea: Author interview with a Tesla manager familiar with the trip.

513 "Traffic is driving": Elon Musk said on Twitter (Dec. 17, 2016), https://twitter.com/elonmusk/status/810108760010043392?s=20.

514 As things dragged into 2018: Bruce Einhorn, et al., "Tesla's China Dream Threatened by Standoff Over Shanghai Factory," Bloomberg News (Feb. 13, 2018), https://www.bloomberg.com/news/articles/2018-02-14/tesla-s-china-dream-threatened-by-standoff-over-shanghai-factory?sref=PRBlrg7S.

515 GM would partner with: Mike Colias, "GM, LG to Spend $2.3 Billion on Venture to Make Electric-Car Batteries," **Wall Street Journal** (Dec. 5, 2019), https://www.wsj.com/articles/gm-lg-to-spend-2-3-billion-on-venture-to-make-electric-car-batteries-11575554432.

515 Volkswagen had committed to: Stephen Wilmot, "Volkswagen Follows Tesla into Battery Business," **Wall Street Journal** (June 13, 2019), https://www.wsj.com/articles/volkswagen-follows-tesla-into-battery-business-11560442193.

516 "Tesla is not niche": Christoph Rauwald, "Tesla Is No Niche Automaker Anymore, Volkswagen's CEO Says," **Bloomberg News** (Oct. 24, 2019), https://www.bloomberg.com/news/articles/2019-10-24/volkswagen-s-ceo-says-tesla-is-no-niche-automaker-anymore.

516 While he might not: Author interviews with Tesla managers over the years.

517 "was very happy with": Dave Michaels and Tim Higgins, "Judge Gives Elon Musk, SEC Two Weeks to Strike Deal on Contempt Claims,"

Wall Street Journal (April 4, 2019), https://www.wsj.com/articles/judge-asks-elon-musk-and-sec-to-hold-talks-over-contempt-claims-11554408620.

523 The automaker was allowed: Wang Zhiyan, Du Chenwei, and Hu Xingyang, "Behind 'Amazing Shanghai Speed'" (translated into English), **Jiefang Ribao** (Jan. 1, 2020), https://www.jfdaily.com/journal/2020-01-08/getArticle.htm?id=285863.

523 The electric grid extended: Luan Xiaona, "The Power Supply Project of Tesla Shanghai Super Factory Will Enter the Sprint Stage Before Production" (translated into English), **The Paper** (Oct. 17, 2019), https://www.thepaper.cn/newsDetail_forward_4700380.

524 "What will our": Tim Higgins and Takashi Mochizuki, "Tesla Needs Its Battery Maker: A Culture Clash Threatens Their Relationship," **The Wall Street Journal** (Oct. 8, 2019), https://www.wsj.com/articles/tesla-needs-its-battery-maker-a-culture-clash-threatens-their-relationship-11570550526.

524 But increasingly: Ibid.

527 By August, when Musk: Description comes from video posted by Jason Yang on YouTube on Oct. 20, 2019: https://youtu.be/bI-My94Ig5k.

527 They were estimated to: Research provided to author by research firm S3 Partners.

EPILOGUE

537 The government seemed: Chunying Zhang and Ying Tian, "How China Bent Over Backward to Help Tesla," **Bloomberg Businessweek** (March 18, 2020), https://www.bloomberg.com/news/articles/2020-03-17/how-china-bent-over-backward-to-help-tesla-when-the-virus-hit?sref=PRBlrg7S.

538 Accustomed to moving: Tim Higgins, "Tesla Cuts Salaries, Furloughs Workers Under Coronavirus Shutdown," **Wall Street Journal** (April 8, 2020), https://www.wsj.com/articles/tesla-cuts-salaries-furloughs-workers-under-coronavirus-shutdown-11586364779.

538 Tesla also began: Tim Higgins and Esther Fung, "Tesla Seeks Rent Savings Amid Coronavirus Crunch," **Wall Street Journal** (April 13, 2020), https://www.wsj.com/articles/tesla-seeks-rent-savings-amid-coronavirus-crunch-11586823630.

539 "Over the past week": Jeremy C. Owens, Claudia Assis, and Max A. Cherney, "Elon Musk vs. Bay Area Officials: These Emails Show What Happened Behind the Scenes in the Tesla Factory Fight," MarketWatch (May 29, 2020), https://www.marketwatch.com/story/elon-musk-vs-bay-area-officials-these-emails-show-what-happened-behind-the-scenes-in-the-tesla-factory-fight-2020-05-29.

539 "Frankly this is the final": Tim Higgins, "Tesla Files Lawsuit in Bid to Reopen Fremont Factory,"

Wall Street Journal (May 10, 2020), https://www.wsj.com/articles/elon-musk-threatens-authorities-over-mandated-tesla-factory-shutdown-11589046681.

541 "I will be on the line": Rebecca Ballhaus and Tim Higgins, "Trump Calls for California to Let Tesla Factory Open," **Wall Street Journal** (May 13, 2020), https://www.wsj.com/articles/trump-calls-for-california-to-let-tesla-factory-open-11589376502.

541 "It is very important": Fred Lambert, "Elon Musk Sends Cryptic Email to Tesla Employees About Going 'All Out,'" **Electrek** (June 23, 2020), https://electrek.co/2020/06/23/elon-musk-cryptic-email-tesla-employees-all-out/.

543 Musk hit his $100 billion: Sebastian Pellejero and Rebecca Elliott, "How Tesla Made It to the Winner's Circle," **The Wall Street Journal,** (Feb. 19, 2021), https://www.wsj.com/articles/how-tesla-made-it-to-the-winners-circle-11613739634.

545 "This is very": Elon Musk told author in an email on May 7, 2020.

548 "I've never met": Scarlet Fu, "Chanos Reduces 'Painful' Tesla Short, Tells Musk 'Job Well Done,'" Bloomberg News (Dec. 3, 2020), https://www.bloomberg.com/news/articles/2020-12-03/tesla-bear-jim-chanos-says-he-d-tell-elon-musk-job-well-done?sref=PRBlrg7S.

INDEX

ABOUT THE AUTHOR

TIM HIGGINS is an automotive and technology reporter for The Wall Street Journal. He appears regularly as a contributor on CNBC. His writing has won several awards from the Society for Advancing Business Editing and Writing, and he is a five-time finalist for the Livingston Awards. After almost a decade reporting on the car business from Detroit, he now lives in San Francisco.